CW01238446

MILIEUS OF MINUTIAE

Milieus of Minutiae

CONTEXTUALIZING THE SMALL IN
LITERATURE, PHILOSOPHY, AND SCIENCE

✦ ✦ ✦

Edited by Elizabeth Brogden
and Christiane Frey

UNIVERSITY OF VIRGINIA PRESS
CHARLOTTESVILLE AND LONDON

The University of Virginia Press is situated on the traditional lands of the Monacan Nation, and the Commonwealth of Virginia was and is home to many other Indigenous people. We pay our respect to all of them, past and present. We also honor the enslaved African and African American people who built the University of Virginia, and we recognize their descendants. We commit to fostering voices from these communities through our publications and to deepening our collective understanding of their histories and contributions.

University of Virginia Press
© 2024 by the Rector and Visitors of the University of Virginia
All rights reserved
Printed in the United States of America on acid-free paper

First published 2024

1 3 5 7 9 8 6 4 2

LIBRARY OF CONGRESS CATALOGING-IN-PUBLICATION DATA

Names: Brogden, Elizabeth, editor. | Frey, Christiane, editor.
Title: Milieus of minutiae : contextualizing the small in literature, philosophy, and science / edited by Elizabeth Brogden and Christiane Frey.
Description: Charlottesville : University of Virginia Press, 2024. | Includes bibliographical references and index.
Identifiers: LCCN 2024017620 (print) | LCCN 2024017621 (ebook) | ISBN 9780813950747 (hardcover) | ISBN 9780813952451 (paperback) | ISBN 9780813950648 (ebook)
Subjects: LCSH: Causation. | BISAC: SCIENCE / Philosophy & Social Aspects | SCIENCE / History | LCGFT: Essays.
Classification: LCC BD591 .M55 2024 (print) | LCC BD591 (ebook) | DDC 122—dc23/eng/20240521
LC record available at https://lccn.loc.gov/2024017620
LC ebook record available at https://lccn.loc.gov/2024017621

The publication of this volume has been supported by *New Literary History*.

Cover art: From *Micrographia*, by Robert Hooke (1665). (Courtesy of the Linda Hall Library of Science, Engineering & Technology)
Cover design: TG Design

CONTENTS

Introduction — 1

Part I. Poetics of Minitude

Particularity and Virtual Witnessing in the Eighteenth-Century British Novel — 21
ROGER MAIOLI

Infinitely Small Differences: The Individual and Its Milieu in Novalis's Natural Philosophy and *Henry of Ofterdingen* — 47
MAREIKE SCHILDMANN

Le menome cose: On the Function of Details in Leopardi's *Zibaldone* — 71
ELENA FABIETTI

Part II. Micrological Biospheres

Landscapes of Disease and Contagion: Putrefaction and Fermentation in Early Modern Corpuscular Medical Theories — 97
CARMEN SCHMECHEL

Confronting the Limits of Optical Interpretation: Some Philosophical Considerations of Microscopic Details — 122
DANIEL LIU

Experimental Environments: Micrologies of Knowledge — 149
HANS-JÖRG RHEINBERGER

Part III. Philological Minima

Donne's Things: Epistemologies of the Small in John Donne's Love Poetry — 165
ANDREAS MAHLER

"Of a Parenthesis" — 181
CYNTHIA WALL

"Love" of Detail: Auerbach, Spitzer, and Curtius   209
  CHRISTOPHER D. JOHNSON

Part IV. Minimilieus of Modernity

Architecture and Ambience: From Walter Benjamin to the
French Avant-Garde   239
  MALTE FABIAN RAUCH

Cell and Cosmos: Siegfried Ebeling's Environmental Architecture   263
  MARGARETA INGRID CHRISTIAN

Literature as a Milieu of the Small: Kleist/Kafka   281
  MARIANNE SCHULLER

Notes on Contributors   299

Index   301

MILIEUS OF MINUTIAE

Introduction

✦ ✦ ✦

THE FIRST two decades of the twenty-first century have made abundantly clear the extent to which the environments we inhabit on a global—and, indeed, planetary—scale are governed by phenomena that, in and of themselves, elude the naked senses. From catastrophic weather and steady warming caused by the accumulation of carbon particles in Earth's atmosphere, to societies brought to a standstill by a microscopic virus, to the multiplication of subliminal microaggressions that can poison an entire social order, the new millennium has reminded us, over and over again, of how the small can have outsized effects on our surroundings, of the intimate and often mysterious links between minutiae and their milieus. In our tendency to exceptionalize the contemporary moment, however, we risk ignoring the longer history of the primary question with which our volume engages: namely, how are minutiae and milieus mutually constitutive of one another? The notion that the smallest of things can fundamentally shape our experiences is no more novel than nonanthropogenic climate change or infectious disease, but it is long overdue for sustained scholarly treatment. In the essays that follow, an interdisciplinary cohort of scholars reflects on the vital importance of various forms of minutiae to a wide range of milieus—a theme that has preoccupied the European and Anglo-American cultural imaginary at least since early modernity. This volume presents a series of original arguments across literature, philosophy, and the history of science in order to explore the epistemological genealogies of the relationship between seemingly trivial matter(s) and the conditions to which they give rise.

In excavating this longer history of milieus of minutiae, we embark from the premise that history is not just that which precedes our own era, nor is it a linear process of successive occurrences. Rather, it is that which surrounds and permeates us at all times: all of our experiences are as much historical as they are residually embedded.[1] Hence parsing the history—or the many intersecting histories—of milieus of minutiae does not imply observing a sequence of so-called historical developments from some fixed (let alone "modern") standpoint but rather suggests the possibility of accommodating both the embeddedness of each historical instance and the inextricable historicity of our own moment. In this regard, history is itself a "milieu": an ambient in which we are always implicated. It is therefore far from something that we can objectify as our past "other." Instead, any historical enterprise transforms the past by projecting into it. When Maurice Merleau-Ponty suggests that history is "a milieu of life," he is also articulating this mutual entanglement of being in history and transforming the past through the very act of telling it.[2] Our volume is methodologically indebted to an approach that keeps this ambient nature of any historical endeavor in mind.

Contours of the Field

OVER THE past few decades, scholarly interest in the category of minuscule entities has mainly arisen in relation to histories of microscopic vision. Catherine Wilson's influential study *The Invisible World* (1995) credits the invention of the microscope with providing the main impetus for the emergence of the modern sciences.[3] In her account, the magnifying lens brought into existence a whole set of new phenomena that had "the intractability of fact."[4] Subsequent contributions to the history of microscopy have since extended Wilson's temporal scope to include the late Enlightenment and the first half of the nineteenth century.[5] Thus far, however, minute particulars have been little historicized in terms of "minutiae" beyond the realm of microscopy.[6] While our volume does attend to this technological revolution and its consequences, it is also interested, more broadly, in what is commonly dismissed as too small or too trivial for scrutiny. Even Gottfried Wilhelm Leibniz, who was deeply invested in expanding the scope of particulars that should be taken into consideration, thought of "minutiae" as meaningless if they fell outside of a general scientific system.[7] At the same time, however, Leibniz himself concedes—in no uncertain terms—that seemingly trivial minutiae may be highly significant in domains as wide-ranging as warfare and politics to economy and medicine. Even more importantly, though, he draws far-reaching conclusions from his encounter with the microscopic and the infinitesimal: that worlds and beings are what they are because of minimal differences, that the organic is animated down to its tiniest parts, that small perceptions make for infinitely many connections, and—last but not least—that merely circumstantial trifles can fundamentally alter whole systems of knowledge.[8]

In keeping with Leibniz, we understand minutiae as all that which, in and of itself, falls below the threshold of our conscious attention but nevertheless constitutes and enables the world(s) of perception. For us, these worlds are embodied as well as imagined, empirical as well as narrative. Atoms and viruses, for instance, despite their subsensory nature, create the material, social, and economic realities we inhabit, whereas practices like punctuation and description make possible the diegetic universes to which they are considered extraneous or subordinate.[9] In this respect, our volume investigates cultural and philosophical forms and appearances of minutiae prior to and beyond the advent of magnification. At the same time, our contributors illuminate connections between the empirical practices and technologies with which minutiae have come to be associated and the broader, more diffuse discourses—from the philosophical to the artistic—that have attended aesthetic, epistemological, and ontological smallness before and after Robert Hooke's *Micrographia*.

Since the middle of the twentieth century, there has been ample investigation of the objecthood or "thingness" of minutiae themselves (fragments, ornaments), the metaphorical potential of smallness, and the sublimity and jouissance adjacency of the detail. Groundbreaking work has been done on the vitalism and transcendence of the seemingly trivial or incidental, from Roland Barthes's "ravish[ing]" *punctum* to Michel Leiris's meditations on fetishism to Daniel Arasse's theory of detail in painting.[10] Susan Stewart's iconic reading of miniatures—in particular the dollhouse—as emblems of interiority has further alerted us to the symbolic power and referential energy of the diminutive.[11] In critical conversation with such work, this volume contributes to ongoing efforts to foreground what is easily overlooked, underappreciated, or dismissed. However, the acts of recuperation in the essays that follow are more attuned to the quietude and subtlety of minutiae than to their capacity for rapture or epiphany. Rather than "prick" or enthrall, the minutiae in these pages—from Athanasius Kircher's animalcules to modern microbiology—engage in more modest modes of assertion: they suffuse, envelop, subtend, and element.

Hence instead of following approaches that stress the extraordinariness and singularity of minute things as an antidote to treating them as negligible, our volume builds on a burgeoning field of research that has developed within German academia over the past two decades: it intends to further explore what could be called an analysis of "microstructures," or, perhaps more fittingly, "micrologies."[12] The central concern of this relatively new research field is how the smallest details and tiniest things; trivia and trifles; and minor, marginal, or peripheral occurrences structure forms of knowledge, scientific research, and artistic creation. How have they been conceived of and coped with (artistically, philosophically, and technically) throughout history via diverse theories and practices? Our volume is meant to deepen these existing lines of inquiry within the field of micrology. At the same time, it seeks to make new incursions into this theoretical terrain. Instead of focusing on the "minute" as such or the "detail" per se, this volume draws our attention, first and foremost, to those aspects of minutiae that concern their embeddedness—which is to say, their tendency to occur in the plural, in series, and in collusion with other minutiae. We are also interested, very literally, in their situatedness and relationality. As Rodolphe Gasché points out, "relations" themselves can be considered "minimal things"—minimal in the sense of not small in scale but rather small in being. Medieval philosophy, he reminds us, "believed in the reality of relations." Instead of considering relation merely as the result of mental connections, scholastic philosophy held that relations are things—that is, they have an ontological status of their own. This status is that of an *ens minimum*, understood "as the superlative of *parvus*"—that is, as "something that is excruciatingly small, the smallest of all entities or things."[13] Of course, while substance exists primarily, the relational exists only accidentally. It is not part of the ontological being of something, but it nonetheless is a matter of not just making connections between substances but taking seriously what *appears* to be merely aleatory and granting it more weight. The relationality of minutiae in this sense is at the core of this volume.

Minutiae: Historical Lenses

NO LESS distinguished a philosopher than Tristram Shandy's father was "too refined a researcher" to weigh things "in common scales"; instead, Shandy père maintains that within the world that surrounds us, as well as in observation and scientific methods, everything depends on the smallest of differences: "To come at the exact weight of things in the scientific steel-yard, the fulcrum, he would say, should be almost invisible, to avoid all friction from popular tenets;—without this the minutiae of philosophy, which should always turn the balance, will have no weight at all."[14] What might seem to others like a circumscribed task depends, for Mr. Shandy, on innumerable details that all need to be taken into account. Any deviation from absolute exactness would falsify the results of scientific inquiry: "Knowledge, like matter, he would affirm, was divisible in infinitum;—that the grains and scruples were as much a part of it, as the gravitation of the whole world.—In a word, he would say, error was error,—no matter where it fell,—whether in a fraction,—or a pound,—'twas alike fatal to truth, and she was kept down at the bottom of her well as inevitably by a mistake in the dust of a butterfly's wing,—as in the disk of the sun, the moon, and all the stars of heaven put together."[15] If we understand Mr. Shandy's scientific approach in Laurence Sterne's *Life and Opinions of Tristram Shandy, Gentleman* as a mere travesty of empirical methods, however, we miss its deeper meaning: it is in fact a sophisticated reflection on both the necessity and the paradox of attending to all the perceptible and imperceptible "minutiae" of the world. The novel both succeeds and fails on the basis of this very approach.[16] At the same time, this midcentury literary confrontation and tension demonstrate how deeply invested the eighteenth century is with the issue of the micrological. Contempt for minutiae is as widespread as ambitious attempts to cope with a world that seems ever more "divisible" and replete with the smallest things and beings. Hence instead of arguing that there is a narrative of "progress" to be told about how the philosophical and scientific fortunes of minutiae increased from the late seventeenth century onward,[17] this volume puts another historical approach to the test: Under what conditions do minutiae proliferate at different historical junctures, and what strategies emerge to manage their overwhelming possibilities? Pursuing these questions, it explores the longer history of what could be called an "epistemic thing"—namely, that of minutiae that, at any point in time, verge on the cusp of becoming major objects of scientific inquiry.[18]

Milieu: Historical Contexts

WITHIN THE realm of Newtonian physics, where the "mechanical idea" of *milieu* originated, it was synonymous with *fluid* or *ether*. The term itself made one of its first appearances in Diderot's and Jean le Rond d'Alembert's *Encyclopédie* in the 1750s before being introduced to biology by Jean-Baptiste Lamarck in 1830 and imported to literature by Honoré de Balzac in his 1842 preface to *La Comédie Humaine*.[19] It was over the course of the long nineteenth century that the relationship between minutiae and milieu became an important focus across a broad range of fields, from public health to meteorology. Understanding the ways in which what is literally invisible actually creates, organizes, or constitutes its surrounding environment became normative practice: germ theory, for instance, emerged as the dominant explanatory model for contagious disease, while the advent of modern genetics illuminated the crucial role that DNA molecules play in determining the character and appearance of biological organisms. This led, in turn, not only to revolutionary breakthroughs in areas such as forensics, criminology, sociology, and psychiatry but to new movements in the arts, above all in literature. In his treatise on the fin de siècle school of literary naturalism, *The*

Experimental Novel, Émile Zola compares the techniques employed in this type of narrative fiction to those used in the medical sciences, citing in particular the work of physiologist Claude Bernard—father of the intracellular *milieu intérieur*[20]—as a catalyst for naturalism's "experimental method."[21] The goal of naturalism, according to Zola, is to "dissect piece by piece this human machinery in order to set it going through the influence of the environment," and he makes pointed reference to the discovery of a cure for scabies as exemplary of the ever-growing attunement to the mutual determination of subject and milieu—most notably, to a heightened awareness of how this determination is often mediated through the minute (mites, in this case).[22]

In our volume, the concept of milieu retains its original fluidity, at least in an intellectual sense. Across our essays, it manifests in a variety of shapes and is largely defined precisely by morphological suppleness and flexibility.[23] Along with Georges Canguilhem, we are interested in the shifting semantics of milieu, depending on which syllable is stressed, and all of the milieus discussed in the essays that follow play on the double entendre inherent in the word, which evokes both *middle* (*mi*-lieu) and *medium* (mi-*lieu*). They all function as transient, membranous, or hemihedral spaces that are simultaneously nuclei of sorts: the writer's notebook, the test tube, the cosmopolitan capital, and the typographic plane of the text. Related to this interest in the milieu's multiplicities is Gaston Bachelard's proposition that minutiae are themselves milieus. In the first chapter of *Atomistic Intuitions*, he describes the mesomorphic constitution of minutiae, as instantiated by dust: "The concept of dust, halfway between that of a solid and a liquid, will . . . furnish a sufficiently mixed proof on which to base atomism."[24] Any attempt to parse the histories of minutiae's varied epistemic and cultural impacts will therefore have to take into account that minutiae are never discrete and fixed but only and always aggregate, composite, and changeling—in other words, of a kind with the milieus they create.

Navigating This Volume

INSTEAD OF proceeding in chronological order, this volume presents an array of different approaches to the history of minutiae within their milieus. The first section draws on the aesthetic richness of minutiae between the eighteenth-century novel and Romantic theory, which is complemented by the second section's focus on medical and biological microdevelopments from early modernity to twentieth-century biology. The third section contributes to ongoing efforts to recover philological "minutiae" and the textual surface for critical attention, while the fourth takes up a long-neglected or overlooked literary form, the "metropolitan miniature," within the context of a milieu—namely, the turn-of-the-century city—whose evolution has so far tended to be thought of in monumental terms.

More specifically, the first section of the volume, "Poetics of Minitude," explores minutiae in literature and aesthetic theory from the early eighteenth century to Romanticism. Not only is artistic production and aesthetic theory at this historical juncture deeply invested in particulars and minute things, but there is also an intimate connection between empirical observation of phenomena in nature and society and, for example, mathematical approaches to infinitesimal differences. This section focuses less on the attempt to "define and defend the *minute particular*" within natural philosophy than on the network of the many connections between different forms of scientific and literary practices.[25] At the same time, new modes and methods of approaching the natural and human environment seem to be called for, and the result is a number of (often new and often just slightly revised) epistemic, literary, and aesthetic approaches to viewing and testing the human habitat, the world that surrounds observers who feel confronted with an increasingly differentiated (down to its smallest dimensions) environment.

Roger Maioli, in his contribution on "Particularity and Virtual Witnessing in the Eighteenth-Century British Novel," opens the field by resituating the emergence of the novel from the perspective of an Aristotelian and Baconian epistemology of particularity. But instead of simply reconstructing how the eighteenth-century novel deals with minutiae, Maioli asks what kind of *epistemological* function particulars and details have in prose fiction: "Novels . . . were milieus of minutiae in the sense that they constituted imaginative environments densely punctuated by a web of details." Hence instead of retelling the story that the eighteenth-century British novel aims at producing effects of liveliness through details, Maioli offers a new approach that shows the multifaceted function of minutiae in this early phase of the novel. In order to be able to better understand the historical relevance of narratively magnifying minutiae, Maioli focuses on the *theory* of the novel. Interestingly enough, this also sheds new light on the practices of novel writing and allows Maioli to harness the unprecedented power thence attributed to narrated minutiae: they are meant to enhance capacities of witnessing and observation and were conceived as "routes to general knowledge"—via the medium of the novel and the faculty of imagination. Here, it is the very milieu of the novel that makes possible what no other form could achieve: "raw data to be observed" for those who have not experienced the original version.

It is noteworthy that toward the end of the century, in a different cultural setting, this rehabilitation of particulars that shaped the very idea of the novel's most salient feature and function (and that had a strong impact on other European literatures) will find remarkably different modulations. Around 1800, as Mareike Schildmann is able to show, the scene of the minute is where major shifts in thinking and aesthetics occur for one of the most philosophical (German) Romantic poets—namely, Novalis. The category of what lies beyond immediate human perception is at the heart of the author's concept of the organism while also being the point at which Novalis goes beyond coeval conceptions of nature. For him, the very idea of nature's innermost dynamic, when seen through the lens of minutiae, changes everything. Nothing less is at stake than the question of how the inner world communicates with its outside: in other words, how individual and particular things interact with their milieus. Novalis imagines new life-forms, which lead to new aesthetic strategies and modulations. Schildmann calls these modulations "the principle of infinitely small differences." It is not the essence but the process—not the ontological status of the minute but its relationality, in other words—that constitutes nature and that makes nature's procedures available to human cognition. Both poetic production and mathematical calculus—which are, for Novalis, intimately related—can function only on the basis of this insight.

Across the Alps, as Elena Fabietti shows in "*Le menome cose*: On the Function of Details in Leopardi's *Zibaldone*," the Italian poet Leopardi is simultaneously preoccupied with remarkably similar concerns. While there is, for Leopardi, a continued obsession with minutiae, they gain not only a new relevance but also a new epistemological function and ontological status. This status "assigns the minute detail a crucial function in shaping the knowledge of a larger, general reality," establishing thereby a new kind of "relation between minutiae and milieus." Interestingly enough, the point for Leopardi is not that the world becomes more graspable and understandable through the experimentation of particulars. It is more that what ineluctably constitutes the world wants to be taken into account, while its observation leads to an estrangement. The result is that cognition is precisely no longer considered to be more accurate or more vivid when occupied with minutiae but rather now based on the fundamental idea that "everything is relative" precisely because it is based on "the tiniest [*menome*], incidental everyday differences." As Fabietti is able to show, this will not lead to any totality of vision: there is no analogical harmony between micro- and macrocosmic phenomena. Rather, in Leopardi, "the dialectic relation between the *minutiae* and its *milieu*" exposes a reality that is defined by its irregularity.

The subsequent section, "Micrological Biospheres," examines developments in medicine, natural philosophy, and microbiology from premodernity to the early twentieth century. While aesthetic explorations of minutiae may find a particularly significant historical momentum in the aftermath of microscopic discoveries and then again around 1800, philosophical and scientific endeavors to uncover the work of micrological entities or the effects of minute beings have a long and complicated history that dates back to antiquity. That the tiniest and subvisible particles—as living or nonliving elements—might have an impact on the development of diseases is an idea with a long history. What is often attributed to seventeenth-century atomism, materialism, or microscopism—to philosophers such as Pierre Gassendi or Leibniz—indeed calls for a fresh profile. This part of our volume therefore turns its attention to underexplored paths in the multitudinous historical trajectories of seemingly irrelevant matter and effects occurring on the level of *subvisibilia*.

Carmen Schmechel's medicohistorical contribution "Landscapes of Disease and Contagion" focuses on putrefaction and fermentation in early modern medicine, drawing on the Aristotelian and Galenic "miasma theory of disease." This theory held that certain milieus with natural occurrences, such as marshes or viscid air, could encourage disease. They are, as it were, hotbeds for illness and infection. In medical corpuscular theory, fermentation and putrefaction were seen as those natural processes happening in these milieus. Now, it is remarkable how theories of contagion from Fracastoro to Kircher negotiate the ancient approach. For the first time, it was assumed that minutiae—in this case, the tiniest microbeings (vermicelli, for example)—were causing and carrying disease. The reason was seen in the propitious milieu favoring putrefaction. Schmechel's contribution not only retraces these important early modern medical developments; it also parses their entanglement with the hybrid ontology of matter and spirit modeled by van Helmont.

In his essay on "Experimental Environments," science historian Hans-Jörg Rheinberger formulates, in continuation of his long-standing work on the modern history of biology,[26] a programmatic and exemplary "micrologies of knowledge"—that is, a historicoepistemical praxeology—of the major impact of "minutiae of method" (Claude Bernard) on both the setting and the performance of any experimental investigation. With a focus on the most consequential stages of in vitro experimentation from the 1930s on, Rheinberger is able to show how, at different moments, the minutest details in the procedures of these experiments eventually lead to major shifts in both modes and results of analysis. With the availability of ultracentrifuges, fractionation of cellular contents via high-speed centrifugation became not only possible but the dominant mode of proceeding. In the 1950s, a hybrid molecule was discovered, again by way of a minor shift in method, which led to a number of further experiments, all trying to repeat the conditions of previous experimental settings. And yet again triggered by minutest and unintended divergences—in Marshall Nirenberg and Heinrich Matthaei's experimentation in the early 1960s—the "first genetic code word, that is, a triplet sequence of ribonucleic acid that codes for a particular amino acid" was identified. A microhistorical close-up is indispensable for these insights, and it remains to be seen how the history of molecular biology—if told from the perspective of minutiae—can serve as an example for research in other fields of knowledge production.

Daniel Liu's essay explores the centrality of what he calls "the metaphysical problem of detail" to the history of microscopy. He focuses on the significance of Ernst Abbe and Hermann Helmholtz's discovery, in 1873, that the microscope functioned differently from the telescope, whose governing principles were magnification and image brightness alone. Abbe and Helmholtz recognized that light diffracts as it penetrates the specimen being observed and moves into the objective lens—in other words, that "the fine details of the specimen itself affects light in complex ways before the objective lens captures it." The implications of this insight were profound: it meant, essentially, that there was a firm threshold beyond which the "realm of the microscopically visible" could not extend. Past the 250 nm Abbe diffraction limit, increased magnification would not produce an image with greater or more vivid detail. Liu's essay considers the ways in which emphasizing the contrast between detail and context (or minutiae and milieu) became the primary technique through which microscopic vision was improved in the postdiffraction limit era. The resolution of detail thus became one of the key strengths of microscopy over the course of the late nineteenth and twentieth centuries—a counterintuitive one, given that the microscope was originally designed to cater to the desire for (semi-)immediate, close-up scrutiny. Liu concludes by touching on the ways in which the digital age has further impacted the ontological and epistemological status of the microscopic detail by making "every resolvable detail in a given image . . . quantified and calculable" (18). Detail has once again become potentially overwhelming, and there is a new urgency around the question of restoring the balance between the infinite minutiae to which modern technology gives us perpetual, unfettered access and the milieus without which they cease to signify.

In our third section, "Philological Minima," we move from the world of the laboratory to that of the literary and artistic studio. As a discipline, classical philology has always been associated with minutiae, and the advent of putatively antithetical methodologies, from poststructuralism to the hermeneutics of suspicion to surface reading, has only continued (if toward amended ends) its passion for close reading and its commitment to redeeming the "bad object" of the detail into a legitimate site of critical and interpretive labor.[27] The essays in this section all, in different ways, engage with questions of scrutinizing and valorizing the textual "skin," whether they deal (metaphorically) with the literal epidermis and the possibility of its penetration (Andreas Mahler) or with the printed page as figurative integument, pierced by colons and dashes (Cynthia Wall). Each explores *milieu* as a space that eludes what could be considered a false dichotomy between depth and surface. From Donne's implicit analogy, always shimmering just below the level of direct enunciation, neither fully buried nor openly exposed, to Defoe's parentheticals, which create a textual alcove (or vulva) of sorts—equivalent neither to full disclosure nor to concealment—to the metaphors, facts, and starting points that Curtius, Spitzer, and Auerbach deploy as textual buoys, all touch in different ways on Nietzsche's assertion, which Werner Hamacher cites in his *Minima Philologica*, that "philology . . . is a goldsmith's art and connoisseurship of the word which has nothing but delicate, cautious work to do and achieves nothing if it does not achieve it lento."[28]

Andreas Mahler begins with John Donne's erotic poems to consider how narrative contexts unleash the hermeneutic power and significance of seemingly inconsequential beings or incidents, notably the minuscule insect at the center of one of Donne's most iconic sonnets. By equating a bug bite with sexual intercourse, the speaker of "The Flea" attempts to rhetorically minimize the act of vaginal penetration in an attempt to seduce a young virgin. But rather than succeed in diminishing the moral and cultural magnitude of extramarital intimacy by comparing it to that shared by two people who have had their blood sucked by the same millimetric hematophage, the speaker in fact showcases the absurdity of reducing coital relations to the mere exchange of bodily fluids. In its three brief stanzas, Mahler argues, "The Flea" "ceaselessly creat[es] imaginary *milieus* in which even trifles or almost invisible *minutiae* . . . can infinitely be invested with new meanings," thereby "explor[ing], and negotiat[ing], the epistemological ways and means—as well as the epistemic conditions—by and under which we all continuously make, and abandon, cultural truths."

Cynthia Wall's essay proceeds to take up a similar question concerning the ontological impact of seeming minutiae—namely, that of punctuation on its rhetorical, grammatical, social, textual, and even architectural surroundings. For Wall, "The parenthesis—both verbal and typographical—is, in fact, a tiny detail carving out precision, modestly cloaking itself as a small or trivial matter, not meaning to interrupt the Important Sentence—but of course its very existence interrupts the Important Sentence." She examines the use of parentheses (colloquially known as *lunulae*, or little moons) over the course of the eighteenth century, from Daniel Defoe's *Robinson Crusoe*, first published in 1719, to Lt. William Bligh's 1789 mutiny on the HMS *Bounty*, arguing that the insertion of a parenthetical leaves a text fundamentally altered, however seemingly meaningless the information it contains. For Wall, "Every sentence has an identity made up of typographic as well as syntactic gestures. The contours of letter, line, spacing, pointing; the grammatical shifts, paratactic accumulation, hypotactic organization, the rising status of the preposition—all provide the landscape of *narrative*," and punctuation in particular "underwent a major landscaping in the eighteenth century." In Wall's account, textual topology is both aesthetic and structural: it instantiates the inextricable nature of surface and depth in a narrative context.

Finally, Christopher D. Johnson reflects on the ways in which the "philological detail" engenders affective ties between text and reader, thus bridging the gap between the role minutiae play within the context of Elizabethan love poetry and the position they occupy within twentieth-century philological practice. The broad scope of Johnson's essay encompasses Ernst Robert "Curtius's collection of topics and metaphors, Leo Spitzer's explications of facts, and Erich Auerbach's 'arbitrary' *Ansatzpunkte*" to argue that all invest minutiae with the power to make sense of—and determine the sensibility of—the whole. Central to Johnson's thesis is the notion that "the *love* etymologically embedded in *philology* . . . proves a bittersweet, if not melancholy one" as these scholars navigated the shifting milieu of postwar literary studies through the meticulous handling of detail.

Our final section, "Minimilieus of Modernity," takes up what Andreas Huyssen calls the "metropolitan miniature" in terms of both content and context. According to Huyssen, a unique (and peculiarly German) tradition of short prose emerged in the early twentieth century, alongside the rise of the modern city. Huyssen focuses not only on the ways in which these innovations revolutionized the setting in which the "miniature" was being produced but also on the dramatic shifts in readerly sensibility and perspective that they occasioned. The formal compactness of the miniature—and the critical negligibility that has attended it—is thus a function, by Huyssen's lights, of the tremendous magnitude of the cultural changes in whose midst it emerged. As opposed to his emphasis on the enormity of the changes that precipitated the miniature's emergence, the authors in this section are interested in exploring the subtleties and gradations of the cityscape's metamorphosis during this period (Malte Rauch) and in ideas about minor-scale domestic dwelling that were, paradoxically, informed by Bauhaus training in harnessing the capacities of mass production to construct minimalist buildings on a grand scale (Margareta Christian). Finally, Marianne Schuller turns to two exemplary literary miniatures—namely, Kleist's essay on the "gradual" and Kafka's late prose text "Josefine the Singer"—to demonstrate the a-chronicity of "modernity" and the elusive affinities by which it coheres.

Malte Rauch starts off by examining "microambiences," paying meticulous attention to the ways in which minutiae underpin the milieu in which the miniature thrives. Rauch's essay focuses on the idea of microambiences as it was developed by the French surrealist Walter Benjamin and subsequent avant-garde artists and theoreticians like Constant and Guy Debord. In different ways, they conceived of extremely subtle nuances—humidity of the air, intensities of light, density of smell, aural textures, the rhythms of specific streetscapes—as defining a city's atmosphere: "The construction of ambiances is thus conceived as the modulation of the most minute elements in (sensorial) experience." While the surrealists explored this stratum of atmospheric minutiae through serendipitous individual encounters and aimless metropolitan strolls, Constant and Debord took the attempt to trace these atmospheric minutiae much further, producing collective maps that became the basis for the construction of an entire ecosystem of new atmospheres through specific architectural forms. Rauch's essay argues that the theorization of microambiences in the 1950s and 1960s foreshadows some of the most exciting contemporary developments around "ecology," "atmosphere," "environment," and "milieu" in aesthetics and media studies, ultimately suggesting that "the interest in the most minute elements of urban space, understood as the constituents of the atmospheric tissue, runs like a guiding thread from the Surrealist aesthetics to Constant and Debord's plans for reconstructing social space."

Margareta Ingrid Christian then turns to the realm of the built environment, looking at how Siegfried Ebeling (1894–1963), a little-known Bauhaus student and architectural utopian, conceived of the ideal dwelling place as one modeled on a cell—and thus premised on the very tension between surface and depth (outside and inside) taken up in the previous section. In his famous tract, *Space as Membrane* (1926), Ebeling argues that a house should function like a miniature environment within the world at large. Drawing on cell biology—in particular, the work of the microbiologist Raoul Heinrich Francé—and theories of the microcosm, Ebeling suggests that architectural space should mimic a biological milieu. His ideal built environment is modeled on the minute foundations of existence itself: cells and their constitutive parts (nucleus, plasma, and membrane). Yet even as *Space as Membrane* borrows from the phenomenon of cellular osmosis, it is animated by a logic of isolation. Ebeling uses biology as a model for engineering, yet he does so in order to emancipate humans from nature. His architecture is paradoxically bent on undoing space; it is therefore a vision of a built environment committed to nullifying the very idea of environment itself.

Finally, Marianne Schuller addresses the "metropolitan miniature" itself, pairing two stories by Kleist and Kafka and concluding the volume with an essay that takes the literary text as a milieu for minutiae. "Literature as a Milieu of the Small: Kleist/Kafka" draws on her field-defining coauthored book *Mikrologien*, which was formative for our project from its inception, to explore new ways of deciphering the micrological. Her contribution links back to the literary-historical concerns that animate the first essays, focusing on texts by two of German literature's most "modernist" authors. Instead of concentrating merely on small forms or the appearance of tiny things in texts, it deals with modern literature *as* a milieu of minutiae. Literature thus becomes an ambience that not only attends to the small but engages with the micrological in the very sphere it thereby becomes. Literature embeds and thrives on minutiae, weaving small occurrences into its very texture while also bringing minutiae and infinitely small distinctions to life. In a transhistorical reading that provocatively juxtaposes Kleist's early nineteenth-century essay, "Gradual Formation of Thoughts" (1805–6) and Franz Kafka's last short story, "Josefine the Singer, or The People of the Mice," from 1924, Schuller is able to show to what extent minute incidents can lead to major events or how infinitely small differences engender extravagant claims of power. Modernism's minima, in this insightful literary-historical constellation, thereby become legible through the *petites perceptions* that undermine the grand narrative of modernity to which we have become so accustomed.

A collected volume is itself a milieu, with its own distinctive contours, tone, and "feel." We want to conclude these opening remarks by acknowledging that each reader will derive unique sustenance from the one we have created here. The breadth of the essays gathered together in these pages is a testament not only to the magnitude of minutiae and milieu as intellectual categories but to the way in which the act of collectively articulating the importance of those categories must inevitably entail their comminution. Like ore refined in a mill, each essay offers a granular perspective on the total deposit. It is our hope that each reader, precisely in finding their own discrete point(s) of salience within the arguments that follow, emerges with an enriched appreciation of our overall topic that serves to energize future explorations of it. As the concept of relevance itself becomes an object of inquiry in the Humanities and Social Sciences, whose institutional survival increasingly depends on making a persuasive case for their own,[29] *Milieus of*

Minutiae provides a valuable contribution to conversations around what is worthy of notice. By showing how we can only make sense of the worlds we inhabit, whether in the flesh or on the page, by seriously attending to what is superficially superfluous or ostensibly insensible, it compels us to reconsider how we might more effectively (or at least more deliberately) distribute the ever-more-endangered resources of our time and attention.

Notes

1. Maurice Merleau-Ponty, *Résumés de cours: Collège de France 1952–1960* (Paris: Gallimard, 1968), 44–50. For Merleau-Ponty, all "instances" one could ever want to oppose to history have a history in and of themselves. Any transformation of the past via later conceptions supposes "une entente ... entre le présent et le passé" (50). At the same time, every present moment remains unrealized (45) and thereby yields itself to the future. This, for Merleau-Ponty, is not just a matter of ideas or events but intrinsically spiritual and material.
2. Ibid., 45.
3. Catherine Wilson, *The Invisible World: Early Modern Philosophy and the Invention of the Microscope* (Princeton: Princeton University Press, 1995).
4. Ibid., 252.
5. Most notably Jutta Schickore, *The Microscope and the Eye: A History of Reflections, 1740–1870* (Chicago: University of Chicago Press, 2007); Marc J. Ratcliff, *The Quest for the Invisible: Microscopy in the Enlightenment* (Oxfordshire, UK: Routledge, 2009); and Tita Chico, "Minute Particulars: Microscopy and Eighteenth-Century Narrative," *Mosaic: A Journal for the Interdisciplinary Study of Literature* 39, no. 2 (2006): 143–61. From a different angle and going beyond microscopy, see also Christiane Frey, "The Art of Observing the Small: On the Borders of the 'Subvisibilia' (from Hooke to Brockes)," *Monatshefte für deutschsprachige Literatur und Kultur* 103, no. 3 (2013): 376–88.
6. Obviously, this realm is vast and encompasses numerous subfields, including rich areas of research from the long history of atomism to quantum physics. See, among many others, Antonio Clericuzio, *Elements, Principles and Corpuscles: A Study of Atomism and Chemistry in the Seventeenth Century* (Berlin: Springer, 2000); for histories and theories of quantum physics, see, e.g., Karen Barad, *Meeting the Universe Halfway: Quantum Physics and the Entanglement of Matter and Meaning* (Durham: Duke University Press, 2007); Jimena Canales, "Modern Physics: From Crisis to Crisis," in *The Cambridge History of Modern European Thought*, vol. 2 (2019); or James Vincent, *Beyond Measure: The Hidden History of Measurement from Cubits to Quantum Constants* (New York: W. W. Norton, 2021).
7. Gottfried Wilhelm Leibniz, *Paraenesis de scientia generali*, 1688 (A VI 4), 979–80. See Enrico Pasini, "Leibniz and Minutiae," in "*Für unser Glück oder*

das Glück anderer": *Vorträge des X. Internationalen Leibniz-Kongresses Hannover, 18.–23. Juli 2016*, ed. Wenchao Li (Hildesheim: Georg Olms Verlag, 2017), 697–706; and Herbert Breger, "On the Grain of Sand and Heaven's Infinity," in Li, *"Für unser Glück oder das Glück anderer,"* 64–79.

8. Pasini, 697. See also Christiane Frey, "Kleinformate und Monadologie: Leibniz, Benjamin," in *Barock en miniature—Kleine literarische Formen in Barock und Moderne*, ed. Matthias Müller, Nils C. Ritter, and Pauline Selbig (New York: De Gruyter, 2021), 135–76.

9. For a concise explanation of punctuation's historical affiliation with style over content, see Elizabeth Bonapfel and Tim Conley, introduction to *Doubtful Points: Joyce and Punctuation* (Amsterdam: Brill, 2014), 1–9. For an account of novel theory's tendency to treat description as inferior to the "main" narrative priorities of plot and characterization, see Dora Zhang, "Toward a Theory of Narrative Description," in *Strange Likeness* (Chicago: University of Chicago Press, 2020), 35–60. For further explorations of description in prose fiction and painting, see Gérard Genette, "Frontiers of Narrative," in *Figures of Literary Discourse*, trans. Alan Sheridan (New York: Columbia University Press, 1982), 127–44; Roland Barthes, "The Reality Effect," in *The Rustle of Language*, trans. Richard Howard (Oxford: Basil Blackwell, 1986), 141–48; Cynthia Wall, *The Prose of Things: Transformations of Description in the Eighteenth Century* (Chicago: University of Chicago Press, 2006); and Svetlana Alpers, *The Art of Describing: Dutch Art in the Seventeenth Century* (Chicago: University of Chicago Press, 1983).

10. See Roland Barthes, *Camera Lucida: Reflections on Photography* (New York: Hill and Wang, 1981); Michel Leiris, *The Ribbon at Olympia's Throat* (Cambridge: MIT Press, 2019); and Daniel Arasse, *Le Détail: Pour Une Histoire Rapprochée de la Peinture* (Paris: Flammarion, 1992).

11. See Susan Stewart, *On Longing: Narratives of the Miniature, the Gigantic, the Souvenir, the Collection* (Baltimore: Johns Hopkins University Press, 1984).

12. Wolfgang Schäffner, Thomas Macho, and Sigrid Weigel, eds., *"Der liebe Gott steckt im Detail": Mikrostrukturen des Wissens* (Munich: Fink, 2003); Marianne Schuller and Gunnar Schmidt, *Mikrologien: Literarische und philosophische Figuren des Kleinen*, 2nd ed. (Bielefeld, Germany: transcript, 2015).

13. Rodolphe Gasché, *Of Minimal Things: Studies on the Notion of Relation* (Stanford, CA: Stanford University Press, 1999), 2.

14. Laurence Sterne, *The Life and Opinions of Tristram Shandy, Gentleman* II.19, Vol. 1 (London, 1761), 162.

15. Ibid., 162.

16. Tita Chico, "'The More I Write, the More I Shall Have to Write': The Many Beginnings of Tristram Shandy," in *Narrative Beginnings: Theories and Practices*, ed. Brian Richardson (Lincoln: University of Nebraska Press, 2008), 83–95. For further inquiries that follow similar trajectories, see also Tita Chico, *The Experimental Imagination: Literary Knowledge and Science in the British*

Enlightenment (Stanford, CA: Stanford University Press, 2018); Roger Maioli, *Empiricism and the Early Theory of the Novel* (London: Palgrave Macmillan, 2016); Jenny Davidson, "The Minute Particular in Life Writing and the Novel," *Eighteenth-Century Studies* 48, no. 3 (2015): 263–81; and Helen Thompson, *Fictional Matter: Empiricism, Corpuscles, and the Novel* (Philadelphia: University of Pennsylvania Press, 2017).

17. Lorraine Daston, "Attention and the Values of Nature in the Enlightenment," in *The Moral Authority of Nature*, ed. Daston/Vidal (Chicago: University of Chicago Press, 2004), 100–126.
18. See, for example, Hans-Jörg Rheinberger, *Toward a History of Epistemic Things* (Stanford, CA: Stanford University Press, 1997).
19. See Georges Canguilhem, "The Living and Its Milieu," trans. John Savage, *Grey Room*, no. 3 (Spring 2001): 7–31. For further references, see also Florian Huber and Christina Wessely, "Milieu: Zirkulation und Transformation eines Begriffs," in *Milieu: Umgebungen des Lebendigen in der Moderne*, ed. Florian Huber and Christina Wessely (Paderborn: Wilhelm Fink, 2017), 7–17.
20. Claude Bernard, *Leçons sur les phénomènes de la vie, communs aux animaux et aux végétaux* (Paris: Baillière, 1878; Paris: J. Vrin, 1966), 113.
21. See Émile Zola, "The Experimental Method," in *Documents of Modern Literary Realism*, ed. Joseph Becker (Princeton: Princeton University Press, 2015): 161–96.
22. Ibid., 176.
23. See also Leo Spitzer, "Milieu and Ambiance: An Essay in Historical Semantics," *Philosophy and Phenomenological Research* 3, no. 1 (1942): 1–42; and no. 2 (1942): 169–218.
24. See Gaston Bachelard, "The Metaphysics of Dust," in *Atomistic Intuitions: An Essay on Classification*, trans. Roch C. Smith (Albany: SUNY Press, 2019), 13–26, here 15.
25. Chico, "Minute Particulars," 144.
26. See, for example, Hans-Jörg Rheinberger, *An Epistemology of the Concrete: Twentieth-Century Histories of Life* (Durham: Duke University Press, 2010).
27. See, for example, Naomi Schor, *Reading in Detail: Aesthetics and the Feminine* (New York: Methuen, 1987); Sharon Marcus and Stephen Best, "Surface Reading: An Introduction," *Representations* 108, no. 1 (Fall 2009): 1–21.
28. Werner Hamacher, *Minima Philologica* (Fordham University Press, 2015), 62.
29. See, for example, Elisa Tamarkin, *Apropos of Something: A History of Irrelevance and Relevance* (Chicago: University of Chicago Press, 2022); Abraham Flexner, *The Usefulness of Useless Knowledge* (Princeton: Princeton University Pres, 2017); Nuccio Ordine, *The Usefulness of the Useless*, trans. Alastair McEwen (Philadelphia: Paul Dry Books, 2017); Martin Savransky, *The Adventure of Relevance: An Ethics of Social Inquiry* (Berlin: Springer, 2016); Helen Small, *The Value of the Humanities* (Oxford: OUP, 2013).

PART ONE

Poetics of Minitude

✦ ✦ ✦

Particularity and Virtual Witnessing in the Eighteenth-Century British Novel

ROGER MAIOLI

✦ ✦ ✦

For a long time in Western aesthetics, particulars were unfavorably compared to their conceptual counterpart, universals. Aristotle, founding a common line of critique, declared that "poetry is a more philosophical and more serious thing than history" on the grounds that "poetry tends to speak of universals, history of particulars."[1] The vast majority of early modern commentators repeated Aristotle's dictum with approval,[2] and we can still hear its powerful echo in one of the most famous statements in English literary criticism. "The business of a poet," says Imlac in Samuel Johnson's *Rasselas*, is "to remark general properties and large appearances; he does not number the streaks of the tulip, or describe the different shades in the verdure of the forest."[3] Particulars, for Imlac, are trivial details that poetry had better gloss over—a common imperative in neoclassical poetics. Johnson himself did not always endorse this principle,[4] and his unsettledness reflects a change in the critical fortunes of particulars. John Hawkesworth, who agreed with Johnson on a lot else, found that "general properties and large appearances" could be rather dull. "An account of ten thousand men perished in a battle," Hawkesworth wrote, "is read ... without the least emotion, by those who feel themselves strongly interested even for Pamela, the imaginary heroine of a novel that is remarkable for the enumeration of particulars."[5] Hawkesworth was not alone in taking the side of particulars. In eighteenth-century Britain, the Aristotelian and neoclassical stance coexisted with a growing appreciation for minutiae and revealing detail, especially as featured in realist fiction and the Dutch school of painting. These two views stood in an uneasy relationship with each other, as best seen in the opposition between the prominent painter and art critic Sir Joshua Reynolds and the then obscure poet William

Blake. For Reynolds, the painter who labors over detail "by regarding minute particularities" will only "pollute his canvas with deformity."[6] Blake, who dismissed Reynolds's views as sheer nonsense, asserted instead that "Art and Science cannot exist but in minutely organized Particulars."[7] Between these opposite poles lay an evolving and complex range of defenses and dismissals of particularity.

This essay provides an account of these transformations in order to achieve two goals. On the one hand, I reconstitute a historical background that is relevant to all essays in this unit. Like Mareike Schildmann's discussion of Novalis and Elena Fabietti's account of Giacomo Leopardi, I am interested in how assessments of particularity ebbed and flowed along with shifting assumptions in epistemology and ontology. The eighteenth century was a crucial stage in such shifts. It constitutes the pre-Romantic background that Novalis will react against, serving as well as the genealogy leading up to Leopardi's pessimistic empiricism. Covering contiguous historical periods in different national contexts, the essays in this unit are all concerned with particulars as iterations of the very small—the infinitesimally small, in Novalis; Leopardi's *menome cose*; or what Blake and Reynolds, as well as a number of other eighteenth-century authors, referred to as "minute particulars." This term registers the historical perception that the shift to particulars is also a shift toward minutiae. Not all particulars are small in scale—London is a particular place but by no means a small one—and yet such usage makes sense given the history of representational genres. Specifying the setting of a novel down to a single city or neighborhood narrows down the almost horizonless settings of older genres such as the allegory, morality plays, epic, or romance. To particularize was to constrain. But the smaller world of the novel added weight to its contents. As Hawkesworth and many others realized, Pamela's domestic drama bestowed value on seemingly inconsequential details. Here, accordingly, is another assumption common to all essays in this unit: that modern narrative modes operated as milieus that qualitatively transformed their contents, however minute. The realist novel is a representational environment densely punctuated by a web of details that gain resonance through ecological juxtaposition.

In retracing the critical fortunes of particularity, I am also pursuing a second goal: to intervene in an unresolved debate concerning the epistemic value of the novel. For eighteenth-century British commentators, particularity in the arts could serve two different purposes, one aesthetic and another epistemic. Those interested in the aesthetic value of particulars gave pride of place to art criticism and neoclassical genres like epic and tragedy, arguing that particularity yielded vividness. "It is by means of circumstances and particulars properly chosen," wrote Hugh Blair, "that a narration becomes interesting and affecting to the Reader."[8] Studies on the aesthetics of particulars have by and large concurred on the essential features of this critical tradition.[9] Conversely, there is less agreement among modern scholars regarding the epistemic value of particularity. One dominant thesis, indebted to Ian Watt, is that the novel's particularity brought fiction closer to the empirical world. On this reading, minutiae regarding locale, furniture, the weather, dates, prices, surnames, incomes, meals, sports, individual idiosyncrasies, minor foibles, facial traits, or physical characteristics increased the novel's faithfulness to social reality. The epistemic value of novels, accordingly, is that they shed light on the empirical world they represent. As Joseph Drury explains, "The increasing legitimacy of the novel as an instrument of philosophical knowledge derived in part . . . from the emerging sense of the basic similarity of works of nature and works of art, which made it newly plausible that the machinery of a romance might reveal something profound about the contrivances of nature."[10] An influential version of this theory, articulated by John Bender, claims that novels enable readers to "virtually witness" imaginary events. Novel readers, for Bender, are "surrogate observers . . . who could not be present at an actual experiment but can come to lend credence to it and to participate in discussion of it through representation."[11]

This assessment of the novel's epistemic value underwent substantive reappraisal in recent years. As Jenny Davidson rightly points out, the eighteenth-century novel is far from being as minutely detailed as its nineteenth-century successor;[12] recognition of such limits has brought renewed attention to the many ways in which eighteenth-century fiction remained invested in pre-empirical narrative modes that prioritized universality over particularity.[13] Conversely, scholars interested in the novel's indebtedness to empiricism have challenged the "virtual witnessing" metaphor by providing a different account of how particulars operated both in empiricism and in the novel. Tita Chico has shown that early experimental philosophers often worried about the uncertain status of minute particulars, as the detection and articulation of particular features of nature involved acts of the imagination that made the description nontrivially different from the real thing.[14] The particulars of empiricist science, for Chico, were understood to be produced as much as witnessed. Building on this insight, Helen Thompson recently challenged Bender's application of virtual witnessing to the novel.[15] In Thompson's view, virtual witnessing implies a questionable detachment between observers and observed, mistakenly assuming that what ends up on the page is "the mimetic imitation of transparently encountered things in the world" (20). The epistemic value of the novel, for Thompson, is not that they represent the observed world in all its minute particularity but that they disclose the mental processes by which empirical facts come into being.

I will have more to say about this debate over the course of this essay. For now, it is enough to conclude that such disagreements leave us with an unresolved question: Can we still speak of the eighteenth-century novel as a surrogate for empirical observation? Not everybody is ready to say no— Anne M. Thell recently gave virtual witnessing a new spin by linking it to digital virtual reality[16]—but Chico and Thompson's challenge is robust enough to give us pause. I propose that to address this impasse, we distinguish between (1) the assumptions of early modern natural philosophy as reconstituted by a recent historiography of science and (2) the popular reception of empiricism by eighteenth-century literary critics and novelists. The theorization of minute particulars, which Chico and Thompson reveal to have been rather complex when coming from microscopists and experimentalists, tended to be more straightforward in eighteenth-century writings on fiction. Keeping in view this distinction between original articulation and popular reception enables us to reconcile these two approaches, accepting Chico and Thompson's reassessment of particularity in science while retaining Bender's application of virtual witnessing to the novel. As will be seen, eighteenth-century commentators did speak of the novel as a tool for observing events in absentia, and they grounded their claims on the novelist's premium on particularity. Hence the double purpose of this essay: to speak of novels as milieus of minutiae is simultaneously to assess their epistemic value as reports on the world of experience.

Before turning to my main task, let me offer two considerations about method. First, I will pay more attention to theories of particularity than to the practice of writing fiction, reflecting the approach I followed in *Empiricism and the Early Theory of the Novel* (2016). The focus on practice has undeniable advantages, since what Aphra Behn or Daniel Defoe said about their novels was usually a far cry from what they actually did as novelists. But it is often difficult to identify, in the absence of theoretical statements, the assumptions guiding a novelist's handling of minutiae. After all, the procedures of the novel were heavily overdetermined, emerging from sources of influence as various as English folklore, Aristotelian poetics, Restoration drama, experimental science, journalism, religious literature, the periodical essay, and the vagaries of buyers' tastes.[17] Because practice can be variously construed, we may risk substituting our understanding of cultural influences for the values and assumptions of historical agents who may have had different motivations or a different conception of their purposes. The advantage of focusing on theories of fiction is that they reveal how practitioners of literary minutiae processed their intellectual inheritances. Defenses of particularity always tell us something about the perceived values of a culture, especially when they happen to conflict with the way authors actually wrote. Secondly, the account that follows combines an attention to fiction and nonfiction, to the eighteenth century and earlier, and to major and minor writers. Even though my ultimate focus is on the novel, the way eighteenth-century novelists conceived of particularity built on debates taking place since the sixteenth century across a wide variety of other genres, and my first task will be to recover that broad background as well as the role of empiricism in it. In the interest of illustrating how diversified these debates were, I prioritize primary sources less frequently featured in the existing scholarship. I will say less about Johnson and Reynolds than about figures like Thomas Taylor, Joseph Priestley, and Mary Hays. While the latter enjoyed a smaller audience than the former, the very fact that their views have been less extensively documented means that they can shed new light on the debate as we currently know it.

Aristotle, Bacon, and the Critical Fortunes of Particulars

I BEGIN with the contrast between two different assessments of narrative particularity: that of neoclassicism, with its Platonic and Aristotelian parentage, and that of the British empiricists. Emerging in the late sixteenth century, this opposition accompanied a shared understanding of the genres of human knowledge. Neoclassical poetics, as well as Baconian empiricism, routinely distinguished between three spheres of learning: philosophy, history, and poetry. Sir Philip Sidney, in his Aristotelian *Defense of Poesy* (1595), explained the difference by saying that philosophers offer precepts, historians offer examples, and only "the peerless poet performs both."[18] Bacon, as later British empiricists and the French *encyclopédistes* led by Diderot and d'Alembert, would map this distinction onto "the three partes of Mans understanding, which is the seate of Learning: HISTORY to his MEMORY, POESIE to his IMAGINATION, and PHILOSOPHY to his REASON."[19] Both Sidney and Bacon agreed that poetry depicts worlds that transcend the world of experience and that there is value in poetical contemplation. Poetry, for Bacon, "doth raise and erect the mind, by submitting the shews of things to the desires of the mind," and its purpose is "to give some shadow of satisfaction to the mind of Man in those points, wherein the nature of things doth deny it."[20] Sidney, analogously, declares that "the poet ... does grow in effect into another nature in making things either better than nature brings forth or, quite anew, forms such as never were in nature."[21] Beyond this agreement, however, lie very different assessments of the epistemological value of philosophy, history, and poetry. For Bacon, with the exception of the "divine poesy" of scripture, the products of the imagination have recreational rather than epistemic value; "imagination hardly produces sciences; poesy ... being to be accounted rather as a pleasure or play of wit than a science."[22] For Sidney, by contrast, poetry is an unsurpassed source of wisdom, a window into the realm of universal truths, while history is a mere account of particulars. And particulars, for Aristotle and his Renaissance commentators, are inferior to universals as a source of knowledge.

For the Aristotelian tradition, there were at least two reasons for preferring universals over particulars. The first had to do with the applicability of narrative examples. In Sidney's reading of Aristotle, those who draw inferences from the particular events of history reason "as if he should argue because it rained yesterday, therefore it should rain today."[23] The uselessness of such inferences was stressed by other Aristotelian critics, such as the French classicist André Dacier:

> History can instruct no further than the Facts it relates give an opportunity, and as those Facts are particular, it very rarely happens that they are suitable to those who read them, and there is not one of a Thousand to whom they agree; and those to whom they do agree, have not perhaps in all their Lives two occasions, on which they can draw any advantage from what they have Read. 'Tis not so with Poetry, that keeps close to Generals, and it is so much the more Instructive and Moral, as General things surpass Particulars; these agree to one only, those to all the World.[24]

Originally published in 1692, Dacier's commentary on the *Poetics* circulated in English translation starting in 1705, but his objection to particularity had already reappeared verbatim, without acknowledgment, in John Dennis's *Remarks on a Book entituled, Prince Arthur* (1696). Like Dacier, Dennis bases his arguments not only on a dismissal of inferences from particulars but also on an additional principle—the Platonic view of the sensible world as an imperfect copy of ideas. This was a second reason for favoring universality. "Particular Men," Dennis writes, "are but Copies and imperfect Copies of the great universal Pattern."[25] Writers who focus on particulars not only offer useless examples that readers will never apply; they also treat distorted instantiations of reality as their original models.

Here, in short, is Plato's well-known view of poetry as a copy of imperfect copies. Defenders of poetry who appealed to universals found ways of avoiding Plato's pessimism by reviving a Neoplatonic argument: that poets represent not the imperfect copies that we know through perception but the perfect ideas we know through intellection.[26] A concise statement of this view appears in Richard Hurd's commentary on Horace (1749): "In deviating from particular and partial, the poet more faithfully imitates universal truth. And thus an answer occurs to that refined argument, which Plato invented and urged . . . against poetry. . . . For, by abstracting from existences all that peculiarly respects and discriminates the *individual*, the poet's conception . . . reflects the divine archetypal idea, and so becomes itself the force of that unusual encomium on poetry by the great critic, *that it is something more severe and philosophical than history.*"[27] This argument reconciles Plato's archetypes and Aristotle's defense of poetic universality; in the process, it fuses two very different theories into a single case against particularity. Taken together, the Aristotelian concern that particular examples reveal little beyond themselves and the Platonic dismissal of particulars as debased copies of the true reality would be championed by generations of critics, forming the backbone of eighteenth-century epistemological objections to particularity in literature.[28]

Meanwhile, the years that separate Sidney from Hurd also witnessed the elaboration of a radically different conception of particularity, one that would invert the Aristotelian hierarchy of genres. These developments initially took place in the domains of law and natural philosophy before they affected literary criticism.[29] They involved an epistemological emphasis on sense experience and reliable testimony as well as a different understanding of how one arrives at general truths about God, humankind, and nature via the premise that everything that exists is particular. The founding figures of this tradition—most notably Bacon—did not invent what we now call empiricism, but earlier empirical traditions such as Aristotle's natural philosophy had assigned a humbler philosophical role to the particulars of experience. As historian of science Peter Dear explains, "An 'experience' in the Aristotelian sense was a statement of *how things happen* in nature, rather than a statement of *how something had happened* on a particular occasion."[30] For the Aristotelian tradition, particular observations (such as that a given body has fallen) were philosophically meaningful to the extent that they confirmed general knowledge claims (bodies fall); conversely, anything seemingly at odds with the way nature generally functioned was treated as a monstrosity or anomaly to be ignored. On this view, Dear concludes, particulars could illustrate how nature functions, but they could not challenge a received understanding of those operations. Bacon's innovation was to regard events that belied the course of nature not as aberrations to be written off but as evidence that existing knowledge was imperfect. A proper philosophy of nature should be able to account for whatever fails to fit within inherited preconceptions, and Bacon advises natural philosophers to start afresh by assembling "a substantial and severe collection of the Heteroclites or Irregulars of nature, well examined and described."[31] These "heteroclites or irregulars of nature"—particular features of the empirical world misaligned with the current state of knowledge—would provide the groundwork for a better natural philosophy. Dear finds here the conceptual beginnings of future experimental science, which would devise singular events to test general knowledge claims.

"This is not to say that Aristotelian philosophy was not empirical," Lorraine Daston clarifies, "only that its empiricism was not that of facts, in the sense of deracinated particulars untethered to any theory or explanation."[32] Whether the facts championed by Bacon and his Royal Society successors were indeed untethered to theory has been questioned by a number of scholars, including Steven Shapin, Mary Poovey, and Tita Chico.[33] But even revisionist studies grant that something of momentous relevance came out of Bacon's program. A "fact," in earlier legal contexts, had referred to an action—"fact" comes from Latin *factum*, "the thing that has been done," or simply "a deed"—but over the course of the seventeenth century, it came to designate not only *actions* but also particular *features* of nature.[34] By subordinating human reason to the authority of these features, Bacon's blueprint for future researchers, the *New Organon* (1620), flipped around the Platonic picture of the empirical world as an impoverished version of reality. That world, with its warts, nooks, and crannies, is all the philosopher has. Rather than "grope and clutch at abstracts with feeble mental tendrils," Bacon advised, "one must travel always through the forests of experience and particular things."[35] The goal is still to arrive at general knowledge claims, but these must be inductively derived from particulars rather than intuited through universal ideas. Bacon did not, as Dear and Poovey point out, confront the problem of induction in quite the same terms as Hume would in the 1730s.[36] But far from taking induction for granted, he shared Aristotle's skepticism about induction by incomplete enumeration, devoting book 2 of *The New Organon* to avoiding its trappings. His "organon" is essentially a new method of induction—or, as Christiane Frey prefers to call it, a "prosthesis"—devised to arrive at general, albeit probabilistic, natural knowledge via the observation of the minute features of nature.[37]

Empiricism, in short, vindicated particulars as sources of knowledge and consequently altered the epistemic value of different types of narrative. Narratives about particular events and individual lives came to be seen not as uselessly specific, as they were for Sidney, but as revealing of larger causal and characterological patterns. History, the lower genre in Aristotle's ranking, directly benefited from these developments. Starting with Degory Wheare in the early seventeenth century, we find commentators defending historical narratives as inventories of facts from which readers could derive useful inductive inferences. By studying history, Wheare noted, we may "foresee the Events of things, perceive their Causes, and by remembring those Evils that are past, provide Remedies against those which are coming upon us."[38] Likewise, Lord Chesterfield writes to his son in 1739 that "the perfect knowledge of History is extremely necessary; because, as it informs us of what was done by other people, in former ages, it instructs us what to do in the like cases."[39] This was more than a revival of the classical view of history as philosophy taught through examples; it reflected a new understanding of general knowledge as emerging from particular facts, including the facts of history. As Locke explains in *Of the Conduct of the Understanding* (1706), natural inquirers should test their empirical conclusions "by what they shall find in history, to confirm or reverse their imperfect observations; which may be established into rules fit to be relied on, when they are justified by a sufficient and wary induction of particulars."[40] Hume similarly regarded the historical record as "many collections of experiments, by which the politician or moral philosopher fixes the principles of his science," just as the natural philosopher learns about plants and minerals "by the experiments, which he forms concerning them."[41] For the empiricists, history expanded the pool of particular data from which to derive inductive generalizations, and this meant, contra Sidney and Dacier, that the particulars recorded by history were not inapplicable to the lives of readers. The power of historical narratives to yield portable insights is lucidly articulated by Dugald Stewart from the vantage point of 1814: "Every fact collected with respect to the past is a foundation of sagacity and of skill with respect to the future. . . . It is from this apprehended analogy between the future and the past, that historical knowledge derives the whole of its value."[42]

In short, by rehabilitating induction and stressing the importance of particulars, empiricism simultaneously raised the status of historical narratives. Accounts of actual events came to be seen by many sectors of British educated opinion as surrogates of direct experience and hence as valuable sources of knowledge. As an anonymous author put it in 1746, history is "our Preceptor in *Wisdom* and *Experience*."[43] History can, in the words of Hester Chapone, "supply the defect of that experience, which is usually attained too late to be of much service to us."[44] For Joseph Priestley, "History presents us with the same objects which we meet with in the business of life" and "may be called anticipated experience."[45] Defenses of history as a supplement for experience are common in the eighteenth-century archive, and the colonial governor Sir William Young spelled out what they meant for the Aristotelian alternative:

"This opinion of Aristotle, that the epic muse was a better and more comprehensive teacher than the historic, I cannot readily adopt.... History may teem with as much philosophic theory as poetry: in the annals of an united people, we find matter for general positions, and the particular examples interspersed assist us in the analysis or composition of our system."[46]

The new prestige of particulars can be attested by the complaints it elicited from the opponents of empiricism. At the end of the seventeenth century, the Cartesian philosopher Thomas Burnet pointed out that modern experimentalists "are apt to distrust every Thing for a Fancy or Fiction that is not the Dictate of Sense, or made out immediately to their Senses. Men of this Humour and Character call such Theories as these philosophick Romances, and think themselves witty in the Expression."[47] In noting that the detractors of wit liked to be thought witty, Burnet is highlighting the mismatch between theory and practice that characterized empiricist rhetoric, as the same authors who belittled the imagination made ample use of its resources.[48] A century later, when empiricism had secured its status as the dominant philosophical framework in Britain, the Neoplatonist Thomas Taylor described the prestige of particulars as a philosophical disaster. For Taylor, "A minute attention to trifles is inconsistent with great genius of every kind," and "the science of universals, permanent and fixt, must be superior to the knowledge of particulars, fleeting and frail." Regrettably, Taylor notes, "to a genius ... with whom the crucible and the air-pump are alone the standards of truth," the attempt to achieve universality "must appear ridiculous in the extreme."[49] To admirers of experimental philosophy, Taylor issues the following challenge: "Where ... is the microscope which can discern what is smallest in nature? Where the telescope, which can see at what point in the universe wisdom first began? Since then there is no portion of matter which may not be the subject of experiments without end, let us betake ourselves to the regions of mind ... where every thing is permanent and beautiful, eternal and divine. Let us quit the study of particulars, for that which is general and comprehensive, and through this, learn to see and recognize whatever exists."[50] This attack bears witness to the philosophical status that particulars had come to enjoy. Taylor's works were rather unfavorably received, suggesting that his uncompromising Neoplatonism had become peripheral by the 1780s. To William Wollaston's objection that "we reason about particulars, or from them; but not by them," John Dove was able to confidently reply, "I ... own myself one of those who Reason about Particulars, in Order to Reason from and by them."[51] Those who did so were the new majority. Despite naysayers like Taylor, it was obvious to a growing number of eighteenth-century observers that particulars held profound philosophical significance.

Fictional Particulars: Witnessing the World in the Novel

EMPIRICIST DEFENSES of history may initially seem promising for the novel, since they anticipate the claims that novelists would make about their own narratives. But while empiricism allowed history to make ambitious claims to knowledge, it only did so by lowering the status of imaginative literature. When Dugald Stewart noted that if it were not for induction, history would "rank with the fictions of poetry," he was declaring the caducity of genres that failed to comply with empirical protocols.[52] That fiction was guilty of this very sin was a widespread view among earlier empiricists. For the prominent Irish physician Sir Edward Barry, we can only understand the principles of nature "if such *Data* are only made use of, which are not the probable Fictions of a luxuriant Imagination, but such as are to be observed by the Senses."[53] Analogously, for Adam Smith, "as the facts [in fiction] are not such as have realy [sic] existed, the end proposed by history will not be answered. The facts must be real, otherwise they will not assist us in our future conduct, by pointing out the means to avoid or produce any event."[54] For Stewart, Barry, and Smith, fiction cannot do the job of history, as it invents its particulars instead of finding them.

Such a suspicion of fiction had become entangled with the empirical worldview since its inception with Bacon. Bacon's view of poetry as a source of pleasure rather than knowledge reappears in Locke, Hume, George Campbell, and James Beattie but also in many eighteenth-century critics not usually described as empiricists. For denouncers of fiction like Jeremy Collier, "Romances ... are a meer Land of Fairies, and lie perfectly out of the Road of History and Life: Thus they furnish no useful Knowledge."[55] Novelists aware of the prestige of factual discourse often practiced and disowned fiction at the same time, with Defoe defining "romance" as a story "put upon the World to cheat the Readers, in the Shape or Appearance of Historical Truth."[56] The main issue, for critics of fiction, is sharply articulated by the divine and naturalist William Jones of Nayland: "When we confine ourselves to real life, and are content with describing facts, with the consequences that actually followed from them, we may be unable to trace the designs of Providence, but we do not misrepresent them.... But when we dare to settle the fate of imaginary characters, we take the providence of God out of his hands, assuming an office for which no man is fit."[57] Because novelists are incapable of replicating the causal patterns that govern events in real life, any inferences drawn from fictional events will fail to apply to the real world. As Lady Mary Wortley Montagu animadverts, novels "encourage young people to hope for ... legacys from unknown Relations, and generous Benefactors to distress'd Virtue, as much out of Nature as Fairy Treasures."[58]

For all these critics, it is a dangerous mistake to seek instruction from the particulars of fiction. "He who seeks in [novels] for a knowledge of real life and manners, must be disappointed," says the "Trifler," a regular contributor to the *Aberdeen Magazine*. For this author, history is "one great source of experience," as it relates "facts recorded for many centuries, with proofs that carry conviction," almost as "if they happened in our own time, and before our eyes." But when it comes to whether novels can put the world before the eyes of readers, the Trifler answers in the negative:

> It is common to refer to Novels for instruction of the kind I allude to. But in vain in such works do we look for representations of real life. The authors of them create men and women such as never existed, and place them in situations such as not one person in ten thousand has any chance to be placed. It was the reproach of false philosophy, before the time of the immortal Lord Bacon, that philosophers laid down a certain position, and adapted their experiments to it, instead of making experiments first, and deducing positions from them. Something of that same philosophy prevails in novel-writing. A story is formed, and all the characters and situations must be adapted to compleat it. Hence the operations of nature are strangely perverted, and the ways of providence absolutely burlesqued. It is not from fiction that experience can be acquired; in the affairs of common life we are sensible that it is from what really happens, and that only, that we learn wisdom.[59]

This passage is strikingly reminiscent of Dacier's objections to the particularity of history, except that it reverts the genres in question: now it is fiction, not history, that fails to yield valid inferences about the lives of its readers. The nod toward Bacon reveals the Trifler's intellectual debts. If novels fail to yield knowledge, it is not because they lack Aristotelian universality but because they tamper with the empirical evidence: they take liberties with the way nature and human societies actually operate.

In Vicesimus Knox's version of a commonplace, "the end of oratory is to persuade, of poetry to please, and of history to instruct by the recital of true events."[60] This division of intellectual labor affirmed the new status of history by demoting poetry from the domain of knowledge to that of leisure. Given such an association between fiction and falsity, how could novels carve out a space in the program of empiricism? They did so in a number of ways, including an attempted compromise between the empirical emphasis on particulars and the traditional principles of neoclassicism. Critics from Dennis's generation used to describe the artist's quest for universality through a mythical paradigm: that of the sculptor Phidias, who produced a statue of Minerva by dispensing with real-life models and consulting instead an eternal idea of beauty. In time, this Platonic paradigm came to share space with a different yet equally venerable one: that of the painter Zeuxis, who painted Helen of Troy by combining the features of five beautiful women. The abbé Charles Batteux, whose *Principes de la littérature* (1753) became popular in English translation, explained that Zeuxis, in pursuing perfect beauty, "gathered together the separate features of several beauties who were then alive"; following Zeuxis, Molière, in portraying a misanthrope, "collected every mark, every stroke of a gloomy temper that he could observe" and "drew a single character, which was not the representation of the true, but of the probable."[61] While Batteux is very much an Aristotelian, he claimed that the representational arts achieved "probable" truth by translating actual particulars into fictional generalizations. As a later critic, James Beattie, would put it, the poet works by "considering each quality as it is found to exist in several individuals of a species, and thence forming an assemblage more or less perfect in its kind."[62] This procedure, involving abstractions from observation and experience, is empirical rather than Platonic. It was recommended by none less than Sir Joshua Reynolds. "This great ideal perfection and beauty," Reynolds wrote, "are not to be sought in the heavens, but upon the earth. They are about us, and upon every side of us."[63]

This way of achieving universality allowed the neoclassical pursuit of universal truths to incorporate the empiricist attention to particulars. It did not take long for defenders of fiction to capitalize on it.[64] A telling example is Henry Fielding's twofold vindication of *Joseph Andrews* (1742). In his preface, Fielding defends *Joseph Andrews* along Aristotelian lines as an epic, and in book 3, he does so along empiricist lines as a true history; he claims that *Joseph Andrews* embodies universal truths (reflecting Aristotle's praise of poetry) but originates in his personal observations (reflecting the empiricist stress on experience). The universal truths of poetry are now firmly grounded in the minute particulars of London life and roadside taverns. In a later essay that owes much to Fielding, William Godwin argues that "the writer of romance collects his materials from all sources, experience, report, and the records of human affairs; then generalises them; and finally selects, from their elements and the various combinations they afford, those instances which he is best qualified to pourtray."[65] The same theory underlies the Reverend Richard Whately's eulogium on Jane Austen in 1821. Novels, for Whately, "being a kind of fictitious biography, bear the same relation to the real, that epic and tragic poetry, according to Aristotle, bear to history: they present us . . . with the general instead of the particular,—the probable, instead of the true; and, by leaving out those accidental irregularities, and exceptions to general rules, which constitute the many improbabilities of real narrative, present us with a clear and abstracted view of the general rules themselves; and thus concentrate, as it were, into a small compass, the net result of wide experience."[66] This insightful passage combines Aristotle's praise of poetry with the generalizations from particulars characteristic of empiricism. Novelists, for Whately, communicate universal truths, but they arrive at them by carefully sieving through raw observational data. For Whately, as for Fielding and Godwin, the particular characters and events in novels constitute instantiations of general truths that, in turn, are inferred inductively from the particulars of experience. To read a novel, consequently, is to access a distilled version of the world of experience, one that reveals the same general patterns discernible, albeit with more dross, in our observations of real life.

Novels, in short, were theorized as containers (or milieus) for imagined minutiae that mediated the gap between the particulars of experience and the universal truths of old. As a genre, the novel straddled the tripartite division separating philosophy, history, and poetry. Being fictional, it falls under the Aristotelian category of poetry, but its narrative procedures, especially its professed faithfulness to the particulars of experience, brought it closer to experimental philosophy or empiricist history.[67] Scholarship on the novel has attended to this double heritage, and in recent decades, the emphasis has fallen on the empiricist side of things. It is Fielding's appeals to experience, rather than his theory of the epic, that led John Bender to choose *Tom Jones* as his prime example of virtual witnessing. According to Bender, "*Tom Jones* explicitly puts its leading character into the laboratory and asks readers to observe his behavior side-by-side with the narrator." Like empiricist history or reports of scientific experiments, the realist novel, for Bender, enables readers to observe events that played out at a remove and then draw inductive inferences from them: "Readers must constantly judge evidence, probability, and the chain of cause and effect as [Fielding] pushes them toward the inductive method of moral philosophy."[68] Novels, on this construal, reflect with accuracy the minute particularity of real life, enabling readers to derive observational conclusions from them just as they would from witnessing actual events.

As I noted in my introduction, this theory has been challenged for its dependence on a debatable interpretation of the history of science. Bender borrows the concept of virtual witnessing from Steven Shapin and Simon Shaffer's influential *Leviathan and the Air-Pump* (1985). The new empirical sciences, for Shapin and Shaffer, faced the problem that experiments performed in private could not be produced as evidence to persuade absent audiences. Divulging science, accordingly, required developing "a literary technology by means of which the phenomena produced were made known to those who were not direct witnesses."[69] This technology, which they called "virtual witnessing," involved particularized reports of scientific experiments to be published and widely distributed. "Through virtual witnessing," Shapin and Shaffer explain, "the multiplication of witnesses could be, in principle, unlimited.... The validation of experiments, and the crediting of their outcomes as matters of fact, necessarily entailed their realization in the laboratory of the mind and the mind's eye."[70] Readers of the reports, in short, observed, with a mental eye, an enactment of the original experiment on a mental stage. The problem with this account, for Helen Thompson, is that it involves an overly straightforward conception of how scientists encountered and described their objects of study. By implying a strict separation between witnesses and the things they witnessed and by downplaying the early modern interest in unobservable entities and processes,[71] Shapin and Shaffer conceal the productive relation between the subjects and objects of observation. Like Chico, Thompson regards minute particulars as both produced and discovered. Experimentalists like Boyle and Hooke were aware of the imaginative processes involved in producing minute particulars, and Thompson claims that such processes went on to play a role in empiricist fiction as well. The novel, similar to chemistry or microscopy, "evokes things that cannot be sensed, things whose power to produce sensation is not mimetically transmitted by sensation itself."[72] What novels do when they build on empirical models is not to transcribe reality for the benefit of absent observers but to "make explicit the *production* of empirical reality as the reader's encounter with forms and powers that enable sensational knowledge."[73] The particulars of novelistic description, on this account, are not meant to enable readers to witness the world by proxy; instead, they make visible the interactional processes through which we cognize the world.

The claim I have been developing in this essay is that eighteenth-century theories of particularity offer a way forward for this debate. They suggest that it is possible to embrace modern reassessments of the Scientific Revolution while at the same time retaining Bender's application of virtual witnessing to the novel. One notable feature of the empiricist praise of history is the absence of misgivings about the status of actual particulars. Most eighteenth-century commentators held a less reticent view of the relationship between observers and observed than we find in Royal Society experimental science. Commentators like Adam Smith and the Trifler were confident that particulars could be observed without interference from the imagination, that writing was able to record the world, and that the problem with fiction is that it failed to live up to that standard. They held, in other words, the naive view of particularity that Shapin and Shaffer attributed to earlier natural philosophers. The upshot is that virtual witnessing is more than a projection of a questionable history of science onto eighteenth-century literary criticism. It is often an accurate description of how observers at the time conceived of the function of texts. The rhetoric of empiricist history already involved recognizable articulations of that principle. Consider, for example, what students at the College of Edinburgh were learning from a midcentury textbook by James or Adam Ferguson:

> Observation is that proceeding of the mind by which we collect facts.
> Facts relate to the existing qualities and operations of different natures.
> Observation terminates in history, or the knowledge of particulars in detail ...
> We are determined by a law of our nature to believe facts to which we ourselves are witnesses, or to which we have the credible testimony of others.[74]

These entries, which students were invited to learn by rote, recapitulate the Baconian program and then formulate the principle of making historical narratives potential sources of factual knowledge: testimonies from credible sources elicit the same type of belief as direct observation, and history is a repository for such testimonies. Nor was the resemblance between history and scientific experiments lost on eighteenth-century observers. Joseph Priestley, a strong defender of circumstantiality in narratives, argued that "real history resembles the experiments made by the air-pump, the condensing engine, or electrical machine, which exhibit the operations of nature, and the God of nature himself."[75] To read history, for Priestley, is like observing the subtle operations of the empirical world. As David Fordyce puts it, while "the Knowledge of Mankind ... depends on long Experience and Observation," it can also be acquired through the study of history, since "history ... re-acts almost every Scene of Life before us."[76] The analogy between reading history and witnessing life as well as the power of description to compel belief are forcefully captured by Hugh Blair, for whom historians endowed with "impartiality, fidelity, gravity, and dignity" are able to "place us, as on an elevated station, whence we may have an extensive prospect of all the causes that cooperate in bringing forward the events which are related."[77] Notice the detachment between the observer and the object of observation:

to read a reliable historical narrative is tantamount to standing on a promontory from which one can survey, via the mediation of writing, the causal revolutions of the human past.

In turn, the same reading protocols informed how eighteenth-century critics and novelists thought of imaginative genres. For critics like Elizabeth Montagu, certain forms of literature were capable of improving upon empiricist history: "If it is the use of history, that it teaches philosophy by experience," then Shakespeare's historical plays "must be allowed to be the best preceptor.... The poet collects, as it were, into a focus those truths, which lie scattered in the diffuse volume of the historian."[78] For Henry Pemberton, a contemporary of Fielding, epic and dramatic poetry "may very justly be compared with the experimental part of natural philosophy," where "artificial experiments are contrived, wherein the powers of nature may discover themselves by acting under less disguise, than in the ordinary course of things."[79] Pemberton denies that novels can do the same, as they "exhibit as false a picture of human affairs, as the knight-errantry and enchantments of romance,"[80] but other critics begged to differ. Charles Dibdin, in the opening of his novel *The Younger Brother* (1793), writes that "a novel ... is a kind of mean betwixt history, which records mere fact, and romance, which holds out acknowledged fiction," and then claims that "there is not a character or circumstance in this whole work which I do not *experimentally* know from a close observation of mankind, to have *virtual* existence."[81] By "virtual" existence, Dibdin seems to mean that his characters are real even though they are not embodied (Johnson defines "virtual" as "having the efficacy without the sensible or material part"[82]), and Dibdin knows such characters "experimentally" in the sense that they originate in his observations, which are now laid out so that readers can have mediated access to them.

Caught between these poles and illustrating the stakes of this debate is Joseph Priestley, who alternated between affirming and denying the value of novels as replacements for observation. In his *Course of Lectures on Oratory and Criticism* (1777), Priestley defends prose fiction by claiming that "reading a romance is nearly the same thing as ... seeing so much of the world, and of mankind." Priestley reserves special praise for narratives that mention "those circumstances of *time, place*, and *person*, which were originally associated with the particulars of the story," on the grounds that "in nature, and real life, we see nothing but *particulars*, and to these ideas alone are the strongest sensations and emotions annexed."[83] To read a richly particularized fictional story, Priestley seems to conclude, may be the same thing as witnessing events in the real world. And yet, having gone this far, Priestley then turns around and makes the following qualifications:

Even the best of our modern romances, which are a much more perfect copy of human life than any of the fictions of the ancients, if they be compared with true history, will be found to fall greatly short of it in their detail of such particulars as, because they have a kind of *arbitrary*, and, as it were, *variable* connexion with real facts, do not easily suggest themselves to those persons who . . . never introduce more persons or things than are necessary to fill [their incidents] up. . . . Whereas a *redundancy of particulars*, which are not necessarily connected, will croud into a relation of real facts.[84]

Fiction, for Priestley as for Godwin and Whately, handles particulars sparingly, subordinating description to the demands of the plot and filtering out that surplus of circumstances that makes up the fabric of real life. But while Godwin and Whately see this as an advantage, Priestley finds that the novel's selectivity lowers its epistemic value compared to history.

Priestley revisits the topic in his *Lectures on History* (1788), where he writes that "true history has a capital advantage over every work of fiction." Whereas true history constitutes "an inexhaustible mine of the most valuable knowledge, . . . works of fiction resemble those machines which we contrive to illustrate the principles of philosophy, such as globes, and orreries, the uses of which extend no farther than the views of human ingenuity."[85] Joseph Drury clarifies Priestley's point by explaining that an orrery illustrates but does not instantiate the operations of gravity; in imitating the orbits of the planets, orreries "moved by clockwork along fixed metal tracks, not by the force of gravity alone." Analogously, novels "provided entertaining illustrations of the workings of nature, without exhibiting those operations directly."[86] For Priestley, consequently, "*historical facts*" are "the only foundation on which men who think . . . will build any conclusions," whereas "all works of *fiction*, both in prose and verse," are mainly designed "to amuse the imagination."[87] Nonetheless, Priestley offers a way to think of novels as modeling reality for the reader's gaze. As Drury notes, novels disseminated knowledge through "compelling simulations of human nature that would abstract from the historical record the fundamental principles of human nature," and this could be true even if they were not offering "accurate records of nature that, like history, simply documented potentially useful experimental data."[88]

Less reticently than Priestley, some defenders of fiction claimed that novels were useful precisely as sources of experimental data. A remarkable example is Mary Hays, who describes her novel *Memoirs of Emma Courtney* (1796) as a contribution to "the science of mind," a science in which "a long train of patient and laborious experiments must precede our deductions and conclusions."[89] In order to make advances in psychology and morals, her heroine notes, "we want a more extensive knowledge of particular facts, on which, in any given circumstance, firmly to establish our data." Hays's ostensible goal is to perform such experiments on the page and "establish" more data through a fictionalized version of her own difficult experiences with unrequited love, and she pursues this goal under the conviction that "the most interesting, and the most useful, fictions, are, perhaps, such, as delineating the progress, and tracing the consequences of one strong, indulged, passion, or prejudice, afford materials, by which the philosopher may calculate the powers of the human mind, and learn the springs which set it in motion." The difference between this conception of fiction and that of someone like Whately is that Hays does not regard fiction as simply illustrating, for the benefit of readers, the knowledge that philosophers already have; instead, she thinks of fiction as leading to new knowledge by providing philosophers with the equivalent of new empirical data. In order to guarantee that her study of the mind is empirically sound, she traces the development over time of a single passion—love—while paying "a sufficient regard ... to the more minute, delicate, and connecting links of the chain."[90] Hays's premise is that such minutiae, however fictional, enact the way mental events causally unfold and accordingly reveal previously invisible features of human psychology—thus opening up the recesses of the human mind to virtual observation.

Anti-Jacobin novelists took note of this ambitious conception of fiction, and Elizabeth Hamilton targeted it in her *Memoirs of Modern Philosophers* (1800). Hamilton satirizes Hays through one of her three protagonists, the extravagant young radical Bridgetina Botherim, who believes, like Hays, that philosophers can learn new truths from witnessing her disappointments in love. "The history of my sensations," Bridgetina tells her friend Julia, "are equally interesting and instructive. You will there see, how sensation generates interest, interest generates passions, passions generate powers; and sensations, passions, powers, all working together, produce associations, and habits, and ideas, and sensibilities." This closely echoes the psychological vocabulary of *Memoirs of Emma Courtney*, but once Bridgetina sets about recovering the minute links in her mental autobiography, everything descends into farce: "The remoter causes of those associations which formed the texture of my character, might, I know, very probably be traced to some transaction in the seraglio of the Great Mogul, or some spirited and noble enterprise of the Cham of Tartary; but as the investigation would be tedious, and, for want of proper data, perhaps impracticable, I shall not go beyond my birth, but content myself with arranging under seven heads (I love to methodise) the seven generating causes of the energies which stamp my individuality."[91]

At the end of Bridgetina's preposterous history of her inner self, Hamilton concludes, "It would be unpardonable to neglect the opportunity that now presents itself of offering a hint to our very much respected friends, the experimental philosophers; to whose serious consideration we would very earnestly recommend a minute investigation of the facts so often recorded in the works of celebrated writers. From these authors sufficient data may be obtained for an exact calculation of the greatest height to which any river was ever known to rise by the fall of a single shower of tears."[92]

Fictional minutiae as data yielding empirical inferences: this may be the last frontier in terms of eighteenth-century conceptions of virtual witnessing. Hays embraced it, but as Hamilton's response shows, it was far from clear, at the turn of the 1800s, that fiction could claim this role for itself. In this essay, however, I have been less concerned with whether novels can indeed enable virtual witnessing than with how eighteenth-century discussions of particularity envisioned that possibility. We have seen that the Aristotelian view of particulars as inferior to universals was reverted under the influence of empiricist epistemology; that literary critics, without abandoning the Aristotelian defense of poetry, often invoked the rising prestige of history on behalf of imaginative genres; and that particulars, whether historical or fictional, and however contentiously, were often described as routes to general knowledge. This elevated the minute particular in epistemic importance, even in genres that laid claim to universality. Defenders of the novel made ample use of these resources, speaking of novels as circumscribed environments populated by lifelike minutiae; for its optimist defenders, novels were empirical like history and were accordingly representations of worlds to be observed or virtual surrogates for experience. "When we cannot see or hear anything ourselves," said the notable English jurist Sir Geoffrey Gilbert, "we must see and hear by report from others," and the novelist's purpose, says Henry Fielding, is "to present the amiable Pictures to those who have not the Happiness of knowing the Originals."[93] With Hays we come to the point where novelistic particulars cease to be replacements for unobserved originals and become instead raw data to be observed—or witnessed—for the first time. Most defenders of fiction did not go this far, but many asserted the power of novels to showcase, for the instruction of young and old, the minute particularity of absent experience.

Notes

1. Aristotle, *Poetics*, trans. Richard Janko, in *The Norton Anthology of Theory and Criticism*, ed. Vincent B. Leitch, 2nd ed. (New York: W. W. Norton, 2010), 1451b, 95.
2. For a survey, see Baxter Hathaway, *The Age of Criticism: The Late Renaissance in Italy* (Ithaca: Cornell University Press, 1962).
3. Samuel Johnson, *The History of Rasselas Prince of Abissinia*, ed. Thomas Keymer (Oxford: OUP, 2009), 28.
4. For a detailed discussion of Johnson's oscillation, see Scott Elledge, "The Background and Development in English Criticism of the Theories of Generality and Particularity," *PMLA* 62, 1 (1947): 147–82.
5. John Hawkesworth, *An Account of the Voyages Undertaken by the Order of His Present Majesty* (Perth: R. Morison, 1789), vol. 1, vi.
6. Sir Joshua Reynolds, "The Idler No. 82," in *Eighteenth-Century Critical Essays*, ed. Scott Elledge (Ithaca: Cornell University Press, 1961), vol. 2, 837.
7. William Blake, "Jerusalem," in *The Complete Works of William Blake*, ed. David V. Erdman (Berkeley: University of California Press, 1982), chap. 2, l. 62, 205.
8. Hugh Blair, *Lectures on Rhetoric and Belles Lettres* (Dublin: Whitestone, 1783), vol. 3, 60.
9. The essential eighteenth-century voices include the usual suspects, Johnson, Reynolds, and Blake, plus midcentury rhetoricians like Lord Kames and George Campbell and Romantic critics such as Coleridge and William Hazlitt. See Houghton W. Taylor, "'Particular Character': An Early Phase of a Literary Evolution," *PMLA*, 60, 1 (1945), 161–74; Scott Elledge, "The Background and Development in English Criticism of the Theories of Generality and Particularity," *PMLA*, 62, 1 (1947), 147–82; and Leo Damrosch, "Generality and Particularity," in *The Cambridge History of Literary Criticism*, vol. 4: *The Eighteenth Century*, ed. H. B. Nisbet and Claude Rawson (Cambridge: Cambridge University Press, 1997), 381–93.
10. Joseph Drury, *Novel Machines: Technology and Narrative Form in Enlightenment Britain* (Oxford: OUP, 2017), 86.
11. John Bender, *Ends of Enlightenment* (Stanford, CA: Stanford University Press, 2012), 66.
12. Jenny Davidson, "The Minute Particular in Life Writing and the Novel," *Eighteenth-Century Studies*, 48, 3 (2015), 264.
13. See, for example, Scott Black's *Without the Novel: Romance and the History of Prose Fiction* (Charlottesville: University of Virginia Press, 2019).
14. Tita Chico, "Minute Particulars: Microscopy and Eighteenth-Century Narrative," *Mosaic: An Interdisciplinary Critical Journal*, 39, 2 (2006), 143–61. Chico vastly expands her argument in *The Experimental Imagination: Literary Knowledge and Science in the British Enlightenment* (Stanford, CA: Stanford University Press, 2018).

15. Helen Thompson, *Fictional Matter: Empiricism, Corpuscles, and the Novel* (Philadelphia: University of Pennsylvania Press, 2017).
16. Anne M. Thell, *Minds in Motion: Imagining Empiricism in Eighteenth-Century British Travel Literature* (Lewisburg: Bucknell University Press, 2017), 111–52.
17. For just how various the sources of the novel's circumstantiality were, see Cynthia Sundberg Wall, *The Prose of Things: Transformations of Description in the Eighteenth Century* (Chicago: University of Chicago Press, 2006). See also Lennard Davis, *Factual Fictions: The Origins of the English Novel* (Philadelphia: University of Pennsylvania Press, 1983); Michael McKeon, *The Origins of the English Novel, 1600–1740* (Baltimore: Johns Hopkins University Press, 1987); and J. Paul Hunter, *Before Novels: The Cultural Contexts of Eighteenth-Century English Fiction* (New York: Norton, 1990).
18. Sir Philip Sidney's *Defense of Poesy*, ed. Lewis Soens (Lincoln: University of Nebraska Press, 1970), 17.
19. Francis Bacon, *The Advancement of Learning*, in *The Major Works*, ed. Brian Vickers (Oxford: Oxford University Press), 175.
20. Bacon, *Advancement*, 186, 187.
21. Sidney, *Defense of Poesy*, 9.
22. Francis Bacon, *De Augmentis Scientiarum*, in *The Works of Francis Bacon*, ed. James Spedding, Robert Ellis, and Douglas Heath (Boston: Brown and Taggard, 1861), vol. 4, 406.
23. Sidney, *Defense of Poesy*, 20.
24. André Dacier, ed., *Aristotle's Art of Poetry* (London: Dan. Browne, 1705), 142.
25. John Dennis, "Reflections Critical and Satyrical, upon a Late Rhapsody, Call'd, an Essay upon Criticism," in *The Critical Works of John Dennis*, ed. Edward N. Hooker (Baltimore: Johns Hopkins University Press, 1939), vol. 1, 418n1.
26. See Erwin Panofsky, *Idea: A Concept in Art Theory*, trans. Joseph J. S. Peake (Columbia: University of South Carolina Press, 1968); and Louis I. Bredvold, "The Tendency toward Platonism in Neo-classical Esthetics," *ELH*, 1, 2 (1934), 91–119.
27. Richard Hurd, *Q. Horatii Flacci Ars Poetica. Epistola ad Pisones: With an English Commentary and Notes* (London: W. Bowyer, 1749), 132.
28. Elledge ("Background") traces aesthetic objections to the English reception of Longinus.
29. I focus on natural philosophy here. For the importance of the legal tradition in the rise of the modern scientific fact, see Barbara J. Shapiro, *A Culture of Fact: England, 1550–1720* (Ithaca: Cornell University Press, 2000).
30. Peter Dear, *Discipline and Experience: The Mathematical Way in the Scientific Revolution* (Chicago: University of Chicago Press, 1995), 4.
31. Bacon, *Advancement*, 176.
32. Lorraine Daston, "Marvelous Facts and Miraculous Evidence in Early Modern Europe," *Critical Inquiry*, 18, 1 (1991), 110.
33. Steven Shapin, *A Social History of Truth: Civility and Science in Seventeenth-Century England* (Chicago: University of Chicago Press, 1994); Mary Poovey,

A History of the Modern Fact: Problems of Knowledge in the Sciences of Wealth and Society (Chicago: University of Chicago Press, 1998); Chico, *Experimental Imagination*.
34. See Shapiro, *Culture of Fact*, 34.
35. Francis Bacon, *The New Organon*, trans. Michael Silverstone, ed. Lisa Jardine (Cambridge: Cambridge University Press, 2000), 220, 10.
36. Dear, *Discipline and Experience*, chap. 1; Poovey, *History of Modern Fact*, xviii.
37. Frey provides a rather different account of Bacon's investment in particulars and in what we call induction. On her account, whereas induction begins with particulars and subsequently articulates general knowledge claims, Bacon's organon begins with both "the observations of an unpracticed eye" and "the opinions of a prematurely conclusive mind" to then *"produce* those particulars and axioms that are needed for the knowledge of natural truth" (159). Christiane Frey, "Bacon's Bee: The Physiognomy of the Singular," in *Exemplarity and Singularity: Thinking through Particulars in Philosophy, Literature, and Law*, ed. Michele Lowrie and Susanne Lüdemann (New York: Routledge, 2015). 151–65.
38. Degory Wheare, *The Method and Order of Reading Both Civil and Ecclesiastical Histories* (London: Charles Brome, 1685), 324. Original Latin published in 1622.
39. From Letter XXV, September 10, 1739, in Lord Chesterfield, *Letters Written by the Late Right Honourable Philip Dormer Stanhope, Earl of Chesterfield, to His Son, Philip Stanhope*, ed. Eugenia Stanhope (Dublin: E. Lynch, 1774), vol. 1, 71.
40. John Locke, *Some Thoughts concerning Education and of the Conduct of the Understanding*, ed. Ruth W. Grant and Nathan Tarcov (Indianapolis: Hackett, 1996), 187, 188.
41. David Hume, *An Enquiry concerning Human Understanding*, ed. Peter Millican (Oxford: OUP, 2008), 60.
42. Dugald Stewart, *Elements of the Philosophy of the Human Mind* (London: A. Strahan and T. Cadell, 1814), 217.
43. Anonymous, *An Essay on the Manner of Writing History* (London: M. Cooper, 1746), 5. The text has been attributed to Peter Whalley (1722–91).
44. Hester Chapone, *Letters on the Improvement of the Mind, Addressed to a Young Lady* (London: H. Hughs, 1773), vol. 2, 125.
45. Joseph Priestley, *Lectures on History, and General Policy* (Dublin: P. Byrne, 1788), 6.
46. William Young, *The History of Athens Politically and Philosophically Considered* (London: J. Robson, 1786), vii.
47. Thomas Burnet, *The Theory of the Earth Containing an Account of the Original of the Earth* (London: R. Norton, 1684), xxii.
48. For the role of literariness in empiricist rhetoric, see Chico, *Experimental Imagination*; Courtney Weiss Smith, *Empiricist Devotions: Science, Religion, and Poetry in Early Eighteenth-Century England* (Charlottesville: University of Virginia Press, 2016); Jules David Law, *The Rhetoric of Empiricism: Language and Perception from Locke to I. A. Richards* (Ithaca: Cornell University Press,

1993); Brian Vickers and Nancy S. Struever, *Rhetoric and the Pursuit of Truth: Language Change in the Seventeenth and Eighteenth Centuries* (Los Angeles: University of California, 1985); and John Richetti, *Philosophical Writing: Locke, Berkeley, Hume* (Cambridge: Harvard University Press, 1983).
49. Thomas Taylor, *Concerning the Beautiful, or, A Paraphrased Translation from the Greek of Plotinus, Ennead I. Book VI* (London: Printed for the author, 1787), xii, xiii, 7–8.
50. Taylor, *Concerning the Beautiful*, 7–8.
51. William Wollaston, *The Religion of Nature Delineated* (London: Sam. Palmer, 1724), 41; John Dove, *A Creed Founded on Truth and Common Sense; with Some Strictures on the Origin of Our Ideas* (London: R. Spavan, 1750), 25.
52. Stewart, *Elements*, 217.
53. Sir Edward Barry, *A Treatise on a Consumption of the Lungs* (Dublin: George Grierson, 1726), 9–10.
54. Adam Smith, *Lectures on Rhetoric and Belles Lettres*, ed. J. C. Bryce (Indianapolis: Liberty Fund, 1985), 90.
55. Jeremy Collier, *Essays upon Several Moral Subjects: Part III* (London: W.B., 1705), 150.
56. Daniel Defoe, *A New Family Instructor; in Familiar Discourses between a Father and His Children, on the Most Essential Points of the Christian Religion* (London: T. Warner, 1727), 55.
57. William Jones, *Letters from a Tutor to His Pupils* (London: G. Robinson, 1780), 28–29. In *Empiricism and the Early Theory of the Novel*, I misidentified Jones as the linguist William Jones. I take this opportunity to rectify that error.
58. Lady Mary Wortley Montagu, *Selected Letters*, ed. Isobel Grundy (London: Penguin Classics, 1997), 402.
59. Anonymous, "The Trifler," no. 32, Thursday, March 26, 1789: "On Self," in *Aberdeen Magazine, Literary Chronicle, and Review* (Aberdeen: J. Chalmers, 1789), vol. 2, 155.
60. Vicesimus Knox, *Essays Moral and Literary* (Dublin: R. Marchbank, 1783), 108.
61. Charles Batteux, *A Course of the Belles Lettres, or The Principles of Literature* (London: B. Law, 1761), 17–18.
62. James Beattie, *Essays on the Nature and Immutability of Truth ... on Poetry and Music, as they Affect the Mind; on Laughter, and Ludicrous Composition; and, on the Utility of Classical Learning* (Dublin: C. Jenkin, 1778), vol. 2, 50.
63. Sir Joshua Reynolds, *Discourses on Art*, ed. Robert R. Wark (New Haven: Yale University Press, 1975), 44.
64. This paragraph sums up arguments I made at greater length in *Empiricism and the Early Theory of the Novel: Fielding to Austen* (Cham: Palgrave, 2016).
65. William Godwin, "Essay of History and Romance," in *Political and Philosophical Writings of William Godwin. Vol. 5: Educational and Literary Writings*, ed. Pamela Clemit (London: William Pickering, 1993), 299.
66. Richard Whately, "Northanger Abbey, and Persuasion," in *Famous Reviews*, ed. R. Brimley Johnson (London: Sir Isaac Pitman & Sons, 1914), 229.

67. This is not to say that empiricism did not influence the formal procedures of other genres. For its impact on poetic description, see, for example, Kevis Goodman, *Georgic Modernity and British Romanticism: Poetry and the Mediation of History* (Cambridge: CUP, 2004); and Melissa Bailes, *Questioning Nature: British Women's Scientific Writing & Literary Originality, 1750–1830* (Charlottesville: University of Virginia Press, 2017).
68. Bender, *Ends of Enlightenment*, 27, 28.
69. Steven Shapin and Simon Shaffer, *Leviathan and the Air-Pump: Hobbes, Boyle, and the Experimental Life* (Princeton: Princeton University Press, 1985), 25.
70. Ibid., 60.
71. These processes include "sublimation, dissolution in aqua regia, and other reactions that prove the physicality of tiny parts as the very result of their sensory disappearance" (13). For the role that invisible entities came to play in empiricist epistemology, see Christiane Frey, "The Art of Observing the Small: On the Borders of the 'Subvisibilia' (from Hooke to Brockes)," *Monatshefte*, 105, 3 (2013), 376–88.
72. Thompson, *Fictional Matter*, 17.
73. Ibid., 20.
74. Ferguson (Adam or James), *Analysis of Pneumatics and Moral Philosophy: For the Use of Students in the College of Edinburgh* (Edinburgh: Sold by A. Kincaid & J. Bell, 1766), 14–15.
75. Priestley, *Lectures on History*, 6.
76. David Fordyce, *Dialogues concerning Education* (London: s.n., 1745), 313.
77. Blair, *Lectures*, vol. 3, 51, 55.
78. Elizabeth Montagu, *An Essay on the Writings and Genius of Shakespeare, Compared with the Greek and French Dramatic Poets* (Dublin: J. Potts, 1778), 48.
79. Henry Pemberton, *Observations on Poetry, Especially the Epic: Occasioned by the Late Poem upon Leonidas* (London: H. Woodfall, 1738), 11, 24.
80. Ibid., 24.
81. Charles Isaac Mungo Dibdin, *The Younger Brother: A Novel, in Three Volumes* (London: printed for the author: 1793), 1, 6. Emphasis mine.
82. Samuel Johnson, "Virtual," *A Dictionary of the English Language: A Digital Edition of the 1755 Classic by Samuel Johnson*, ed. Brandi Besalke, last modified February 15, 2014, https://johnsonsdictionaryonline.com/virtual/.
83. Joseph Priestley, *A Course of Lectures on Oratory and Criticism* (London: J. Johnson, 1777), 83, 85, 84.
84. Ibid., 85.
85. Priestley, *Lectures on History*, 4–5.
86. Drury, *Novel Machines*, 40.
87. Priestley, *Lectures on History*, 12, 4.
88. Drury, *Novel Machines*, 41.
89. Mary Hays, *Memoirs of Emma Courtney*, ed. Eleanor Ty (Oxford: OUP, 2009), 9.
90. Ibid., 8, 3, 4.

91. Elizabeth Hamilton, *Memoirs of Modern Philosophers*, ed. Claire Grogan (Peterborough: Broadview, 2000), 174.
92. Ibid., 197.
93. Sir Geoffrey Gilbert, *The Law of Evidence* (London: Henry Lintot, 1756), 4; Henry Fielding, *Joseph Andrews and Shamela*, ed. Thomas Keymer (Oxford: OUP, 2008), 15.

Infinitely Small Differences
The Individual and Its Milieu in Novalis's Natural Philosophy and *Henry of Ofterdingen*

MAREIKE SCHILDMANN

✦ ✦ ✦

SMALL THINGS have long exerted a strong pull on the human mind. The minute can be understood not only as embodying "the whole" in miniature, thereby revealing its structure, but also as manifesting a special kind of potency—namely, a marked potential for growth and expansion. The Romantic poet Friedrich von Hardenberg, better known as Novalis, programmatically expresses this double quality in his *Teplitz Fragments* (1798) when he writes, "Where is the primal seed [*Urkeim*]—the type embodying the entirety of nature—to be found? The nature of nature etc.?" (II, 386; my translation).[1] On the one hand, as a "type," the seed, nucleus, or germ (in German, *Keim*[2]) is assigned a systematic function within the order of nature. Whoever was searching for the primordial seed was also looking for the structural principle, the blueprint of nature par excellence, which had to be both discovered and deciphered. On the other hand, the attribute "primordial"—expressed in German with the prefix *Ur*—includes a temporal element and invokes the idea of a primal origin: of life, man, the individual, precisely the questions that in the eighteenth century had increasingly become the focus of science, philosophy, and literature.

The status of the minute and its importance for Novalis's natural philosophy and aesthetics have received scant critical attention. I will argue, though, that the category of the smallest thing, or what one might want to call *minutiae*[3]—up to the level of the (infinitely) small—not only is central to Novalis's concept of life but also marks the crucial point at which Novalis transgresses contemporary discourses of nature. His negotiation of minutiae, as the first part of this essay shows, has as its starting point the question of the origin of life and his conceptualization of the organism. It then proceeds to develop around how one conceives of the relationship between inner and outer worlds: between the individual organism and its environment, or milieu. Novalis therefore brought up crucial questions of the milieu before its later theoretical elaboration in the nineteenth century by deploying and conceiving the related terms of "surrounding" (*Umgebung*) and "outside world" (*Außenwelt*).[4] At the same time, he transcended the scientific theory of the organism and its environment with his utopian project of "infinitesimal medicine." The epistemic condition of this anthropological revaluation of the (infinitesimally) small, according to my thesis, consists in its desubstantialization and proceduralization: in Novalis's work, the minute is transformed into a concept of relation in both temporal and spatial terms. This ultimately leads to the suggestion of new forms of existence—and the establishment of a new form of poetics—that obeys what I call the principle of infinitesimally small differences.

In the second part, I consider the poetological consequences of (what I will call) Novalis's infinitesimal anthropology upon his own writing. I will show that Novalis's aesthetic preoccupation with minutiae is not limited to the "small form" as discussed with regards to his aphorisms and fragments;[5] it also defined Novalis's approach to the novel, which Georg Lukács has theorized as the prototypical organic genre. By offering a close reading of Novalis's novel fragment *Henry of Ofterdingen*, I aim to demonstrate how the principle of infinitesimally small differentiation is asserted in the form of the novel itself, where it takes on the form of poetry. Poetry becomes, in Novalis's writing, a fluid medium of mediation between the inner and outer worlds; it constitutes the inner milieu of the organism that is the "novel."

Small Things: Germs, Atoms, and Pollen

SINCE THE invention of the microscope in Europe in the seventeenth century, the theoretical and experimental exploration of minutiae has gained new legitimacy and importance within scientific disputes.[6] In the eighteenth century, germs, seeds, cells, atoms, and corpuscles all formed partly complementary but also conflicting models of the small. Since they were often not clearly defined in terms of concepts, they testified to how epistemological boundaries—and the boundaries of the objects of scientific study—were constructed but also confused as new disciplines like chemistry, physiology, and embryology evolved.

Novalis's work and life manifest the paradigmatic entanglement of early Romantic concepts with the scientific debates of his period like hardly any other Romantic author. After completing his law degree in Jena—at this time, a major hotbed of idealistic philosophy—Novalis began studying mining engineering at the prestigious Freiberg Academy in 1797. In the intellectual constellation of Freiberg (which included thinkers like Johann Ritter, Franz Baader, and Abraham Werner, as well as indirectly Friedrich Schelling, Alexander von Humboldt, and others), Novalis was introduced to some of the most cutting-edge developments in natural science of his time, which tended to center on questions about the generative force of life and the emergence of life-forms.[7]

Novalis's view of the minute and its role within the order of knowledge has an eye to the *longue dureé*: he connects the specificity of the small with universal types of thinking that—in his perspective—had organized the epistemic access to the world since antiquity. Thus Novalis emphasizes in an entry of the *Logological Fragments* in 1798, "The crude, discursive thinker is the scholastic. The true scholastic is given to mystical subtleties. He builds his universe out of logical atoms. He destroys all living nature in order to put a mental trick in its place—his goal is an infinite automaton" (L I, no. 13, 49). In the subtle thinking of the *micrologue*, nature is segmented and reduced to a conglomerate of atoms. Life is, Novalis therefore criticizes, treated as a dead form and automaton, organized by the laws of mechanics.[8] The "myopic" thinker contrasts with the view of the antimechanistic and "far-sighted" (II, 383) *macrologue*, whose broad gaze is directed at the random movements and dynamic forces that permeate life as a whole and, hence, the whole of nature: "He hates rules and fixed form. Wild, violent life reigns in nature—everything is vivified. There is no law—but only the arbitrary and the miraculous everywhere. He is purely dynamic" (L I, no. 13, 49). By juxtaposing the smallest particles to a concept of unified dynamics, Novalis refers back to a natural-philosophical discussion dating back to antiquity, which he calls elsewhere "the atomistic and dynamic sectarian strife" (AB, no. 648, 118). According to the atomistic doctrine that was closely connected to mechanistic models of life, matter was believed to consist of inseparable, imperishable, tiny particles with solid nuclei, which—often somewhat vaguely—were referred to as either atoms, corpuscles, or particles.[9] This model was questioned in the seventeenth and eighteenth centuries by dynamists following Leibniz, who, in accordance with the newly developed infinitesimal calculus, emphasized the principle of continuity within nature (*natura non facit saltus*).[10] By defining the nature of matter by a duality of forces—that is, the "moving forces of attraction and repulsion originally inherent in them"[11]—it was Immanuel Kant who laid the foundation for a Romantic, antimechanistic, dynamic notion of nature that was taken up and expanded upon by thinkers like Schelling, Ritter, and Georg Wilhelm Friedrich Hegel.[12]

At first glance, Novalis seems to have committed himself to the dynamic concept of nature. Not only does he repeatedly emphasize the infinite divisibility of matter and thus adopt the central argument against atomism,[13] but he also refuses to understand the originary moment of creation (or creative beginnings) as a purely mechanical process: "The beginning of all life must be antimechanical—opposition to mechanism" (II, no. 230, 364; my translation).

Yet the "merely dynamic" conception of nature is also confronted with a predicament when it comes to the explanation of life: if everything is merely active dynamics and disorderly movement, then no "fixed form" (L I, no.13, 49)—that is, no delimited, contoured forms; no microcosm; no individuality—is possible.[14] Unsurprisingly, for the Romantic author, it is the figure of the ingenious artist who, in the *Logological Fragments* cited above, succeeds in mediating by thinking of both concepts as being "united in a common principle" (L I, no. 13, 50). Because the "artist" is able to grasp both atomism and dynamism, matter and force, as organic "limbs" (*Glieder*) of one being, he offers a unifying model that functions as both an epistemic model for the explanation of life *and* a creative model for the work of the artist: "Here that *living* reflection comes into being, which with careful tending afterwards extends itself into an infinitely formed spiritual universe—the kernel or germ of an all-encompassing organism. It is the beginning of a true *self-penetration of the spirit* which never ends" (L I, no. 13, 50). Atomism and dynamism combine in a new model of the organism: the whole is neither the sum of its (atomistic) parts nor a random, encompassing power but a constantly growing conglomeration that self-organizes; it becomes, in other words, a self-organizing process. By referring to the common principle, Novalis shows himself committed to an idea of a vital power of formation (*Bildungskraft*) that at the end of the eighteenth century not only was posited (and elaborated by Kant) as the key principle responsible for the generation and self-organization of living organisms[15] but also entered into aesthetic discourse: conceptualized as *Einbildungskraft*—that is, the power of imagination—and "*living* reflection," it constitutes, for Novalis, the "germ" of an "all-encompassing organism" (L I, no. 13, 50).[16]

Within this framework, Novalis's fragment implicitly introduces a metonymic logic of substitution: the organic microform, the germ, replaces the indivisible, solid, unchangeable atom as the smallest basic element of nature and cipher of the beginning. Seeds, germs, and pollen henceforth function in Novalis's work as the epitome of an organic potentiality. As a miniature embodiment of the organism, they form a new small *and* at the same time large world. They are continuously evolving—though they are already complete—and open to the future; they are dynamic but delimited, autonomous but not anarchical. The small germ becomes and contains the promise of a new beginning, of organic life, but, in a metaphoric sense, also of its artificial equivalent, the aesthetic piece of art. *Blütenstaub*, or "pollen," Novalis calls his 1798 collection of fragments. These collected small germs constitute not only his first publication but also the first edition of the leading organ of early Romanticism, the journal *Athenaeum*. Quite aptly, they are a work of art and an "origin" point for Novalis and the early Romantics.

Media of Excitement: The Organism and Its Environment

From a historical-epistemological view, it is the perspective on the individual organism Novalis develops in his discussion of Schelling and John Brown that rebalances the relationship between part and whole, small and large, individual and general, in an innovative way: as a question not of similarity or mise en abyme but of interaction and mutual dependence of the individual organism—the human being—and its other, the outside world.[17]

In a fragment dated from April 1800, Novalis describes his resistance to the analytical method that "isolates" its objects "from their surroundings, the history, the soil" and thereby "comes too close to the nature of things and by focusing on this single viewpoint forgets its value as an element of the larger whole" (II, 813; my translation). On the one hand, this interjection seems to reflect the holistic worldview that informed the Romantic critique of the Enlightenment and its rational-scientific dogma. At the same time, however, Novalis's criticism of an epistemological method predicated upon "isolation" contains a thoroughly materialistic punch line: the individual phenomenon is implicitly declared to be dependent on "surroundings, the history, the soil," influenced and shaped by the contingent circumstances of its existence. In other words, "Everything that surrounds us, daily incidents, ordinary circumstances, the habits of our way of life, exercises an uninterrupted influence on us, which for just that reason is imperceptible but extremely important" (L I, no. 27, 54).

With the concept of a determinant "surrounding" or environment (*Umgebung*), Novalis calls forth a prehistory of the concept of "milieu." Playing a pioneering role in conceptualizing the interrelations between internal orders and their external worlds since the nineteenth century, this concept can be traced back to the antique tradition of anthropogeography, which gained new popularity in (early) modern Europe, when the influence of environmental conditions (soil, temperature, climate, history) on the human constitution were discussed in philosophical and political writings from Jean Bodin and Montesquieu to Johann Gottfried Herder.[18] While historically this concept of "surroundings" has primarily focused on the matter of geographical influences, Novalis expands its purview to include the panoply of external and internal factors that constitute and condition the individual. When Novalis connects the question of the "effect of warm air on the chest" to the insight that "everything is a stimulus—relations of stimuli" (AB, no. 371, 55), he commits to a physiological model of the organism that became popular in medicine in the late eighteenth century. Based on the experimental physiological studies of Albrecht Haller in the 1770s, it was disseminated by the Scottish physician John Brown, whose main work, *Elementa Medicinae* (1780), quickly received broad attention in German intellectual circles.[19]

Novalis, whose notebooks demonstrate an intense preoccupation with Brown,[20] was fascinated by his works but also criticized its "tendency toward mechanics."[21] He therefore shifted the "Brownian system" in two decisive ways. On the one hand, Novalis assigned two different forms of excitability to the stimuli. While "sensibility" controls the reactions to internal stimuli, "irritability" is responsible for processing external stimuli: "Excitability," Novalis notes in *Das Allgemeine Brouillon*, "consists of elastic sensibility and elastic irritability" (AB, no. 440, 69).[22] On the other hand, Novalis expands the concept of "sensibility," which he wants to understand not only as a purely physiological state but also as an attribute of the spiritual. Consequently, the inner stimuli of the human organism consist not only of—as in Brown—nerves, brain activities, and bodily fluids but also of spiritual forces: "The internal inciting potentials themselves are a composition—out of soul and body—in different relations" (AB, no. 504, 90). Instead of a one-sided relation of determination of soul and body, inner world and outer world, the individual and their environment enter into a "perfect reciprocal relation to one another" (LG I, no. 70, 61). Both systems mutually depend on each other.

But how is this "reciprocal relation" (*Wechselverhältnis*) of inner and outer worlds possible? How can the individual organism be thought of as both—conditioned by its environment and, at the same time, autonomous?[23] Novalis's solution is not only characteristic of the dialectical thinking of early Romanticism but also represents—as Nelly Tsouyopoulos has worked out for Schelling—an important stage in the modern concept of milieu. He installs an instance of mediation, an instance that belongs to both worlds: the inner and the outer worlds. In Novalis's work, the body and the senses function as this instance of "touch" (*Berührung*),[24] as "intermediary" and "medium," connecting the inside and outside:[25] "The body is the intermediary—the product, as it were, of both these infinitely variable quantities [between the inner and outer stimuli—between the soul and world], the irritant, or better still, the medium of stimulation. The body is at once the product and the modification of the stimulation" (AB, no. 399, 61). The individual body, the organism, is, on the one hand, the product of the separation of inner and outer worlds, of inner and outer stimuli. On the other hand, the body and the senses—Novalis especially refers to the tactile sense—operate as the medium of their interaction. The function of the body is thereby rendered potent: it mediates the stimuli of the external world (*Außenwelt*) but is also an external stimulus itself.[26] What Novalis ultimately suggests is an ontological doubling of the "external world" into an "external world outside us," which cannot be directly perceived, and an "external world within us":[27] "We now behold the true bindings connecting subject and object—behold that there is also an external world within us [*Außenwelt in uns*], united in an analogous manner with our internal being, just as the external world outside us [*Außenwelt außer uns*] is united with our external being; and hence the former and latter are joined, like our internal and external realms" (AB, no. 820, 151). The fundamental separation between inner and outer worlds is therefore repeated in the organism itself—and it is exactly this duplicity that functions as the condition of possibility of the organism's autonomy and the *Eigensinn* (obstinacy) of life.

"Individual Calculus": Infinitely Small Differences

IN DISCUSSING the work of Brown and Schelling, Novalis develops an innovative physiological model of the organism in which the complex mutual dependence between the organism and its environment is mediated by the body itself. However, with his project of "infinitesimal medicine" (AB, no. 399, 62), anticipated in the *Allgemeine Brouillon*, Novalis seems to transcend the physiological discourse of life as it flourished in his time. The epistemic description of the organic body transforms into a utopian doctrine of its perfection.[28]

As Brown, who identified the cause of all illnesses as a disturbed relationship between the organism and its environment, also Novalis's thoughts on the organism revolve around the organism's health (and illness).[29] At the same time, Novalis criticizes Brown, whose therapeutic measures were only aimed at regulating *external* stimuli (such as alcohol, certain foods, etc.) and the restitution of a disturbed balance (cf. AB, no. 399, 62). The new art of medicine Novalis has in his mind is instead intended to be a "theory of the ordering of life" (*Lebensordnungslehre*) with a higher destination: "The artist of immortality practices higher medicine—infinitesimal medicine. . . . He practices medicine as a higher art—as a synthetic art" (AB, no. 399, 62). The "art of the formation and improvement of the constitution" (62) does not aim at restitution but rather at the improvement and reconfiguration of the individual organism that ultimately leads to infinite health—that is, immortality. This requires an adjustment of how the organism reacts to stimuli. The perfect constitution, according to Novalis, is the one that combines highest irritability (the receptivity to large stimuli) with highest sensitivity (the receptivity to the minutest stimuli): "Receptivity toward large—and small stimuli—receptivity for both together.... The greater the excitability (if we wish to describe the synthesis by this name) ... the more perfect the constitution" (AB, no. 437, 68).

Since, according to Novalis, the external stimuli are already large and the outer milieu under control of the individual, the main challenge of the "artist of immortality" is the stimulation of sensitivity—that is, "the gradual increase in the inner stimulus" (AB, no. 399, 62).

The process of infinitesimal medicine, in other words, implies the manipulation of the organism's inner world that produces and processes inner and external stimuli. Only a sensitive organism with finely tuned, easily excitable inner circulation is able to absorb and process the smallest "imperceptible tensions and motions": "If the highest irritability manifests itself in vigorous motions and tensions, then in contrast the highest sensibility manifests itself in imperceptible tensions and motions. Irritability reveals itself via large variations and effects—Sensibility via small variations and effects—Infinite irritability via infinitely large variations—infinite sensibility via infinitely small variations" (AB, no. 409, 64).[30] In strengthening the sensitivity for the inner, infinitely small stimuli, Novalis implements a model that privileges the power of the minute over and against the physical-sensual superiority of the sublime. The micrologist, who *sees and recognizes* the smallest things, is replaced by the microphenomenologist, who *feels or perceives* the infinitely small, imperceptible forces, tensions, and dynamics.

This power of the infinitely small is by no means to be understood only metaphorically, for according to Novalis, it is precisely the strengthening of the inner sphere of stimulation by exercise that promises an increase in the independent activity and self-regulation of the organism. The organism can now not only select, locate, dissect, and distribute the irritability of its organs but can also make "modifications of the real world as he will." The organism thus creates its own external world; that is, the individual becomes the artist and creator of the reality and milieu around them: "The artist has vivified the germ of self-formative life in his sense organs—he has raised the excitability of these for the spirit and is thereby able to allow ideas to flow out of them at will—without external prompting—to use them as tools for such modifications of the real world as he will" (L II, no. 17, 72). It is no coincidence that Novalis's attempt to integrate the infinite into the finite organism is centered around the anthropological-physiological adaptation of the "infinitely small."[31] Especially during his time in Freiberg, Novalis was intensively preoccupied with the infinitesimal calculus,[32] which already Leibniz understood as the final refutation of atomism. For Novalis's own physiological model of the organism, the infinitesimal calculus entailed two important shifts: instead of designating a determinable, substantial unit, the minute becomes a concept of relation, which aims at the ever-decreasing distances *between* two entities.[33] At the same time, the inherent principle of continuity challenges the model of duality: entities that are polarized in the classical model of organism—the infinitely large versus the infinitely small, irritability versus sensitivity, outer versus inner, soul versus body, health versus illness—appear instead as two extremes on a continuous scale, whose gradual shifts take place perceptibly, without being distinguishable as distinct, punctual units.[34]

In terms of a theory of the subject, the ability of the sensitive organism to perceive infinitely small stimuli and gradual tensions corresponds to its own infinite divisibility: the individual's constitution, assets, and dispositions. Novalis rejects the idea of the individual as being by definition "indivisible," a concept that was closely connected with the conceptual history of the atom since antiquity.[35] Instead, the individual emerges as the result of many infinitesimally small differences: "Leibniz," Novalis writes in his notebook, "also terms the infinitesimal calculus: analysis indivisibilium. (Constant quantities—constant transitive quantities.) Infinitesimal calculus really means calculation, division or measurement of the nondivisible—noncomparable—immeasurable. Analysis indivisibilium = analysis of an individuum—individual calculus—genuinely physical calculus" (AB, no. 645, 118).[36] One can recognize the nonreductionist potential of Novalis's reformulated theory of the organism and its milieu in this focus on the minute as an entity determined purely by relation: it substitutes the interaction between a dichotomously defined inside and outside by an associative, dynamic, minutely differentiable network of references. In this manner, the individual is placed in a relationship of infinite variation and minuscule distances not only to its environment but to other individuals.[37] As the object and occasion of infinitesimally multiplying microdifferentiations and microcharacterizations, the individual ironically multiplies (i.e., stretches and grows) into infinity: "An infinitely characterized individual is a member of an infinitinomium" (AB, no. 113, 19). The promise of a successive formation of the massive from the small, whose paradigmatic image was the germ, is inverted: the potential to multiply or grow ad infinitum corresponds to a potential for infinite division into increasingly smaller units.[38] "The common or the smaller character can be infinitely developed right into the smallest details. Likewise the larger character" (AB, no. 239, 44). This is the surprisingly egalitarian side of a psychology of minute differences in Novalis: not only the artist but even the most ordinary person can participate in infinity on a small scale.[39] Everyone is "the seed of an infinite genius."[40]

The Novel in the Medium of Poetry: *Henry of Ofterdingen*

THE FIGURE of the artist features prominently in Novalis's writings. On the one hand, it is the artist who, as previously discussed, succeeds in uniting a dynamic and atomistic worldview in the organic model of the "germ": the imagination (*Einbildungskraft*) of the artist is conceived analogously to the inherent power of formation (*Bildungskraft*), which, in the natural sciences of the era, was posited as the organizational principle of life par excellence. On the other hand, Novalis's "transcendental medicine" characterizes the type of sensitive existence necessary to perceive the infinitely small and minute (and thus to emancipate itself from the outside world) as a genuinely *artistic* one.

In this manner, his work suggests that the laws of life are intricately interwoven with the laws of art. This interweaving becomes even more explicit when Novalis's attention turns to the function of poetry: the order of literature and the order of the world, according to Novalis, are structurally related to each other. Life, he argues, can be understood as a novel in which the subject is placed in a distinct relationship with the surrounding circumstances: "We live in a colossal novel (writ large and small). Contemplation of surrounding events" (AB, no. 853, 155). Yet if the world is a huge novel, the "real poet" and the poet's work are also a "real world in miniature" (II, 381). To understand the work of art as a world—this is a common topos of Romantic thought—means to understand it as a living organism. Accordingly, and contrary to Schlegel's well-known commentary on Novalis's "atomistic" thinking,[41] the central metaphor that Novalis himself employs for his writing is the germ, the seed. "Everything is seed," Novalis concludes in a fragment from 1798 (L, 100, 66). His aphorisms are germs, implying the hope for further development and advancement, for manifold life: "Friends, the soil is poor, we must sow abundant seeds / So that even modest harvests will flourish."[42]

It is the artist's task, he intimates, to shape the fragile organism of the art and its environment. First, Novalis formulates the conditions of an extraliterary environment that encounters the artwork in the mode of critique: a better harvest, or a greater aesthetic impact, requires a greater receptivity of the "soil," or a larger and better-instructed readership, that knows "how the poem must be read—under what circumstances [*Umständen*] alone it can please"

(F, no. 414, 133).[43] Second, however, the work of art contains within itself its own environment that forms, in turn, the relationship to the reader: "Every poem has its relations to a variety of readers and numerous circumstances—It has its own surroundings, its own world, its own god" (133). The work of art, one is given to understand, is just like the organism; it has a double external world: an internal and external one.[44] It stands in a productive and receptive conditional relationship to its extra-aesthetic reality, which is determined by the coordinates of time, place, and the disposition of the reader. At the same time, it creates an innerliterary, aesthetic outside world, which operates as the external condition of the narrated characters, scenes, and actions that, in turn, affects the reader.

This complex organization of the art organism, one might assume, is especially true for the novel, a genre that, as Georg Lukács has observed, was constituted precisely through the will to form an organic, living totality.[45] The novel, as Novalis reads it, therefore finds its fulfillment in the "harmonious mood," "where everything finds its proper aspect—everything finds an accompaniment and surroundings that suit it" (LF, no. 3, 153). As is well known, in his *Theory of the Novel*, Lukács reproached the Romantic novel for its "immoderateness," writing that "the inner wealth of pure soul-experience is seen immoderately as the only essential thing."[46] Since in a modern age, the outside world is no longer in harmony with the inner world but is instead perceived as "completely atomised or amorphous," as "heterogeneous, brittle and fragmentary parts,"[47] the Romantic novel withdraws to the promise of poetry and therefore to a "subjectivity that is uninterrupted by any outside factor or even."[48] In this way, however, the Romantic novel refuses, according to Lukács, to shape the intricate relationship between the outer and inner world, between the individual's self-activity and its determination by its environment, by the "prosaic vulgarity of outward life."[49]

Lukács's analysis of the Romantic "triumph of poetry"[50] refers explicitly to Novalis's novel *Henry of Ofterdingen*, which prompts us to take a closer look at the relationship between natural philosophy, organism theory, and the novel in Novalis. The significance of poetry for this particular novel, I would like to argue, does not imply a mere retreat into subjective inwardness. Instead, Novalis's physiological model of the organism instantiates poetry as the fluid medium of the most subtle and fine stimuli *between* or at the threshold of the inner and outer worlds. The novel corresponds in a certain sense to his project of a "synthetic poetry" that wants to be both: "analysis of the external and the internal at the same time" (GD, no. 25, 129).

Novalis interrupted the work on his first novel fragment, *The Novices of Sais*, and began to write his novel *Henry of Ofterdingen* in 1799. He worked on it until shortly before his death in 1801. *Henry of Ofterdingen* is conceptualized not only as a *Bildungsroman* but even as an explicit counterproject to the most famous *Bildungsroman* of his time, Goethe's *Wilhelm Meister's Apprenticeship*. In contrast to Goethe's novel, where inside and outside worlds are mediated by an omniscient narrator, the narrative point of view in *Henry of Ofterdingen* can be read as evidence for the author's disinterest in the outside world: the focus is internal but variable. The reader learns about the external circumstances of Henry's family background and everyday life, his outer appearance, his encounters, and his experiences over the course of the protagonist's journey, above all from Henry's inner perspective or that of his companions.[51] It is, however, precisely the protagonist's inwardness, his penchant for interiority, that is addressed from the very beginning as a problem: the fact that Henry was "far more quiet and inclined to brood than formerly" (*HO*, 24) is occasion for motherly concerns and motivates the journey south. There, the new and unfamiliar landscape, the confrontation with foreign cultures, the encounters with new types of people, and, according to the "secret" hope of the mother, the "charms of a girl in her native Augsburg" ("charms" is the translation of German *Reize*, which means also "stimuli"; *HO*, 24) serve the goal of reviving the young man's constitution and of brightening his "gloomy mood" (24). The mother apparently plays on the physiological register Novalis learned from Brown; the educational journey appears as a school of excitability, the *Bildungsroman* as the logbook of a stimulating cure aiming to restore the fragile equilibrium of organism.

In fact, it is the journey itself that brings Henry to the banal but essential realization that he *has* a specific outside world. The displacement from his milieu, the experience of separation elicited by a new environment, allows Henry to perceive for the first time the border between inner and outer worlds, between the organism that the individual represents and the world that surrounds him.[52] Only by leaving his familiar surroundings again, when descending into the narrow world of the caves, does Henry become aware of the permeability of this border. The separation enables him to recognize the manifold relationships and connections with the world: "Now he surveyed at a glance all his relations to the wide world around him, felt what he had become through it and what it would become to him, and grasped all the strange concepts and impulses he had often felt in contemplating it" (*HO*, 77). Realizing the close connections to the "environment" (*Umgebung*) and "region" (*Gegend*) through which the individual "becomes" what he is—that is, realizing the external conditionality of one's own existence—is presented as a precondition for the possibility of overcoming these bonds. Henry's reflection is accompanied by insight into a hierarchy of species that reaches from inorganic to organic beings and differentiates between their various degrees of dependence:

> I have only got to know my home region [*Gegend*] fully ... since leaving it and seeing many other regions. Every plant, every tree, every hill and mountain has its particular setting [*Gesichtskreis*], its peculiar territory [*Gegend*]. This belongs to it, and its structure and its whole make-up are explained by it. Only animals and man can move from place to place; all regions are theirs. So all of them together make up a world environment [*Weltgegend*] and a limitless setting which in turn influence animals and man as plainly as the more restricted environment [*Umgebung*] affects the plant. (HO, 162)

Humans and, in a certain sense, animals are able to live in different milieus and are able not only to adapt to a given environment but to appropriate it. The novel presents various examples of such flexible existences: the miner and the hermit but also the merchant and the crusader.[53] While the crusaders and merchants adapt to the different "settings" of the world only at the price of their hardening (*Abhärtung*)—that is, through sensory desensitization—the miner and hermit stand for the opposite process:[54] that of an infinite sensitization. This is the path, it may be recalled, of transcendental health and the perfect constitution, and it is also the path that is laid out for the future artist Henry.

On his educational journey, the protagonist is, on the one hand, exposed to a multitude of external stimuli that appeal to his senses; the entries corresponding to these in the draft sheets for the novel are "sheen, smell, colors, and aridity—exciting [*reizende*] figures"[55] and "travel—acquaintance with people of many kinds—opinions of many kinds." The notes further suggest that the author planned even more entanglements of his hero with the lowlands of the prosaic world, the world of "bourgeois trades" (*Händel*), wars, and military conquests.[56]

In order for Henry to be united with the coveted blue flower at the destination of his journey, Novalis's preliminary notes, on the other hand, indicate that the protagonist must first "be made susceptible" (I, 393). What Henry experiences over the course of the narrative is a refinement and strengthening of his inner senses that transpires through the awakening of a stimulus that not only was unknown to Henry until then but also is, according to Novalis, the strongest of all stimuli. This stimulus is love.[57] Henry thereby becomes that prototypical figure of the artist whose seemingly "passive nature" (I, 389) imbues him with a "multisided susceptibility" to interface with the phenomena of the external and internal world. His "calm, attentive mind" prevents him from forming overly close attachments to "objects," "insignificant [*kleinliche*] affairs," and "earthly business" (I, 384f.; my translation) and instead enables him to perceive the smallest stimuli, relations, and connections that develop between him and the world around him. This act of ever-more-minute differentiation and characterization, as Novalis's doctrine of infinitesimal medicine implies, is an infinite one, and so, it can also be assumed, is the novel, which itself is devoted to this infinite (if not unfinishable) project.[58]

With the perspective of the smallest dynamic relations, stimuli, and tensions, the novel aims to both manifest and produce a multilayered and variated perceptivity, which seems to dissolve the border between inside and outside in the vibrating and elastic state of the in-between. It is no coincidence that the temporal-spatial coordinates of the novel are also situated in a transition, in the middle. It is this condition between two states that simultaneously functions as a condition of possibility for both the protagonist and his narrative: "During every period of transition higher spiritual powers appear to want to break through as in a sort of interregnum [*Zwischenreich*]. And just as on the surface of our dwelling place the districts richest in natural resources ... lie between the wild and inhospitable primeval mountains and the boundless plains, so between the rough and crude times of barbarism and the modern age abounding in wealth, art, and knowledge" (*HO*, 25). This spatial-temporal "in-between" is the location where the real gems can be found (as the gold miner knows),[59] and it is in this milieu that the individual is constituted; the individual, we remember, receives his individuality precisely through those minutest differences that it processes in relation to himself and to the other. The individual is itself a being in-between (*Zwischenwesen*). In the "ardent, living state between two worlds"—between a "transparent" "outer world" and a "varied" "inner world"[60]—the individual and the novel are equally in the middle, "au milieu" of their own formation.

The anthropological order is shifted from a topical-static order to a process-oriented one. The middle and the transition become the locus of mutual contact between the individual with himself and his other, and this midpoint is associated with an omnipresent, oscillating, floating condition. Novalis had described already this dynamic state of the in-between earlier in his first novel fragment, *The Novices of Sais*, as new mode of (self-)perception and (self-)recognition:[61] "New kinds of perceptions [arise], which appear to be ... strange contractions and figurations of an elastic liquid. From the point where he has transfixed the impression, they spread in all directions with a living mobility and carry his self with them. Often he can stop this movement at the outset by dividing his attention or letting it wander at random, for thoughts seem to be nothing other than emanations and effects which the self induces all around in that elastic medium" (*NS*, 75). In Novalis's concepts of an "elastic medium," regulating the relationship between the inside and outside of the individual, the old idea of a fluid, generative milieu reemerges. Circulating in the concept history of the "milieu" since antiquity, it was updated through Newton's concept of an ethereal *ambient medium* before it became crucial to Claude Bernard's concept of the inner milieu in the nineteenth century.[62] For Novalis, however, it is less the physiological than the poetological consequence that seems to be of interest: it is poetry that takes over the function of this elastic medium and inner, fluid milieu that mediates and connects the inner and outer worlds. "Poetry [*Die Poesie*]," Novalis writes to August Wilhelm Schlegel, "is fluid by nature—completely malleable [*allbildsam*]—every stimulus moves in all directions. . . . It becomes a kind of organic being—whose whole construction reveals its emergence out of fluidity, its originally elastic nature" (I, 656f.; my translation).

Understanding the novel as an organism inevitably means equipping it with the medium of poetry. Only poetry, with its multiple and all-around receptivity to infinitesimally small stimuli, guarantees mediation and continuity without harshly exposing the work, the individual, to the outside world. Poetry as a fluid medium within the organism itself functions as the "inner milieu" of the work of art. Novalis asserts accordingly in the notes to the *Ofterdingen* novel, "Poetry speaks between each chapter" (I, 392). He thereby addresses the many poems, songs, and fairy tales that are inserted in the prose text, not least the prominently located *Klingsohrs Märchen*, which he places in the middle of the novel fragment.[63] It is only in this fragile space of in-between where the duality of inner and outer worlds seems to be suspended in favor of a kind of radiation spreading in all directions. The mobile elasticity of poetry does not destroy the form in process—a risk that unquestionably the novel fragment is constantly demonstrating—but on the contrary serves as its condition of possibility. Poetry, the fluid medium, makes the life of the organism possible in the first place as well as the novel.

Novalis's conception of the organism as a model of both life and art is, to conclude, centered around the complex relationship between the inner and outer worlds, between the organism and its environment. Novalis thus participates in a multifaceted prehistory of the concept of milieu that increasingly came to inform biological and sociological theories in the nineteenth century. However, it is precisely by interrogating the function of the small that he develops a genuine, personal commitment. While Novalis's reflections on the organic miniature form of the germ already testify to a differentiated approach to the life science discourse of his time, it is above all his preoccupation with the relational concept of the minutely small—as infinitesimally small difference and distance—that makes his theory of the organism such an original contribution within the history of a poetic knowledge of life. It initiates the transformation of a static-topological order into a processual-relational one, in which the relationship between the organism and its outside appears as one of multiple, multidirectional microrelationships and effects. In the elastic, mobile space between inside and outside, the principle of duality is superseded by a state of infinitely small differences, and it is precisely this place where the individual can emerge as an autonomous entity—or its aesthetic counterpart, the novel.

Notes

1. Novalis, *Werke*, is cited in text by the volume number (I, II, III) and page number (my translation). Except for a few untranslated fragments and notes (my translations), I use the English translations. AB = Allgemeines Brouillon (cited by Novalis, *Notes for a Romantic Encyclopaedia*); L I and L II = *Logological Fragments* I and II; FL = Faith and Love; T = Teplitzer Fragments; LF = Last fragments (all cited by Novalis, *Philosophical Writings*); FS = Novalis, *Fichte Studies*; HO = Novalis, *Henry of Ofterdingen*; NS =Novalis, *Novices of Sais*.

    I want to thank Sasha Rossman for his excellent proofreading and Christiane Frey and Elizabeth Brogden for their many helpful comments and careful editing of this essay.
2. The German term *Keim* invokes connotations of the English terms *germ* and *seed* and will be translated in the following essay depending on the context.
3. In this essay, I use the terms *minute* and *minutiae* in a neutral, nonpejorative sense, referring to minimal details, the smallest elements, and the finest differences (as elaborated by Frey and Brodgen in their introduction) and as a specification of the general attribute *small*. The term *small* will be used exclusively in the context of the (originally) mathematical concept of the "infinitely small," as adapted by Novalis.
4. To my knowledge, there is only one article that looks more closely at Novalis's concept of environment: Wanning, "Poet and Philosopher," 43–62.
5. See, for example, Neumann, *Ideenparadiese*; Stadler, "Kleines Kunstwerk."
6. See Frey, "Observing the Small," 381.
7. Biareishyk, therefore, proposes the term "Freiberg Romanticism" (see Biareishyk, "Rethinking Romanticism," 272).
8. In his aphorism collection *Faith and Love*, Novalis suggests that the atomistic-mechanical idea of life also informs political structures. On Novalis's criticism of the "state factory" as presented in the state of Friedrich Wilhelm I, see Matala, *Der verfaßte Körper*, 131–72.
9. Cf. Snelders, "Atomismus und Dynamismus," esp. 187.
10. Snelders, 187.
11. Kant, *Metaphysical Foundations of Natural Science*, 72.

12. Cf. Snelders, "Atomismus und Dynamismus," 189–91.
13. Cf. Snelders, 190.
14. Tsouyopoulos referred to this problem in her groundbreaking studies on the concept of organism in Schelling's work (Tsouyopoulos, "Schellings Konstruktion," 593).
15. The biological principle of self-organization became known under the terms of *vis vitalis* (Friedrich Wolf) and *Bildungstrieb* (Johann Blumenbach). On the success story of the concept of *Bildungstrieb*, which was closely related to an epigenetic concept of development, see Müller-Sievers, *Self-Generation*.
16. For theories of creation in Novalis, see Holland, *Procreative Poetics*; and Lehleiter, "Novalis' Denken im Kontext der zeitgenössischen Biologie."
17. On the natural, aesthetic, and political status of the organism in Novalis as well as its epistemicohistorical roots in contemporary physiology and natural science, see Neubauer, *Bifocal Vision*; Uerlings, *Friedrich von Hardenberg*; Matala, *Der verfaßte Körper*; Holland, *German Romanticism and Science*.
18. Cf. Canguilhem, "Living and Its Milieu," 9; Wessely, "Milieu: Zirkulation und Transformation," 10f. Novalis was probably familiar with these ideas through Herder, who discussed his project of a "climate history of all human thought and sensory powers" (*"Klimatologie aller menschlichen Denk- und Empfindungskräfte"*) in his *Ideas toward a History of Humanity*.
19. Brown, *Elements of Medicine*. On Brown's theory of the organism, cf. Cheung, *Organismen*, 29–42, as well as Henkelmann, *John Brown*.
20. Novalis probably read the German translation (1796) of Brown by Christoph Heinrich Pfaff that was based on the extended English version. He also definitely read Andreas Röschlaubs *Untersuchungen über Pathogenie* (1798–1800), where the German physician refers extensively to Brown. Cf. Uerlings, *Hardenberg*, 169.
21. "The . . . falseness of the Brownian system is its tendency toward mechanics" (AB, no. 721, 133). Novalis ignores that Brown already provides a self-regulating moment in his concept of the organism (cf. Henkelmann, *John Brown*, 31).
22. Novalis intensively read Kielmeyer's book *On the Relationships of Organic Forces* (1793), where he distinguishes five forces: sensitivity, irritability, reproduction, secretion, and propulsion. Sensitivity forms the highest force (cf. Saul, "Mitteilen," 159f.).
23. Schelling described this problem in his lectures on nature philosophy (1799) explicitly: "If the external world could determine the organism as subject then it would cease to be excitable. Only the organism as object is determinable through external influences, the organism as subject must be unreachable by them" (Schelling, *System of the Philosophy of Nature*, 106; cf. Tsouyopoulos, "Schellings Konstruktion," 597f.).
24. On the concept of "touch" as the gateway to a materialistic tradition following Spinoza, see Biareishyk, "Rethinking Romanticism," 276–84. On Novalis's anthropology of the senses, see Wellmon, *Becoming Human*, 213–35.

25. Novalis strives for a solution similar to that which Schelling developed in his *System of the Philosophy of Nature* (1799). Instead of the body, however, Schelling sees the "organism (taken as a whole)" as the "medium through which external influences act upon it"; Schelling, *System*, 107; cf. Tsouyopoulos, "Schellings Konstruktion," 597f.
26. Already in the *Fichte Studies* (1795–96), Novalis writes on this process of communication: "That is, at the same time the body serves to communicate outer objects to the soul via the senses, and insofar as it is itself an outer object, it acts on itself as such, through [the effect of] the senses upon the soul" (F, no. 568,170).
27. Here, too, Novalis corresponds with Schelling's idea for an "original duplicity" of the organism: "'The organism should itself be the medium, etc.' means (expressed more generally) nothing other than: *there must be an original duplicity in the organism itself*. . . . 'There must be in the organism an original duplicity' means, therefore, . . . precisely *that the organism must have a dual external world*" (Schelling, *System*, 107). See also Tsouyopoulos, "Schellings Konstruktion," 597f. and Cheung, *Organismen*, 116f.
28. Novalis's project of "infinitesimal medicine" is mostly conceived in terms of therapeutical-medical and metaphysical implications (cf., for example, Krell, *Sexuality, Disease, and Death*; Pethes, "Novalis' 'Infinitesimalmedizin'"). So far, to my knowledge, only Holland dealt with the implications of the infinitely small within Novalis's "infinitesimal medicine," connecting it with the "Law of Minimum" in homeopathy; cf. Kuzniar, *Birth of Homeopathy*.
29. According to Brown, the excitability of the body is either too great (*sthenia*) or too small (*asthenia*); Brown, *Elements of Medicine*, 52.
30. Haller's and Schelling's assumption of a reverse reciprocity of the two stimuli is thus explicitly invalidated (AB, no. 409, 64).
31. The infinitesimal calculus informed the interest of the early Romanticist for the infinite; see Smith, "Friedrich Schlegel's Romantic Calculus."
32. Novalis deals with the infinitesimal calculus in the "Freiberg Natural Scientific Studies," the "Teplitz Fragments," and the "Allgemeine Brouillon." On Novalis's occupation with the infinitesimal calculus, see Pollack, "Novalis and Mathematics Revisited"; Pollack-Milgate, "Mathematical Infinite"; Bomski, *Die Mathematik*.
33. The infinitesimal calculus, which Leibniz and Newton developed independently of each other in the seventeenth century, provided a mathematical formulation for the infinite differentiation of various magnitudes tending to zero. In this process of infinite division, a series of endlessly tiny intervals are created that are continuously connected without being resolved into independent parts or entities. The infinitely small thus approaches in a certain sense the scholastic understanding of the relation as a nonsubstantial, tiny entity (cf. Gasché, *Of Minimal Things*, 2f.).
34. Cf. AB, no. 479, 82; AB, no. 554, 98. In Novalis's case, the transformation of the supposed oppositions into different degrees affects almost all anthropological

categories, from excitability, health, and soul to the will, the human, etc. (Senckel, *Individualität und Totalität*, 55).

35. Since the Greek term *atom* was translated in Latin as "individuality," the two principles were repeatedly crossed over in the history of concepts (Kaulbach, "Individuum, Atom," 299f.).
36. By asserting the mathematical-physical infinitesimal calculus (which Leibniz called analysis indivisibilium—i.e., the analysis of the nondivided and immeasurable) as a method for the analysis of the individual, Novalis differs from Leibniz: while Leibniz rejected an atomistic view of the material world and insisted on the infinite divisibility and continuity of matter, he regarded the individual as the only "true," indivisible atom and "monad" (cf. Kaulbach, "Individuum, Atom," 299f.).
37. "We catch a glimpse of ourselves as an element in the system—and consequently, in an ascending and descending line, from the infinitely small to the infinitely large—human beings of infinite variations" (AB, no. 820, 51).
38. Novalis takes up an idea that Schelling formulated in the *Weltseele*: "In each organization the individuality (of the parts) goes to infinity" (Schelling, *Weltseele*, 223).
39. "Even the most ordinary character can be infinitely developed. The differentials of the infinitely large behave like the integrals of infinitely small—because they are also one" (AB, no. 290, 43). Passavant has discussed the surprising mediation of the ordinary and the eccentric in Novalis's idiosyncratic poetics: Passavant, *Nachromantische Exzentrik*.
40. "Every person is the seed of an infinite genius. They may be divided into numerous people and yet still be one. The true analysis of the person as such brings forth people" (AB, no. 63, 10).
41. See Neumann, *Ideenparadiese*, 282.
42. Novalis, *Philosophical Writings*, 4.
43. According to Novalis, however, the work of art is conditioned by the external world not only in terms of its reception but already in terms of its production: "Insofar as we are contemporaries at a certain time or members of a specific body, we are nevertheless hindered by it in the higher development of our nature. Divinatory, magical, truly poetic people cannot come into being under circumstances such as ours" (L I, no. 27, 54f.).
44. Cf. Pethes, "Milieu," 153–55.
45. Lukács, *Die Theorie des Romans*.
46. Lukács, 118.
47. Lukács, 113, 124.
48. Lukács, 118.
49. Lukács, 104. By transferring the modern concept of milieu to the aesthetic work of art, I refer to Pethes (Pethes, "Milieu"), who deals in particular with Hegel's lectures on aesthetics. According to Pethes, Hegel assigns the concept of prose to the everyday, contingent environments of the living—i.e., the "Außenwelt"—while poetry is associated with the concept of beauty of a living freedom and signifies the "Innenwelt."

50. Lukács, *Die Theorie des Romans*, 140.
51. See Stadler, "Novalis und Lavater," 186f.
52. "Henry left his father and his native city with sorrow [*in wehmütiger Stimmung*] in his heart. Now for the first time it became clear to him what separation means. His preconceptions of the journey had not been accompanied by the strange feelings he now had when first his familiar world was torn from him and he was washed up as it were on foreign shore" (*HO*, 26).
53. At the same time, however, Henry is confronted with tragic fates that result precisely from the impossibility of "adapting" to culturally and geographically foreign milieus. Thus for Zulima, the Palestinian prisoner of war, separation from her own "setting" comes at the price of loneliness and melancholy, while the young children of the old hermit who grew up abroad didn't adapt to the "harsher climate of the West" (*HO*, 89) and died shortly after their arrival in Europe.
54. "People born to carry on trade and business ... have to take a hand in a great many things and hurry through a host of details; they have to steel their minds, as it were, against the impressions of a new situation, against the distractions of many and diverse objects, and accustom themselves even through the press of great events to hold fast the thread of their purpose and skillfully bring it to a conclusion. Their soul may not indulge in introspective reverie; it must be steadily directed outward" (*HO*, 93).
55. The German *reizend* means both "charming" and "stimulating."
56. The German *Händel* means both "trade" and "quarrel." According to Novalis's notes, the "unity ... of the novel" should result precisely from the "fight of poetry with non-poetry": "Henry of Af[terdingen] is meddling with civil trade [*Händel*] in Switzerland ... Italian trade [*Händel*]. Here Henry becomes a commander. Description of a combat" (I, 389; my translation).
57. "All stimuli are relative—they are quantities—except for one which is absolute.... The most perfect constitution arises through stimulation and absolute union with this stimulus. This stimulus is—*absolute love*" (FL, no. 53, 96). On poetry and love as the strongest inner stimulus in Novalis, see Matala, *Der verfaßte Körper*, 154–57.
58. In his notes on Novalis's unfinished novel, Tieck reports that it was "the poet's intention to write six more novels after the completion of Ofterdingen, in which he wanted to set down his views on physics, bourgeois life, plot [*Handlung*], history, politics and love, and, as in the Ofterdingen, of poetry" (I, 405; my translation).
59. "With what veneration," the miner reports, "I saw for the first time in my life ... the king of metals in delicate flakes embedded in the rock" (*HO*, 67) (original: "*in zarten Blättchen zwischen den Spalten des Gesteins*"; *HO*, I, 289).
60. This state of the in-between is described in *The Novices of Sais* as the state of the sensitive human being in nature: "The outer world [*Außenwelt*] becomes transparent and the inner world [*Innenwelt*] becomes varied and

meaningful; thus man finds himself in an ardent, living state between two worlds" (NS, 77). This state of the in-between, on a vibrating border between inner and outer world, is, according to Novalis, also the state in which the soul is constituted (II, 233).

61. Pethes points out that the term *elasticity* was introduced by Lavoisier for describing the state of matter between liquid and solid in his *Traité élémentaire au chimie* of 1789 (Pethes, "Der Topos 'Prozessualität,'" 141).
62. See Spitzer, "Milieu and Ambiance," part 1, 6, 34–42. Pethes relates the "elastic medium" in Novalis to the concept of a liquid inner milieu in Bernard's work, without, however, reflecting on the implications for the concept of milieu in Novalis himself.
63. In *Klingsohrs Märchen*, the crossing and connection of borders, areas, and persons are significantly conducted by the permanently circulating and mediating "fabula." See Althaus, *Strategien*, 19f., who argues though that the poetic attempt of an overarching connectivity is the condition not of possibility of the subject but of its dissolution.

Bibliography

Althaus, Thomas. *Strategien enger Lebensführung: Das endliche Subjekt und seine Möglichkeiten im Roman des 19. Jahrhunderts*. Hildesheim, Germany: Olms, 2003.

Biareishyk, Siarhei. "Rethinking Romanticism with Spinoza: Encounter and Individuation in Novalis, Ritter, and Baader." *Germanic Review* 94, no. 4 (2019): 271–98.

Bomski, Franziska. *Die Mathematik im Denken und Dichten von Novalis: Zum Verhältnis von Literatur und Wissen um 1800*. Berlin: Akademie Verlag, De Gruyter, 2014.

Brown, John. *The Elements of Medicine, or A Translation of the Elementa Medicinae Brunonis*. 2 vols. London: Johnson, 1788.

Canguilhem, Georges. "The Living and Its Milieu." Translated by John Savage. *Grey Room*, no. 3 (2001): 6–31.

Cheung, Tobias. *Organismen: Agenten zwischen Innen- und Außenwelten 1780–1860*. Bielefeld, Germany: transcript, 2014.

Daiber, Jürgen. *Experimentalphysik des Geistes: Novalis und das romantische Experiment*. Göttingen: Vandenhoeck & Ruprecht, 2001.

Frey, Christiane. "The Art of Observing the Small: On the Borders of the 'Subvisibilia' (from Hooke to Brockes)." *Monatshefte für deutschsprachige Literatur und Kultur* 103, no. 3 (2013): 376–88.

Gasché, Rodolphe. *Of Minimal Things: Studies on the Notion of Relation*. Stanford, CA: Stanford University Press, 1999.

Henkelmann, Thomas. *Zur Geschichte des pathophysiologischen Denkens: John Brown (1735–1788) und sein System der Medizin*. Berlin: Springer, 1981.

Holland, Jocelyn. *German Romanticism and Science: The Procreative Poetics of Goethe, Novalis, and Ritter.* New York: Routledge, 2009.

Huber, Florian, and Wessely, Christina. "Milieu: Zirkulation und Transformation eines Begriffs." In *Milieu: Umgebungen des Lebendigen in der Moderne,* edited by Florian Huber and Christina Wessely, 7–17. Paderborn: Wilhelm Fink, 2017.

Kant, Immanuel. *Metaphysical Foundations of Natural Science.* Translated and edited by Michael Friedman. Cambridge: Cambridge University Press, 2004.

Kaulbach, Friedrich. "Individuum, Atom." In *Historisches Wörterbuch der Philosophie,* edited by Joachim Ritter, vol. 4, 299f. Basel: Schwabe 1976.

Köchy, Kristian. *Ganzheit und Wissenschaft: Das historische Fallbeispiel der romantischen Naturforschung.* Würzburg: Königshausen und Neumann, 1997.

Krell, David Farrell. *Contagion: Sexuality, Disease, and Death in German Idealism and Romanticism.* Bloomington: Indiana University Press, 1998.

Kuzniar, Alice. *The Birth of Homeopathy out of the Spirit of Romanticism.* Toronto: University of Toronto Press, 2017.

Lehleiter, Christine. "'Wer weiß, ... welch wunderaren Generationen uns noch ... bevorstehen': Novalis' Denken im Kontext der zeitgenössischen Biologie." *Blütenstaub: Jahrbuch für Frühromantik* 5 (2019): 137–52.

Lukács, Georg. *The Theory of the Novel: A Historico-philosophical Essay on the Forms of Great Epic Literature.* Cambridge, MA: MIT Press, 1971.

Matala de Mazza, Ethel. *Der verfaßte Körper: Zum Projekt einer organischen Gemeinschaft in der politischen Romantik.* Freiburg im Breisgau: Rombach, 1999.

Müller-Sievers, Helmut. *Self-Generation: Biology, Philosophy, and Literature around 1800.* Stanford, CA: Stanford University Press, 1997.

Neubauer, John. *Bifocal Vision: Novalis' Philosophy of Nature and Disease.* Chapel Hill: University of North Carolina Press, 1971.

Neumann, Gerhard. *Ideenparadiese: Untersuchungen zur Aphoristik von Lichtenberg, Novalis, Friedrich Schlegel und Goethe.* München: Fink, 1976.

Novalis. *Fichte Studies.* Edited by Jane Kneller. New York: Cambridge University Press, 2003.

Novalis. *Henry of Ofterdingen.* Translated by Palmer Hilty. Long Grove, IL: Waveland Press, 1992.

Novalis. *Notes for a Romantic Encyclopaedia: Das Allgemeine Brouillon.* Translated, edited, and with an introduction by David W. Wood. New York: State University of New York Press, 2011.

Novalis. *The Novices of Sais.* Translated by Ralph Manheim. Brooklyn: Archipelago Books, 2005.

Novalis. *Philosophical Writings.* Translated and edited by Margaret Mahony Stoljar. New York: State University of New York Press, 1997.

Novalis. *Werke, Tagebücher und Briefe Friedrich von Hardenbergs.* Edited by Hans-Joachim Mähl and Richard Samuel. 3 vols. München/Wien: Hanser, 1987.

Passavant, Nicolas von. *Nachromantische Exzentrik: Literarische Konfigurationen des Gewöhnlichen.* Göttingen: Wallstein Verlag, 2019.

Pethes, Nicolas. "'In jenem elastischen Medium': Der Topos 'Prozessualität' in der Rhetorik der Wissenschaften seit 1800 (Novalis, Goethe, Bernard)." In *Rhetorik: Germanistische Symposien*, edited by Jürgen Fohrmann, 131–51. Stuttgart: J. B. Metzler, 2004.

Pethes, Nicolas. "Milieu. Die Exploration selbstgenerierter Umwelten in Wissenschaft und Ästhetik des 19. Jahrhunderts." *Archiv für Begriffsgeschichte* 59 (2017): 139–56.

Pethes, Nicolas. "Vom 'Erfindungsgeist neuer Experimente' zur 'transcendentalen Gesundheit': Novalis' 'Infinitesimalmedicin' als Modell einer Poetik der Prosa." *Blütenstaub: Jahrbuch für Frühromantik* 5 (2019): 59–71.

Pollack, Howard. "Novalis and Mathematics Revisited: Paradoxes of the Infinite in the Allgemeine Brouillon." *Athenäum* 7 (1997): 113–40.

Pollack-Milgate, Howard. "'Gott ist bald $1 \cdot \infty$—bald $1/\infty$—bald 0': The Mathematical Infinite and the Absolute in Novalis." *Seminar: A Journal of Germanic Studies* 51, no. 1 (2015): 50–70.

Ritter, Johann Wilhelm. *Beweis, dass ein bestaendiger Galvanismus den Lebensprocess in dem Thierreich begleite: Nebst neuen Versuchen und Bemerkungen über den Galvanismus*. Weimar: Verl. des Industrie-Comptoirs, 1798.

Saul, Nicholas. "Blütenstaub. Leben und Mitteilen. Zum Kommunikationsbegriff der Romantik." *Blütenstaub: Jahrbuch für Frühromantik* 5 (2019): 153–70.

Schelling, Friedrich Wilhelm Joseph von. *First Outline of a System of the Philosophy of Nature*. New York: State University of New York Press, 2004.

Schelling, Friedrich Wilhelm Joseph von. *Von der Weltseele, eine Hypothese der höhern Physik zur Erklärung des allgemeinen Organismus*. Hamburg: Perthes, 1798.

Senckel, Barbara. *Individualität und Totalität: Aspekte zu einer Anthropologie des Novalis*. Tübingen: Niemeyer, 1983.

Smith, John H. "Friedrich Schlegel's Romantic Calculus: Reflections on the Mathematical Infinite around 1800." In *The Relevance of Romanticism: Essays on German Romantic Philosophy*, edited by Dalia Nasser, 239–57. New York: Oxford University Press, 2014.

Snelders, Henricus Adrianus Marie. "Atomismus und Dynamismus im Zeitalter der Deutschen Romantischen Naturphilosophie." In *Romantik in Deutschland: Germanistische Symposien*, edited by Richard Brinkmann, 187–201. Stuttgart: J. B. Metzler, 1978.

Spitzer, Leo. "Milieu and Ambiance: An Essay in Historical Semantics." *Philosophy and Phenomenological Research* 3, no. 1 (1942): 1–42; and no. 2 (1942): 169–218.

Stadler, Ulrich. "Kleines Kunstwerk, kleines Buch und kleine Form: Kürze bei Lichtenberg, Novalis und Friedrich Schlegel." In *Die kleinen Formen in der Moderne*, edited by Elmar Locher, 15–36. Studien-Verlag: Innsbruck 2001.

Stadler, Ulrich. "Novalis und Lavater: Hardenbergs höhere 'Physiognomie' im Heinrich von Ofterdingen." In *Physiognomie und Pathognomie: Zur literarischen Darstellung von Individualität*, edited by Wolfram Groddeck, 186–201. Berlin: de Gruyter, 1994.

Tsouyopoulos, Nelly. "Schellings Konstruktion des Organismus und das innere Milieu." In *Philosophie der Subjektivität? Zur Bestimmung des neuzeitlichen Philosophierens*, 2 vols., edited by Hans Michael Baumgartner and Wilhelm G. Jacobs, 591–600. Stuttgart: Frommann-Holzboog, 1993.

Uerlings, Herbert. *Friedrich von Hardenberg, genannt Novalis: Werk und Forschung*. Stuttgart: Metzer, 1991.

Wanning, Berbeli. "Poet and Philosopher: Novalis and Schelling on Nature and Matter." In *Ecological Thought in German Literature and Culture*, edited by Gabriele Dürbeck, 43–62. Lanham: Lexington Books 2017.

Wellmon, Chad. *Becoming Human: Romantic Anthropology and the Embodiment of Freedom*. Pennsylvania: Pennsylvania State University Press, 2010.

Le menome cose
On the Function of Details in Leopardi's *Zibaldone*

ELENA FABIETTI

✦ ✦ ✦

Minutus, minuto ec. from minuo, for piccolo [small].

Here and further in this essay, I will reference the English translation of the Zibaldone, edited by Michael Caesar and Franco D'Intino (2013). The first number (e.g., Z4246) provides the original page number of Zibaldone, which can be traced in any Italian version too, and is followed by the page number of the English edition. The Italian expressions from Zibaldone that I provide in brackets when deemed relevant are taken from the critical edition edited by Pacella (Leopardi 1991). The essay epigraph is from Z4246, 1880.

GIACOMO LEOPARDI's *Zibaldone di pensieri* is a work that was never intended for publication and was published only posthumously, sixty years after the death of its author.[1] Giacomo Leopardi (1798–1837) kept writing this personal intellectual journal over sixteen years of his life (1817–32), a wide time frame inevitably filled with a variable intensity of the writing and an active development of his thought. In the more than four thousand pages of this work, Leopardi jotted down, annotated, copied from other texts, working on sharpening his concepts and theories, as well as on pinning down the exact, etymological meanings of words. The *Zibaldone*—a name that designates a varied mix, or hodgepodge—has been variously defined as *journal philosophique*[2] or as a writer's workshop: a place for tentative investigations but also a technical laboratory partly destined to merge into his published work.[3] Leopardi created indices to navigate through the thousand pages, proposing thematic paths as well as drafts for possible future works.[4] By alternating explorations of language, emotions, natural history, social history, and metaphysics, Leopardi collects and weaves them together into a philosophical and poetic fabric that will serve, in part, as material for his other writings, both poetry and prose. From this pragmatic scope, the *Zibaldone* draws—together with the repetitive, contradictory, and open quality of its observations—a special concentration on the detail in various forms: the single example, the individual case, the singular aspect or form of things and words. This thematic focus reflects epistemological concerns circulating at the busy juncture between European Enlightenment and Romantic culture, including the reflections on empirical particulars by eighteenth-century novelists discussed in Roger Maioli's essay or Novalis's aesthetics of the infinitely small in Mareike Schildmann's. In the workshop set up by Leopardi's extraordinarily capacious work, these threads are recorded, sifted, and originally elaborated.

In this essay, I want to show how this general dedication to details is sustained, beyond the praxis of writing, by specific gnoseological intentions that assign the minute detail a crucial function in enabling and shaping the knowledge of a larger, general reality, thereby establishing a necessary relation between *minutiae* and *milieus*. The inexhaustible philosophical drive that guides Leopardi's poetics—which has been captured in the happy critical formula of "pensiero poetante"[5]—is key to understanding not just the relevance of the technical work on details conducted there but also how a particular concentration on details can help disclose philosophically urgent truths.[6] The minute detail, in all its possible forms, is invested with the gnoseological function of disclosing aspects of reality otherwise concealed, which, by becoming visible, transform the apprehension of reality entirely.

I will begin by showing the heuristic function of the detail in Leopardi's gnoseological reflections throughout the *Zibaldone*, thereby identifying the epistemic potential assigned to the *minutia* in relation to a *milieu*, which in this case coincides with an operative system of knowledge. I will then turn to the central place in *Zibaldone* where I believe Leopardi invests the detail with an ontological function proper. Written in 1826 and one of the most famous passages from this work, the description of a lovely garden turns this environment, through a skillful visualization of some details of the botanical *milieu*, into a place of creaturely suffering. I will interpret this passage from the standpoint of an ontology of the detail that reconfigures the ontological status of its *milieu*; in order to do this, I will refer to Leopardi's reception of eighteenth-century philosophical as well as scientific thought, weaving into this selective account at least some of the threads of Leopardi's thought concerning nature, his theory of pleasure, and the notion of the infinite. Lastly, I will address the function of the detail in Leopardi's *Zibaldone* from the point of view of his poetics proper, which is a decisive lens through which to look at his overall thought. The interconnection between poetics and philosophy in Leopardi's "poetic thought" is the reason why the detail across Leopardi's reflections can be defined as poetic, even when the focus lies on other intellectual operations. Parallel and coextensive to the gnoseological, as well as the ontological, power Leopardi assigns to the details is his emphasis on the *minutia* integrated into an aesthetic whole (*milieu*) through poetic imagination.

To summarize, in the *Zibaldone*, the role of the detail—which Leopardi, in different contexts with different emphases and interchangeably, also refers to as *minutia* (in Italian: *minuzia*), minimal thing (*menoma cosa*), very small thing (*piccolissima cosa*), or again, the particular (*il particolare*)—is diversified and spread across multiple epistemic levels, ranging from the heuristic and gnoseological through the ontological to the poetic. In my concluding paragraph, I will look at the question of scale as one fundamental structure of thought and thematic concern in Leopardi's *Zibaldone*, a theme that could be envisioned as the primary generator of his concern for the minute and the small in all their variations. The question of scale, as I will show, fastens the philosophical dimension to the poetic by inserting them into a structure of knowledge (as well as of aesthetic experience) that doesn't sacrifice aesthetic pleasure for analytical truth.

"Fingernails": A Gnoseology of Details

"IF I may be permitted an observation regarding a trifling matter [*una minuzia*]," pleads Leopardi at the beginning of a passage from 1821 (Z1307, 1, 625), in which he seeks to exemplify the notion of habituation (*assuefazione*) as "cause and norm" of aesthetic sense (Z1306, 624). The *minutiae* he wishes to present turn out to consist of the "minute parts of the human body that man is only able to observe with difficulty . . . in others," such as "the nails of the hand." Humans can conveniently observe such bodily details upon themselves, becoming completely used to their exact features and proportions. The effect of this familiarity with the minute details of one's own body is that other people's bodies appear to us ugly and deformed, often triggering a sense of "revulsion." By observing one's own corporeal minutiae, then, Leopardi consolidates his claim that "habituation" is the fundamental factor in regulating aesthetic taste. This sketch of aesthetic theory yields to a broader epistemological reflection just a few lines below: "These observations are trifling, yet it is only by unraveling, investigating, unmasking, pondering, and observing trivial matters [*menome cose*], and by resolving what are actually great matters into their trivial parts [*menome parti*] that a philosopher arrives at great truths" (Z1310, 626). Observational and speculative procedures based on minute details, ranging from particular cases to singular forms and details of objects, are not just corollaries and exemplifications but heuristic turning points in the understanding of reality.

This passage shows Leopardi's sensualist heritage and his discursive use of the body as the fundamental groundwork for philosophical knowledge.[7] Leopardi's interest in the body has been read on multiple levels. Michael Caesar critically addresses interpretations hinging on Leopardi's own diseased and nonconforming body.[8] Caesar moves beyond biographism to show how Leopardi's thinking about the body was shaped by eighteenth-century empiricism as well as the physiognomical and moral literature available in his library (including Lavater, Cabanis, and d'Holbach).[9] Leopardi's "human being" is, from this vantage point, not only a biological "body" but—as Andrea Campana has put it—the *homme physique-moral* forged by Buffon, Cabanis, and the other *idéologues*.[10] The ubiquity of the corporeal dimension in Leopardi, based on his empiricist-sensualist formation, becomes a topic in his historical thinking and a building block of his poetics. The juxtaposition between ancients and moderns, for example, plays out in Leopardi as a history of loss of the bodily dimension: "ancient" poetry being conceived as a language interpreted through the body, which became, with "modernity," abstract and incorporeal.[11]

Beyond expounding the bodily dimension, the passage on fingernails can be seen as one possible articulation, through corporeal semantics, of Leopardi's broader gnoseology of the detail. The observation of the minute detail here enables the apprehending of philosophical truths, thereby presenting one form of the relation between the particular and the universal within an empiricist gnoseology. The philosophical roots of the sensist discourse on the body and the gnoseology of the empiricist detail do in fact coincide. Already starting around 1819, Leopardi had embraced Locke's critique of innate ideas and his attendant theory of experience and knowledge; in an entry from 1821, he writes that "everything is taught to us by our sensations alone" (Z1340, 638). Later, playing on an established philosophical simile, he writes, "Each man is like a soft dough [*pasta molle*], susceptible to every possible shape, impression, etc. It hardens over time, and at first it is difficult, and finally it is impossible to give it a new shape" (Z14552, 683).[12] The gnoseology Leopardi borrows and then works into his thought operates through single sensations, which can be analyzed only by paring things down to their irreducible components. The operativity of an empiricist gnoseology is, in other words, that of the minimal parts. And in fact, Leopardi uses the passage on fingernails and habituation to affirm that philosophers can access truth only by investigating the "minimal things" and by resolving big things in their "minimal parts" through minute observation (*osservazione minuziosa*; Z481, 1, 269). All this bears heavily on the epistemological level. The first of its consequences is a fundamental relativism: "everything is relative" precisely because it is based on "the tiniest [*menome*], incidental everyday differences and ways" (Z1259, 1; Z452, 256). Furthermore, Leopardi's gnoseology of minute details leads to a more general epistemology of "unlearning" (Z4190, 1832) and endless possibilities ("all things are possible"; Z1341, 639).[13]

As we are going to see with increasing evidence in the next pages, the power of the minute detail is heuristic only in relation to the greater complex of knowledge it is supposed to disclose. This structure of knowledge rests on the Lockean theory of ideas, in particular with regard to their combination, which is conceptualized as a synthetic operation captured by the metaphor of the "glance" (coup d'oeil):

> Minute, refined analysis [*la minuta e squisita analisi*] is not the same as seeing at a glance [*colpo d'occhio*] and never discovers a major point of nature, the center of a great system, the key, the mainspring, the entire workings of a great machine (Z1852, 831).... When you set out to compose a great whole out of the most minutely [*più minutamente*] but separately considered parts, you run into a thousand difficulties..., a sure sign and necessary consequence of the lack of the ability to take things in at a glance that discovers the things contained in a vast field, and their reciprocal relations. (Z1852, 832)[14]

This synthetic function is later assigned to the faculty of imagination, which once more Leopardi borrows from English sensualism, mediated through the *Encyclopédie* and the French *philosophes*.[15] Analytic and imaginative operations have the potential to make moments of knowledge coextensive with those of aesthetic experience, guiding the poet as much as the philosopher: "The lyric poet, when inspired, the philosopher in the sublimity of speculation, the man of imagination and sentiment in the throes of enthusiasm..., sees and looks at things as though from a high place" (Z3269, 1341). A bird's-eye view seems to be a prerequisite for creative achievements. In the next paragraph, I will look at the ontological effects of this gnoseological model in one of *Zibaldone*'s most renowned passages.

The Garden in Pain: Toward an Ontology of Details

IN BOLOGNA between 1825 and 1826, Leopardi was exposed to new social and intellectual interactions that left a profound trace on the *Zibaldone*.[16] The following passage attests to the literary, philosophical, and scientific echoes that Leopardi weaves together and engages with:

> Go into a garden of plants, grass, flowers. No matter how lovely it seems. Even in the mildest season of the year. You will not be able to look anywhere and not find suffering. That whole family of vegetation is in a state of souffrance, each in its own way to some degree. Here a rose is attacked by the sun, which has given it life; it withers, languishes, wilts. There a lily is sucked cruelly by a bee, in its most sensitive, most life-giving parts. Sweet honey is not produced by industrious, patient, good, virtuous bees without unspeakable torment for those most delicate fibers, without the pitiless massacre of flowerets. That tree is infested by an ant colony, that other one by caterpillars, flies, snails, mosquitoes; this one is injured in its bark and afflicted by the air or by the sun penetrating the wound; that other one has a damaged trunk, or roots; that other has many dry leaves; that other one has its flowers gnawed at, nibbled; that other one has its fruits pierced, eaten away. That plant is too warm, this one too cold; too much light, too much shade; too wet, too dry. One cannot grow or spread easily because there are obstacles and obstructions; another finds nowhere to lean, or has trouble and struggles to reach any support. In the whole garden you will not find a single plant in a state of perfect health. Here a branch is broken by the wind or by its own weight; there a gentle breeze is tearing a flower apart, and carries away a piece, a filament, a leaf, a living part of this or that plant, which has broken or been torn off. Meanwhile you torture the grass by stepping on it; you grind it down, crush it, squeeze out its blood, break it, kill it. A sensitive and gentle young maiden goes sweetly cutting and breaking off stems. A gardener expertly chops down trunks, breaking off sensitive limbs, with his nails, with his tools (Bologna, 19 April 1826). . . . The spectacle of such abundance of life when you first go into this garden lifts your spirits, and that is why you think it is a joyful place. But in truth this life is wretched and unhappy, every garden is like a vast hospital (a place much more deplorable than a cemetery), and if these beings feel, or rather, were to feel, surely not being would be better for them than being (Bologna, 22 April 1826). (Z4175–76, 1823)

The passage opens with a glance at an apparently lovely garden, which is, however, almost immediately exposed as a place of suffering. The description of the garden unfolds by zooming in on single details of its botanic environment: the rose, the lily, bees, ants, caterpillars, flies, snails, and mosquitoes, followed by textural details of trunk barks, leaves, flowers, and grass. The different actions taking place on the scale of these small elements and their outcomes show a pervasive condition of torment and suffering among the inhabitants of this environment. It is, in particular, the texture of the natural objects (the injured, wounded bark; the dry leaves; the nibbled flowers; the pierced fruits) that shows the suffering most visibly, revealing brokenness, corruption, and corrosion.

The description of the garden employs a technique of close-ups that achieves visually powerful results. The transition from the wide-angle shot of the first delightful garden frame to the close-ups of the single suffering elements is so drastic that it becomes impossible afterward to look at the garden the same way as at the beginning. The initial glance at a garden of delights transforms, by focusing on its details, into the vision of something entirely different: a garden of extensive suffering, which Leopardi fittingly calls a "vast hospital." What is crucial in this transition is the transformative function played by the minute detail as such, a transformation that inheres in gnoseological experience and that I will henceforth call "ontological" because it reorganizes and thereby establishes anew the ontological *milieu* of the garden.[17] By focusing on the details of the garden, it becomes possible to unveil its truth: that is, to grasp the ontological dimension of its beings, which for Leopardi is a universal condition of infelicity. The relation between truth and infelicity is a theme that runs through Leopardi's thought all along: already in 1821, he had written, "The true is unhappy through and through" (Z1974, 874).

The garden passage, with its marked descriptive literary quality and its pace of philosophical experiment, entails obvious literary and philosophical echoes. The passage was written in April 1826, following an essential development in Leopardi's thought that is generally dated around 1824. Until then, Leopardi had been largely shaped by the influence of Rousseau, which is reflected in the recurrent antinomy nature versus reason, formulated, early on, in comparative terms of scale: "Reason is the enemy of all greatness: reason is the enemy of nature: nature is great [*grande*], reason is small [*piccola*]" (Z14, 15).[18] By 1826, Leopardi had shifted from a Rousseauian view of nature as fundamentally benevolent (only corrupted by civilization) toward an assessment of general (not just creatural but also universal) unhappiness (*infelicità*), or what could be called "an anthropology of evil."[19] The *Zibaldone*'s garden passage is significantly preceded by a more abstract and generalizing passage about the evil nature of the universe, introduced by the apothegmatic statement "everything is evil" (Z4174, 1822), which the description of the garden is supposed to exemplify through a sort of empirical and ostensive demonstration.[20]

The shift in emphasis from a first Rousseauian phase has been variously defined as a transition, a turn, a caesura, or even an implosion in Leopardi's thought.[21] For Sergio Solmi, however, this transition is rather a critical fiction that tries to freeze Leopardi's everchanging, intrinsically contradictory view of nature. The rhythm of this thought "in movement" is paced by open reflections following a model of writing closer to the essay than systematic philosophy, which privileges a circumstantial argumentative pattern.[22]

The main rhetorical model in the garden passage is that of the *locus amoenus*, whose structure Leopardi overturns, attaining a sort of "reversed Eden."[23] The garden as locus amoenus offers a special topography of security as well as the aesthetic guarantee of curated botanical beauty. All this is challenged by Leopardi and ultimately invalidated through a skillful aesthetic inversion achieved through the aforementioned visualizing technique of close-ups. The reversal of the topos of the Garden of Delights wasn't new either, and Leopardi was surely familiar with other literary passages playing with an anti-idyllic description of nature, such as famous passages from Goethe's *Werther* and lines by Vincenzo Monti.[24]

Beyond these literary echoes, a complex of philosophical and scientific themes finds its way into the passage. As scholars have extensively demonstrated, Leopardi's intellectual thinking is profoundly shaped by experimental sciences.[25] Leopardi had educated himself in the main scientific fields of astronomy, mechanics, optics, chemistry, botany, zoology, physiology, and, as already noted, physiognomy. An early reader of scientific authors and manuals, Leopardi was also an extremely precocious compiler of scientific compendia, including the *Dissertazioni fisiche* and the *Trattato dell'astronomia*, written when he was just fourteen and fifteen, respectively.[26] His access to scientific knowledge, however, was limited to what made its way to the provincial town of Recanati, and his formation was, for the most part, bookish, the most current developments of his time beyond his reach.[27]

A connoisseur of botany, Leopardi had read Linnaeus, which was present in his father Monaldo's extensive family library.[28] In the garden passage, Leopardi shows a stark anti-Linnean position as to the "economy of nature"—one that will only intensify with time: in 1829, he writes that "there are an infinite number of disorders in the course of things.... They are clearly disorders, and they cannot be attributed to the intention of nature" (Z4461–62, 2019).[29] Leopardi was also very familiar with Buffon's thirty-six volumes of the *Histoire naturelle*, included in his father's collection and available to him in the Italian translation from 1782.[30] Buffon begins his *Natural*

History with a methodological premise that seems to have left a fundamental trace on Leopardi's thought: "For it can be said that the love of the study of nature supposes two qualities of mind which are apparently in opposition to each other: the grand view of the ardent genius who takes in everything at a glance [coup d'oeil], and the detailed attention [*petites attentions*] of an instinct which concentrates laboriously on a single minute detail [*un seul point*]."[31] As to the question of the economy of nature, however, Buffon held the view of nature as "grand design," as he wrote in his chapter "General Views of Nature."[32] In that section of his text, Buffon also described nature as a potentially destructive system, if left to its own devices.[33] He imagined a man staring at this desolate nature and declaring, "Uncultivated nature is hideous and unflourishing; it is I alone who can render it agreeable and vivacious."[34] Leopardi is also therefore distancing himself from Buffon: while the French scientist claimed that human intervention could lift nature up to a state of comfort, beauty, and organization, Leopardi no longer grants this redeeming task to human activity. The undoing of the organizational principles of an economy of nature, which manifests some pre-Darwinian tensions,[35] can be best understood in line with the shift in Leopardi's thinking about nature that took place around 1824. In *Dialogo della Natura e di un Islandese*, one of the best-known philosophical dialogues constituting the *Operette morali* (published in 1827), Leopardi assigns to the figure of the Icelander the task of condemning an undomesticated, destructive, and always threatening nature.[36]

Furthermore, the passage shows the possible influence of the debate over plant sensibility and the materialism of Brownian medicine through its Italian reception.[37] But beyond a skillful re-elaboration of preexisting themes and models, the passage acquires its originality from the function it assigns to details.

Leopardi's eclectic scientific formation may have indeed provided fundamental tools to appreciate the ontological function of the minute, minimal things (*menome cose*) in relation to their greater, more general environment.

Andrea Campana argues that Leopardi's broad, general scientific knowledge enabled him to build sociophysical and sociomoral analogies using the vocabulary of science, mobilizing a metaphorical language with a gnoseological rather than literary function.[38] Campana's excellent study has, for example, reconstructed, in Leopardi's work, the role played by the scientific element of the molecule, which Leopardi calls a "globetto"[39]—merging the Galileian name for celestial bodies and the Cartesian name for the simplest unit of matter while also showing familiarity with microscopic biology, derived mainly from the tables and descriptions of Robert Hooke's *Micrographia*

(1665).⁴ᶜ And it is another author of microscopy, Antony van Leeuwenhoek, author of the *Arcana naturae* (1715–22), who calls the invisible parts made visible by the microscope *globuli*. It is precisely the discourse of microscopy that is woven into the garden passage as one of its potentially constitutive dimensions. In the Recanati family library, furthermore, Leopardi could use an actual microscope, which was one of several scientific instruments his father made available to his children.[41] Microscopy is one of the scientific fields that has shown to be most productive for the literary imagination.[42] Barbara Stafford, looking at the works of some of the great microscopy divulgators of the seventeenth century, has defined the advent of microscopy as a decidedly aesthetic "enterprise," effectively "conveying the vivid impression that life down under the lens was even more bloody and rapacious than existence above it. Before Darwin, they showed nature to be red in miniature tooth and claw."[43]

Microscopy enters eighteenth-century literature by not just providing metaphors for old things but uncovering an unknown universe, as one can infer from Henry Baker's dedication in *The Microscope Made Easy* (1743), where he assigns this instrument "a farther discovery of the minute wonders of the creation; which may not, perhaps, improve our knowledge less than the grander parts thereof."[44] Beyond the promissory wonders to be discovered in nature's minimal elements, however, these works also opened up the vision of horrific actions occurring on the smallest scale of reality. This dimension is particularly productive precisely because it breaks the line of expectations (the wonders of nature) stemming from the classic understanding of nature's economy. In Laurence Sterne's *Tristram Shandy*, of which Leopardi owned a French translation from 1784,[45] the imaginative resources of microscopy are active in at least one famous passage, which describes *ad absurdum* an imaginative "optical beehive" as the inner space of a transparent human being; the passage unravels the image of the beehive by crowding it with crawling maggots.[46] This satirical passage offers the space to represent interiority in terms of a disturbing motion, dismantling the utopian fantasy of a transparent human and thereby expounding a new expressive dimension that the visual potential of Hooke's *Micrographia* helps shape. In this visually powerful potential of reversing apprehensions and expectations lies the ontological function of the detail that Leopardi might have refined under the suggestion or influence of his knowledge of microscopy.

The *minutiae* in the passage of the garden thus become something more than a gnoseological tool, attaining an ontological function: when the detail of the whole is analyzed—when the minute part of the system, by becoming visible, is subject to scrutiny—it may well reveal a completely different ontological composition of the system itself. Familiar with the images and discourse of microscopy, Leopardi seems to have embraced its lessons and exploited its aesthetic resources in the garden passage. The *minutiae* under the microscopic gaze of the poet unveil an unexpectedly new (and in this case horrific) *milieu*.

A Poetics of Minute Things and Vast Imagination

ITALIAN STYLISTIC critic Vincenzo Mengaldo has insisted on locating Leopardi's poetics at a safe distance from European Romanticism, to which it has nonetheless often been associated in a number of ways. For Mengaldo it is clear how Leopardi, indebted to the materialist and rationalist roots of the illuministic European tradition, can only end up rejecting the myth of a nurturing and mysterious nature through its transcription in symbolic language.[47] Leopardi's anti-Romantic inclination consists, according to Mengaldo, of a stylistic dimension characterized by attention to the single, concrete realist detail.[48] Departing from a cumulative, expansive representation, Leopardi privileges a concise and condensed representation of things: a style of "continence."[49] It is impossible here to reconstruct even summarily the linguistic and stylistic elements in Leopardi's capacious oeuvre, which ranges from lyric to prose and encompasses multiple registers of both.[50] My limited scope here pertains to the poetic dimension expounded in *Zibaldone* and only with regard to that ontology of the detail I set up to investigate.

A key passage in the reflections on poetics from 1823 can be read as one variation on the pervasive discussion on ancients and moderns, which iterates the well-known *querelle*: "In Greek tragedians (as in the other ancient poets or writers, too), we do not find the same minute details, the same particular and distinct description and development of passions and characters so typical of modern dramas ... because the ancients were not strong on detail, and did not care for it greatly, indeed they scorned it and avoided it, and precision and minute detail [*l'esattezza e la minutezza*] were as untypical for the ancients as they are typical and characteristic of the moderns" (Z3482, 1423–24). The ancients' lack of precision makes them "far inferior to the moderns in terms of their knowledge of the human heart." This poetic and psychological annotation places the power of literature in the domain of minute details. While here precision and minuteness seem to define the poetic function, other passages in *Zibaldone* seem to go in a very different direction. In an earlier passage from 1820, Leopardi addresses the degree of lexical precision. The more a language becomes specialized and technical, the more its "terms" grow, and the less rich it will be in imagination, which is conveyed and evoked through a different linguistic unit: that of the "word" (Z110, 99–100). This stark antithesis of *terms* versus *words*, which seems to contradict the previous reflections on concrete details in literature, leaves its traces in the peculiar oscillation between the imaginative, poetic pole and that of reason and analysis.[51]

A possible way to navigate this issue is to look at Leopardi's theory of pleasure. The passage immediately following that of the garden considers Anacreon's poetry and its fruition: "The beauty lies in the whole, in such a way that it is not in the parts at all. The pleasure only comes from it altogether, from the sudden and indefinable impression of the whole [*intero*]" (Z4177, 1824). Pleasure is the result of a broad, comprehensive gaze that does not stop at details. Only the *milieu* of the details grants the experience of pleasure, which is impossible through the mere analytic observation of minute components. If we think about the garden once more, the delight derives indeed from the first general apprehension of that environment. The analytic gaze, on the contrary, which focuses on details and single parts, recovers a fragmented experience of pain that is only meaningful—exposing pain and infelicity as universal ontological conditions—when referred back to the whole. While pleasure becomes impossible in the garden after the analytic operations of minute observation, pain as a dimension of meaning retains a synthetic and aesthetic quality that is made possible only by returning to the whole. Pleasure, then, is a fictional operation of suspension and interruption of analytical operations, and in this sense, it precedes the intervention of analytic reason. While this early nucleus of Leopardi's theory of pleasure (1820) still operates within a view of nature as fundamentally benevolent, the later shift toward a view of nature as inherently evil—rather than change the structure of this theory—only reinforces the hiatus between the operations of imagination and the contents of truth.[52]

From the superior poetic value of the detail in modern literature, we moved to the opposite and ostensibly contradictory disparagement of the technical word and the respective analytic gaze as an impoverishment for the imagination. To understand this oscillation, I briefly introduced the theory of pleasure, based on the imaginative faculty as opposed to the analytic faculty of reason. Taking for granted that fluctuation is the very rhythm of Leopardi's "moving thought," the point is that the two moments described here (words and imagination vs. terms and analytic reason) characterize an oscillation that tries to account for both pleasure and pain in aesthetic terms. This movement, in other words, is not just gnoseological but also foundational for poetics. In the next and last section, I will show how this epistemic network, hinging on the relation between *minutiae* and *milieu*, finds a further formulation in the conceptual frame of scale.

A Matter of Scale

THROUGHOUT THE *Zibaldone*, the reader finds many observations concerning the scale of things. Scale, already a Pascalian theme, seems to be a highly productive dimension of reflection for Leopardi in thinking about history, society, and morals.[53] In a passage from 1823, for example, he states, "No one thing shows the greatness and power of the human intellect or the loftiness and nobility of man more than his ability to know and to understand fully and feel strongly his own smallness" (Z3171, 1302). Expounding on the semantic field of the universal immensity that man faces ("he loses himself in the immensity of things"), Leopardi concludes that true (that is, metaphorical) greatness coincides with the recognition of one's own smallness.

Speaking of the world and the individual, he repeatedly enlists the lexicon of vastness and smallness, showing a very early taste for "cosmic resizing."[54] The small and the big are the stations of a perspectival shift, rarely charged with absolute meaning. In fact, nothing is small or big in itself, and it is the perception of smallness in relation to greatness that generates gnoseological experiences. Let's take the following passage from 1820 as an example of the heuristic function assigned to the scale:

> It makes no difference. Things are not small in themselves. The world is not small [*piccola cosa*] but vast, especially with respect to man. Even the organization of the most minute, invisible creatures [*più minuti e invisibili animaluzzi*] is something amazing.... But as soon as man has gained the measure of something hitherto immeasurable, as soon as he becomes familiar with the parts of it or is able to conjecture about them in accordance with the laws of reason, then that thing immediately seems very small [*piccolissima*], it no longer satisfies, and leaves him feeling very discontent. (Z246–47, 171)

If "things are not small in themselves," and smallness is only a matter of scale and proportion, then the minute detail has nothing really "trifling" about it; on the contrary, it enjoys the same ontological completeness of the "greater" element, only on a different scale. It is therefore only logical that the analysis of the small detail on an enlarged scale should provide relevant information for the assessment of the greater object (the system, or *milieu*, as in the garden). The above excerpt, however, combines the assessment of the relativity of scale with a reflection upon habituation and its effects on perception: smallness is one of the effects of familiarity with measurements, once these become available and known. Leopardi is probably also thinking of the way our world becomes small as soon as we know its extent and are able to move comfortably within its limits—a practical truth that today has the air of prophecy. In a passage from 1821, Leopardi comes back to the question of scale with regard to the world and the individual, endowing it with a historical dimension: "The more the world grows with respect to the individual, the more the individual shrinks [*impiccolisce*]. Our forebears, though they knew only a very small part of the world, and were in contact with a still smaller part [*piccola parte*] of it, and very often only with their own homeland, were very great [*grandissimi*]. We, though we know the whole world, and are in contact with the whole world, are very small [*piccolissimi*]. . . . Man and his faculties shrink [*impiccoliscono*] as the world grows in relation to them" (Z1175–76, 561–62). Colored with a moral and ambiguously metaphorical sense of "greatness" ascribed to the ancients, the passage consolidates the idea that scale grows alongside knowledge, which is another way to state the heuristic value of smallness, in this case. In fact, the only apparent existence of smallness (its relative status) is at the heart of one of Leopardi's main poetic ideas and one that his lyric poetry is best known for: the notion of the infinite. The perception of smallness becomes, in Leopardi's poetics, the catalyst for constructing the imaginative space of the infinite, which has no ontological basis in itself ("infinity is a dream, an idea, not a reality"; Z4178, 1824) but is nonetheless fundamental to fictional and poetic operations. In 1826, around the same time as the garden passage, he declared, "Nothing in nature actually announces infinity, the existence of anything infinite. Infinity is a product of our imagination, and at the same time of our smallness [*piccolezza*] and our pride. We have seen things inconceivably greater than we are, than our world, etc., forces inconceivably greater than ours, worlds greater than ours, etc. That does not mean that they are great, but that we are tiny [*minimi*] in respect to them. . . . What was incomparably greater than us and greater than our things which are tiny [*minime*], we have thought infinite" (Z4177–78, 1824). The relative impression of smallness pushes human beings to strive for the limitlessly great, the infinite, which works as a fictional projection of desire. We can see how the theory of pleasure outlined above is coextensive with the idea of the infinite.

The passages on scale seen here, with their logics based on perspective and proportion, bind together the dimension of the small and particular and that of the great and general across multiple levels: moral, psychological, gnoseological, aesthetic. The structure of scale appears as ubiquitous as it is general: "It is very important to note how the smallest effects [*menomi effetti*] derive from great causes, how large and small things harmonize together, how the nature of the age influences even the most trifling [*menome*] customs, how from the smallest [*piccolissime*], everyday observations one may make one's way back to the largest [*grandissime*] and most general" (Z1608, 742). The "harmonizing" moment in the passage should be considered carefully, as it does not commend the idea of an inherently natural harmony of all things. Rather, we have seen how, by discovering the unseen detail and investing it with the ontological function of unveiling the true being of things, Leopardi was able to endorse a reconfiguration of knowledge based on minute observations, linking together the particular and the general, the *minutiae* and the *milieu*, of knowledge and experience. The relationship between the two can be in fact utterly inharmonious, as in the garden passage, but nonetheless necessary.

We are circling back to the gnoseological quest that runs through *Zibaldone*. In a passage from 1824 that deals with Newton's system, Leopardi writes, "Great minds in physics and in other sciences and in every search after the truth and in every intellectual endeavor, have had recourse to the examination of particulars (without which it is impossible to generalize with any truth and profit)" (Z4057, 1729). In any operation of knowledge, the particular grants truth to the general, which in turn is the ultimate aim of knowledge. The attention to the particular is, therefore, aimed at generalization, which—in Leopardi's broad understanding of the "search after the truth"— means the need to always return to a more general vision, be it a gnoseological systemic assessment or an aesthetic apprehension. In another passage from 1823, this structure is described even more explicitly and with concrete reference to German philosophy, with which Leopardi didn't have an easy relation because of both prejudices and a lack of deeper knowledge:[55] "Whoever examines the nature of things using pure reason, and without the help of the imagination or feeling, or without affording either of them any scope, which is the procedure adopted by many Germans in philosophy . . . will certainly be capable of doing what the meaning of the word to analyze involves, that is, to resolve and undo nature, but they will never be able to recompose it" (Z3237–38, 1329). The passage goes on to develop this line of argument and the antithesis between analytical reason and synthetic imagination. Those who limit themselves to the first, even if successful in resolving nature "in its smallest and least elements" (*menomi ed ultimi elementi*), will always miss the meaning of the whole (*l'intero*). Providing a vivid example of this structure, Leopardi imagines the dissection of a human body operated by animals of an unknown yet intelligent species, into "its smallest parts" (*le più menome parti*; Z3239–40, 1329–30). From the exclusive knowledge of this anatomical dissection, argues Leopardi, it would be impossible to infer the form of life this dead human body ever possessed and the vital relations between the parts in light of their function in the system. Subtly channeling an underlying critique of anatomical medicine from the standpoint of vitalist medicine, Leopardi unravels his analogy in order to build an argument about poetical operations: "Nature, by which we mean the universe of things, is made up of, fashioned and ordered toward a poetic effect, [and] nothing poetic can be found in its parts by separating them one from the other ... nothing poetic in nature decompounded, broken down, and as though cold, dead, bloodless, motionless, lying, so to speak, under the anatomist's knife, or fed into the chemical fire" (Z3241, 1330–31). Of course, the discussion about the gnoseological power of imagination had been crucial to eighteenth-century aesthetics, but what is unique to Leopardi's thought is how gnoseological investigation and poetic quest are intertwined in one and the same pattern, where the knowledge of the general and aesthetic experience begin with the smallest parts but can never stop there.[56] As seen in the garden passage, these dimensions are not just analogous but coextensive: in the garden, a gnoseological process unravels, which enables an ontological epiphany that unfolds into an aesthetic experience. Across these epistemic fields, the general system is the endpoint of the investigative operations conducted through details.

This return to the general, systemic reality must be understood, once more, as distinct from the Romantic aspirations toward totality.[57] Leopardi envisions a relation between the small element and the system it participates in that resists the logics of analogy entailed in the idea of correspondences within nature or that of the relation between microcosm and macrocosm. In Leopardi, the dialectic relation between the *minutiae* and its *milieu* reveals a reality at the cost of upsetting any understanding or expectation of regularity and order. In the garden, this pattern unravels as the discovery of the disorder and destructiveness of existence.

In this dialectic movement, philosophy, science, and poetics converge: "All faculties of a great poet, and all contained in and deriving from the ability to discover relations between things, even the most minimal [*menomi*], and distant, even between things that appear the least analogous, etc. Now this is the philosopher through and through: the faculty of discovering and recognizing relations, of binding particulars together, and of generalizing" (Z1650, 758). The poet, the scientist, and the philosopher share a methodology and a task, a rhythmic sequence of minute and synthetic analysis that culminates in imaginatively productive reorganization.

Notes

1. The work was redacted on separate sheets, which Leopardi used to carry in a wooden box (see Cacciapuoti 2020, xvii). The first published edition of

Zibaldone (1898–1900) was edited by the poet Giosuè Carducci, but with the title of *Pensieri di varia filosofia e di bella letterature*. The title of *Zibaldone di pensieri* is assigned by Francesco Flora in his edition from 1937 (see Cacciapuoti 2010, 39).

2. See Dolfi 2000, 187.
3. On the meanings and uses of the term, see the useful review by Cacciapuoti 2010, 27ff.; and D'Intino's introduction to the English translation, xviff. For the phrase "l'officina dello Zibaldone," see Luporini 1992, 127; and Cacciapuoti 2010, 3. Luporini warns about seeing in the *Zibaldone* an auxiliary text only and claims back a certain autonomy for it (1992, 129). Similarly, Pacella argues that the text cannot be seen as a draft, "brogliaccio," because its formal discontinuity rather corresponds to the inner development of Leopardi's thought, including its contradicting moments (159–60).
4. On the writing patterns of *Zibaldone* and its meticulous internal references as well as its indicization, see Cacciapuoti 2010, in particular 159ff.; and Pacella 1989, 151–53. D'Intino suggests the notion of "hypertext" to define the transversal indicization as well as the "reticular structure" that is observable in the original manuscript (D'intino 2013, xxi).
5. The expression is the title of Antonio Prete's study, devoted to the exploration of the poetical shaping of philosophical questions in Leopardi's work as a whole (Prete 1980).
6. On Leopardi's use of the notions of "philosophy," see Campana 2008, 92–93.
7. In Franco D'Intino's words, "For Leopardi, experiencing one's body is a privileged point of access to knowledge" (D'Intino 2013, xxxiv). See also Caesaer 1992, 27: "For Leopardi our whole sense of ourselves and our whole sense of the world is coextensive with our bodies, our bodies are matter, which in turn is the limit of everything that we know, think and feel."
8. Caesar summarizes the discussion by looking at the materialist reading by Sebastiano Timpanaro, which he paraphrases so: "The heightened awareness of the body afforded by sickness allows privileged insight into the material nature of the world" (Caesar 1992, 23).
9. On Leopardi as reader of Locke, see D'Intino 2013, xxxv; Cacciapuoti 2010, 12–13; and Martinelli 2003. Locke was available in translated versions in Monaldo's library. Leopardi also shows familiarity with Lavater—for example, in Z3200–3201, 1314–15. On Lavater in Leopardi, see Cacciapuoti 2010, 128; Campana 2008, 228–30. As for Cabanis, his work *Rapports du physique et du moral de l'homme* is registered on one of Leopardi's lists of readings reproduced in Cacciapuoti 2010, 187. On Cabanis and the *idéologues*, see Campana 2008, 231–37. On d'Holbach's influence on Leopardi, see Campana 2008, 50–52.
10. Campana 2008, 80–81. On the tradition of "self-observation" of Montaigne as model for Leopardi, see D'Intino 2013, lxvi–lxvii.
11. See Caesar 1992, 28: "The body, therefore, becomes, in this account, that which modernity has lost." In *Zibaldone*, see, for example, Z3181 on civilization and "deterioration" of the human body.

12. See Caesar 1992, 25–27. On the sources for the image of the dough, see Campana 2008, 76.
13. This aspect has been read as anticipatory of contemporary scientific developments by Frattini 1978, 35–37; see also Polizzi 2001, 117–23.
14. For a thorough review of Leopardi's "epistemological method" with regard to the structure of the "glance," see Campana 2008, 91ff., esp. 101.
15. Campana 2008, 133ff.; on Leopardi's theory of imagination, see Gensini 2000.
16. On Leopardi in Bologna as well as on the influence of Voltaire, Volney, and La Bruyère, see Benvenuti 1999, 174–75, 183.
17. The "ontological" reading of the garden passage is also suggested by Martinelli, 2002, 16, 20, who compares it with the gnoseological analytical operations of Condillac.
18. One of the index entries prepared by Leopardi to peruse the *Zibaldone* collects notes from 1821 devoted to "Civiltà. Incivilimento"; these entries closely follow the philosophy of Rousseau's *Second Discourse*, arguing the progressive loss of innocence and decay through the development and specialization of social structures (some examples: Z402, 237; Z579, 307). See Cacciapuoti 2010, 136ff. On Rousseau, see Polizzi 2001, 76. On the antinomy in particular, see Luporini 2018, 113; and Berardi 2001.
19. On the chronology of this development in his thought, see Martinelli 2002, 7–8; 2001, 815. "Antropologia del male" is in Prete 2010, 7. On Leopardi's form of theodicy, see Girolami 2001, 214.
20. See Martinelli 2001, 805 ("verifica empirica") and 815 ("experimentum crucis"). For Martinelli, it is precisely because the passage is based on experience and is ostensive that it cannot constitute a demonstration strictly speaking (Martinelli 2002, 27). On the extensive logical "weaknesses" of the passage, see Martinelli 2001, 811ff., 823.
21. Mengaldo speaks of "a point of no return" and "caesura" in the *Operette* from 1824 (Mengaldo 2012, 60). On implosion, see Luporini 2018, 115. The starkness of this "turn" has been variously questioned. For a problematic restating of its pertinence, see Folin 1996, 18ff.
22. Solmi 1987, 102 (original essay from 1967); and Solmi 1987, 117 (original essay from 1970): In the latter, Solmi speaks of a "critical mythology" from the twentieth century. The expression "pensiero in movimento" is from Solmi's 1965 essay also reprinted in Solmi 1987, 61. Solmi loosens the contradictory tension between a vision of nature as benevolent, finalistic, and oriented toward the conservation of species (the Rousseauian view but with far wider grips in the scientific tradition) and one of nature as intrinsically evil by introducing a double distinction in the use of the term, one signifying nature proper and the other the entire cosmos and vicious order of things, the latter dimension being confusingly called by Leopardi "nature" as well, thereby generating a problem in nomenclature (Solmi 1987, 109). On the polysemic notion of nature in Leopardi, see Casini 2001, 71. I employ this notion of "circumstantial" following Polizzi 2001. On Leopardi and systems, see Pacella 1989, 158.

23. See Martinelli 2001, 817 and 819.
24. Martinelli 2002, 13 and 20, as well as Janowski 2001, 549n59. The passage from Werther describes a *Spaziergang* that costs thousands of poor worms their life, destroys ants' buildings, and reduces everything to a grave. The anti-idyllic destruction is therefore the work of the *promeneur*, of the human presence in nature, while in Leopardi it is the work of nature itself (and of humans too). See Goethe 2013, 40.
25. See in particular the two excellent overviews in Casini 2001 and Campana 2008.
26. On the *Storia dell'astronomia*, see also Polizzi 2001, 65–68.
27. See Casini 2001, 73, for a list of what doesn't enter his orbit.
28. Campana, 2008, 16; 20.
29. On the economy of nature in Linnaeus, see Müller-Wille 2012. For more passages on this point, see also Z4467–68, 2024–25.
30. On the influence of Buffon on Leopardi, see Cacciapuoti 2010, 12n22; Cacciapuoti 2012, 69; Martinelli 2001, 83–835; D'Intino 2013, lxiv, lxvi. Buffon appears on one of the lists that Leopardi wrote down to keep track of his readings (Pacella 1966).
31. Buffon 1976, 145. The translation of Buffon's "first discourse," omitted in the English translations of the work, is published for the first by Lyon in his article from 1976.
32. Buffon 1749, 327.
33. Buffon 1749, 336–37.
34. Buffon 1749, 338.
35. Gill Beer calls tensions the struggles of an intrinsically competitive natural world expounded by Darwin, after the pattern of "life, making and destroying itself," in her study on the "Darwinian imagination" (Beer 2004, xix).
36. Leopardi 1982. However, attesting the continuing development of Leopardi's thought, one year after the declaration that "everything is evil," Leopardi will argue, in the pages of the *Zibaldone*, with more restrained attitude, that the nature of the universe entails "as much bad as good" (Z4257, 1889), thereby proposing a way out of his own impasse in the passage "everything is evil." On this development, see Martinelli 2001, 806.
37. See Forlini 1999, 136, 153ff. In Bologna, Leopardi befriended the Brownian doctor Giacomo Tommasini.
38. Campana 2008, 23–24, 30.
39. Campana 2008, 35ff.
40. Campana 2008, 41.
41. Miniati 2001, 184.
42. See the classic study of Wilson for the impact of microscopy on philosophical thought, in particular chap. 6 and 8 (Wilson 1997).
43. Stafford 1993, 105–6.
44. Baker 1743, iv.
45. The record is on one of Leopardi's reading lists, in Cacciapuoti 2010, 188.

46. Sterne 1978, 82–84. The passage has been analyzed in relation to microscopy by Tunstall 2016, 208, and in Fabietti 2022, where I show which tables of Hooke's *Micrographia* Sterne might have been looking at.
47. The anti-Romanticism of Leopardi is evident in his work if we look, for example, at his early *Discorso di un italiano intorno alla poesia romantica* (1818), but parts of his aesthetics have been also in fact associated with Romantic poetics (see Campana 2008, 142n147).
48. Mengaldo 2012, 32.
49. Mengaldo 2012, 40, 54. See also Campana 2008, 173ff.
50. On the different registers of Leopardi's prose, see Blasucci 2001. Blasucci claims a polyphonic quality for the text while rejecting an aphoristic one (122–23).
51. According to Polizzi, this distinction must be understood in relation to Leopardi's "polemics against mathematics" and therefore should not be extended to science as such (Polizzi 2001, 93).
52. On Leopardi's theory of pleasure, see Folin 2001, in particular 28–29.
53. On Pascal and "disproportion," see also Wilson 1997, 190.
54. "Ridimensionamento cosmico" (Campana 2008, 35).
55. For a harsh critique of German philosophy, see, for example, Z1850–57, 830–33.
56. See Campana 2008, 93ff.
57. See Campana 2008, 142ff.; for a review of the discussion on Leopardi and Romanticism, see 142n147, 167–73. See also Casini 2001, 72.

Bibliography

Bacon, Francis, *Francis Bacon: A Critical Edition of the Major Works*, ed. Brian Vickers, Oxford, Oxford University Press 1996.

Baker, Henry, *The Microscope Made Easy*, London 1743.

Beer, Gillian, *Darwin's Plots: Evolutionary Narrative in Darwin, George Eliot and Nineteenth-Century Fiction*, 2nd ed., Oxford, Oxford University Press 2004.

Benvenuti, Giuliana, "Gli appunti bolognesi nello *Zibaldone*," in *Leopardi e Bologna*, ed. Marco A. Bazzocchi, Firenze, L. S. Olschki 1999, 171–86.

Berardi, Luigi, "Natura e Ragione lungo lo Zibaldone," in *Zibaldone cento anni dopo: Composizione, edizioni, temi: Atti del X Convegno internazionale di studi leopardiani: Recanati, Portorecanati, 14–19 settembre 1998*, 729–35.

Blasucci, Luigi, "I registri della prosa: Zibaldone, Operette, Pensieri," in *Zibaldone cento anni dopo: Composizione, edizioni, temi: Atti del X Convegno internazionale di studi leopardiani: Recanati, Portorecanati, 14–19 settembre 1998*, Firenze, L. S. Olschki 2001, 17–35.

Buffon, Georges Louis Leclerc, *Buffon's Natural History*, trans. James Smith Barr, vol. 10, London 1797.

Cacciapuoti, Fabiana, *Dentro lo Zibaldone: Il tempo circolare della scrittura di Leopardi*, Roma, Donzelli 2010.

Cacciapuoti, Fabiana, ed., *Giacomo dei libri: La biblioteca Leopardi come spazio delle idee*, Recanati 2012.
Cacciapuoti, Fabiana, "Introduzione," in *Zibaldone di pensieri: Nuova edizione tematica Condotta sugli Indici leopardiani*, by Giacomo Leopardi, ed. Fabiana Cacciapuoti, 3rd ed., Feltrinelli, Milano 2020.
Caesar, Michael, "Leopardi and the Knowledge of the Body," in *Romance Studies* 10 (1), 1992, 21–36.
Campana, Andrea, *Leopardi e le metafore scientifiche*, Bologna, Bononia University Press 2008.
Casini, Paolo, "L'iniziazione di Leopardi: Filosofia dei Lumi e scienza newtoniana," in *Giacomo Leopardi: Il pensiero scientifico*, Roma, Edizioni Fahrenheit 451 2001, 59–77.
Cori, Paola, *Forms of Thinking in Leopardi's Zibaldone: Religion, Science and Everyday Life in an Age of Disenchantment*, Legenda, Oxford 2019.
Dolfi, Anna, "Tipologie del pensiero e forme della soggettività nello Zibaldone," in *Ragione e passione: Fondamenti e forme del pensare leopardiano*, Roma, Bulzoni 2000, 187–200.
Fabietti, Elena, "*A Dioptrical Beehive*: Die Repräsentation von Innerlichkeit zwischen Sichtbarkeit und Verblendung in der Literatur des 18. Jahrhunderts," in *Alles Verblendung? Was wir nicht wahrnehmen können, sollen, wollen*, ed. Sebastian Donat, Beate Eder-Jordan, Alena Heinritz, Magdalena Leichter, Martin Sexl, Bielefeld, Aisthesis Verlag 2022, 51–59.
Folin, Alberto, "Il pensiero e il desiderio: Note sulla 'teoria del piacere' di Leopardi," in *Leopardi e la filosofia*, ed. Gaspare Polizzi, Firenze, Edizioni Polistampa 2001, 17–35.
Folin, Alberto, *Pensare per affetti: Leopardi, la natura, l'immagine*, Venezia, Marsilio 1996.
Forlini, Adolfo, "Stratone e l'"ospitale" dei viventi. Tradizioni filosofiche e contesti scientifici per il Leopardi Bolognese," in *Leopardi e Bologna*, ed. Marco A. Bazzocchi, Firenze, L. S. Olschki 1999, 133–70.
Frattini, Alberto, *Letteratura e scienza in Leopardi e altri studi leopardiani*. Milano, Marzorati 1978.
Gensini, Stefano, "Osservazioni sulla teoria leopardiana dell'immaginazione," in *Leopardi e lo spettacolo della natura*, ed. Vincenzo Placella, Napoli, Centro Nazionale di Studi Leopardiani 2000, 25–46.
Girolami, Patrizia, "L'antiteodicea leopardiana. Leopardi, Leibniz e il problema del male nello 'Zibaldone,'" in *Ripensando Leopardi: L'eredità del poeta e del filosofo alle soglie del terzo millennio*, ed. Alberto Frattini, Giancarlo Galeazzi e Sergio Sconocchia, Roma, Edizioni Studium 2001, 213–40.
Goethe, Johann Wolfgang von, *The Sufferings of Young Werther*, trans. and ed. Stanley Corngold, New York, Norton 2013.
Janowski, Franca, "Figure del negativo nello Zibaldone leopardiano: La tristezza della natura," *Lo Zibaldone cento anni dopo: Composizione, edizioni, temi,*

Atti del X Convegno internazionale di studi leopardiani, Firenze, Olschki 2001, II, 525–52.
Leopardi, Giacomo, *Operette morali: Essays and Dialogues*, trans. G. Cecchetti, Berkeley, University of California 1982.
Leopardi, Giacomo, *Zibaldone*, ed. Giuseppe Pacella, Bologna, Zanichelli 1991.
Leopardi, Giacomo, *Zibaldone*, ed. Michael Caesar and Franco D'Intino, New York, Farrar, Straus and Giroux 2013.
Lyon, John, "The 'Initial Discourse' to Buffon's *Histoire naturelle*: The First Complete English Translation," *Journal of the History of Biology* 9 (1), 1976, 133–81.
Martinelli, Bortolo, "Entrate in un giardino: Leopardi e la fine del mito dell'Eden," in *Lo Zibaldone cento anni dopo: Composizione, edizioni, temi: Atti del X Convegno internazionale di studi leopardiani*, Firenze, Olschki 2001, II, 801–45.
Martinelli, Bortolo, "Leopardi: La prova del giardino," in *Otto/Novecento: Rivista quadrimestrale di critica e storia letteraria* 26 (2), 2002, 5–28.
Martinelli, Bortolo, *Leopardi tra Leibniz e Locke: Alla ricerca di un orientamento e di un fondamento*, Roma, Carocci 2003.
Mengaldo, Pier Vincenzo, *Leopardi antiromantico e altri saggi sui "Canti,"* Bologna, il Mulino 2012.
Miniati, Mara, "Libri di scienza e strumenti in casa Leopardi," in *Giacomo Leopardi: Il pensiero scientifico*, Roma, Edizioni Fahrenheit 451 2001, 177–86.
Müller-Wille, Staffan, "The Economy of Nature in Classical Natural History." In *Studies in the History of Biology* 4 (4), 2012, 38–48.
Pacella, Giuseppe, "Elenchi di letture leopardiane," *Giornale storico della letteratura italiana* 143 (444), 1966, 557–77.
Pacella, Giuseppe, "Lo Zibaldone: brogliaccio o opera sistematica?," in *Leopardi e il pensiero moderno*, ed. Carlo Ferrucci, Milano, Feltrinelli 1989, 151–62.
Polizzi, Gaspare, "Filosofia delle circostanze e immagini della scienza nello Zibaldone," in *Leopardi e la filosofia*, ed. Gaspare Polizzi, Firenze, Edizioni Polistampa 2001, 61–168.
Polizzi, Gaspare, *Leopardi e le "ragioni della verità": Scienze e filosofie della natura negli scritti leopardiani*, Roma, Carocci 2003.
Prete, Antonio, *Il pensiero poetante*, Milano, Feltrinelli 1980.
Prete, Antonio, "Sull'antropologia poetica di Leopardi," in *La prospettiva antropologica nel pensiero e nella poesia di Giacomo Leopardi: Atti del XII Convegno internazionale di studi leopardiani, Recanati 23–26 settembre 2008*, Firenze, L. S. Olschki 2010.
Solmi, Sergio, "Ancora sulle due 'ideologie' di Leopardi," in *Studi leopardiani*, Milano, Adelphi [1970] 1987, 113–17.
Solmi, Sergio, "Le due ideologie di Leopardi," in *Studi leopardiani*, Milano, Adelphi [1967] 1987, 99–110.
Stafford, Barbara Maria, "Voyeur or Observer? Enlightenment Thoughts on the Dilemmas of Display," *Configurations* 1 (1), 1993, 95–128.
Sterne, Laurence, *The Florida Edition of the Works of Laurence Sterne*, ed. Melvyn New and Joan New, Gainesville, University Presses of Florida 1978.

Tunstall, Kate E., "The Early Modern Embodied Mind and the Entomological Imaginary," in *Mind, Body, Motion, Matter: Eighteenth-Century British and French Literary Perspectives*, ed. Mary Helen McMurran and Alison Conway, Toronto, University of Toronto Press 2016.

Wilson, Catherine, *The Invisible World: Early Modern Philosophy and the Invention of the Microscope*, Princeton, Princeton University Press 1997.

PART TWO

Micrological Biospheres

✦✦✦

Landscapes of Disease and Contagion
Putrefaction and Fermentation in Early Modern Corpuscular Medical Theories

CARMEN SCHMECHEL

✦ ✦ ✦

"MAN WAS to dwell in the Air," wrote Joan Baptista van Helmont in 1648. Such an aerial milieu would not only enable and sustain life but also carry the "seeds" of disease and contagion. This essay explores the environmental dimension of contagion in sixteenth- and seventeenth-century writings, including the effects of miasma ("bad air") alongside an emerging germ theory of disease. In this context, fermentation and putrefaction played a central role, as these natural processes circumscribed the generation and propagation of such disease-inducing minutiae. Hence the essay attempts to relate the cluster of fermentation, putrefaction, and their role in disease in early modern medicine with the concepts of milieu and minutiae. In particular, I look at how the views on contagion evolved from Fracastoro (*De contagione*, 1546) to Athanasius Kircher (*Scrutinium pestis*, 1658) due in large part to the mediation of Jan Baptista van Helmont's reworking of the philosophy of ferments (*Ortus medicinae*, 1648), which affected how the "seeds of disease" were perceived and negotiated within coeval medical thought. In traditional miasma theory, the milieu itself was understood as the immediate (efficient) cause of disease, since putrefaction on its own sufficed to produce disease in a predisposed body. With Fracastoro, however—thanks in part to a Renaissance revival of ancient atomism—the locus of disease causation shifts to "imperceptible particles," minute entities—minutiae, one might say—which emerge from the milieu and reside predominantly within (but are ontologically separate from) it. Van Helmont in turn reframed disease as the result of the action of "ferments" and fermentation as a ubiquitous process in nature, thus laying the basis for a reworking of the "seeds of disease" that proved very influential not only for adepts of the chemical philosophy.[1] Even Athanasius Kircher, deeply religious and deferential to ancient authorities, seems to rely on chemical theories to a small but significant extent, such as in this respect.[2] Kircher sees the venomous minutiae as engendered by putrefaction, yet as he relates them to spontaneous generation, his minutiae come closer to small animals such as worms (*vermes*). These multiply and propagate in the body of the host, especially in the blood, in the manner of a "fermentation" not unlike how van Helmont had described it. With his theory of seeds of contagion, Kircher had landed on a strand of medical thought that, while rather eccentric in his own time, would subsequently be adopted with great enthusiasm by iatrochemistry, incorporating Kircher's contribution. Later physicians who understood themselves as Helmontian and even radically anti-Galenist, such as Marchamont Nedham, relied heavily not only on van Helmont but also on Kircher's account of infection being caused by worms in the blood. Thus, both Kircherian and Helmontian ideas about fermentation, putrefaction, and disease would later greatly influence much of English iatrochemical thought.

The relevance of this story lies in the dynamics between the concurrent influences of minute corpuscles and their milieus—the putrefying air and the human body itself—in generating and propagating disease. In other words, it is relevant for an epidemiological history in which the issue of *virulence*[3] (as a vector result of environment, individual constitution, and disease-causing germs) is constantly being negotiated within a framework of multifactorial disease causation. Indeed, it has never quite ceased to be a point of contention throughout history to what extent a disease is due to the environment, to what extent the damage it inflicts depends on characteristics of the receiving body, and which aspects of the disease are predetermined by the germs themselves. In the early modern age, debates surrounding the issue of disease causation and transmission are often focused more pointedly on either of several concurring factors. Yet in an age long before the bacteriological shift, a focus on one aspect did not come at the exclusion of any others.[4] Neither Fracastoro, nor van Helmont, nor Kircher would have thought to leave out the roles that environment, on the one hand, or individual constitution, on the other hand, play, even though material "seeds" indeed constituted for them the immediate cause of disease. Early modern approaches to disease causation may be instead best understood with reference to ancient frameworks of pluralist causation. While the seeds of disease may play the role of the immediate cause triggering the disease (an efficient cause in Aristotelian parlance),[5] the milieu acts as a Galenic antecedent (procatarctic) cause.[6]

Thus, this pluralist causal framework of explorations of virulence serves to highlight the dynamics between pathogenic[7] minutiae (be it Fracastoro's particles or Kircher's seeds) and their milieu—the latter either external to the human body or constituted by the body itself.

Environment, Miasma, and Contagion from Antiquity to the Early Modern Age

THE REFERENCE work that, within the Western tradition, has set the tone regarding thinking of humans as highly dependent on their environment for their health is *Airs, Waters, Places*, an ancient Greek treatise that is part of the Hippocratic corpus. This foundational text, composed around 450–400 BCE, explains how the environment shapes human physiology and pathology: the nature and qualities of waters, how those waters affect those who drink them, the qualities of airs and winds, and so on. According to the Hippocratic author, a good physician must take into account a multitude of environmental factors—such as geography, climate, and the quality of waters—when assessing a patient's affliction as well as prescribing the proper course of treatment.[8] Particularly important for local health, waters "must be brighter, sweet-smelling and light; while all that are salt, harsh and hard are not good to drink."[9] The unwholesomeness of waters was related to potential stagnation: "Such [waters] as are marshy, standing and stagnant must in summer be hot, thick and stinking, because there is no outflow; and as fresh rainwater is always flowing in and the sun heats them, they must be of bad colour, unhealthy and bilious."[10] Further, the author states that those people who drink such waters exhibit various afflictions and "need more powerful drugs."

It becomes apparent that waters become a vector of disease whenever they turn into a milieu that favors putrefaction. Rainwater, in particular, is damaging to health because it "grows foul quicker than any other, and has a bad smell; being a mixture gathered from very many sources it grows foul very quickly."[11]

As a natural phenomenon, putrefaction was later given a framework of explanation in Aristotle's *Meteorology*, book 4. There, Aristotle differentiated between internal heat, which causes concoction (*pepsis*, πέψις), and external heat, which causes putrefaction (*sepsis*, σῆψισ). While concoction was a teleological process of maturation,[12] which included the phenomena we now call fermentations, sepsis was essentially decay and decomposition. Decay was hence seen as an effect of forces that are inherent in the environment: "Decay is the destruction of a moist body's own natural heat by heat external to it, that is, the heat of its environment."[13] The cause of decay (external heat) was therefore conceived as belonging structurally to the milieu in which the body exists rather than being engendered within the body itself.

Putrefaction also plays an essential role in ancient etiologies of disease. In Galen of Pergamon's writings, "pestilence and putrefaction are constantly entwined,"[14] with putrefaction being a source of contagion and disease.[15] In the treatise *On Differences between Fevers*, Galen writes,

> In pestilential constitutions, the inhalation (of air) is the most important cause. For, if the fever is sometimes caused by the humours in the body that are susceptible to causing putrefaction, when the living being receives a slight impetus from the ambient air for the beginning of the fever, most often it is following inhalation that the fever starts, inhalation of the surrounding air which is polluted by putrefied odours. The origin of putrefaction is either a mass of cadavers that have not been cremated, as normally happens during combat, or fumes from swamps or lakes during the summer.[16]

Galen himself draws on another work from the Hippocratic corpus, *On the Nature of Man*, where the author had stated that putrefied air was produced through excretions from sick bodies. This was a common theme in both Roman and Greek antiquity. The historian Diodorus of Sicily, active in the first century BCE and commenting on the "plague" of Athens, also mentions "thick and foul-smelling 'vapours' (ἀτμίδας) which, once 'exhaled' (ἀναθυμιωμένας), pollute (διαφθείρειν) the surrounding air."[17] Such knowledge would have been the result of observation, in a world where, for instance, dead bodies left unburied on combat sites were common experience. Rotting corpses and other putrefying matter, vegetable or animal, causing widespread contamination, including that sort of contamination embodied by disease,[18] formed the ancient background of what came to be called the miasmatic theory of disease, according to which the inhalation of vitiated air from a milieu dominated by processes of decomposition would generate illness.[19]

What exactly was transmitted through the inhalation, however, was not straightforward. In other words, the ontological status of the minutiae remained unclear. They could be tiny bits of matter, such as had been argued by the Greek atomists (such as Democritus) or by Lucretius in his *De rerum natura*, or they could be of a more spiritual nature, understood more as principles than as particles. In antiquity, miasmatic contagion had a wider semantic range, including a moral or religious dimension that might have been primary (with the medical sense being secondary). Referring to ancient Greece in particular, Robert Parker explains that "the basic sense of the *mia-* words is that of defilement, the impairment of a thing's form or integrity."[20] Such defilement may be "deriving from things that are not dirty in themselves, or not deriving from matter at all,"[21] such as "the pollution of a reputation through unworthy deeds, or of truth through dishonesty."[22] This type of contamination, historically laced with strong religious overtones, is an element of cultural context that must be taken into account when considering the phenomenon of putrefaction as a source of disease; putrefied air or water is understood as a physical milieu as much as a moral one, associated with occult evil forces. Even in the early modern age, as experimental science pushed back on this kind of more traditional notions of disease and contagion, putrefaction as a source of contamination retained its nefarious aura.

Putrefaction, Spontaneous Generation, and Vermin

PART OF that nefarious aura came from a correlation between putrefaction and insects. Toward the end of the seventeenth century, this correlation would play an important role in the development of the idea of disease caused by tiny worms or other animalcules. For many centuries if not millennia before that, however, putrefying matter was a milieu held responsible for the generation of various insects and small animals, which were believed to arise from decaying matter through spontaneous generation, as opposed to a generation *ex ovo* or by mammalian birth. Originating in antiquity, the theory gained particular exposure with Virgil[23] and held sway across many centuries. "Living things," Aristotle had written, "are generated in decaying matter because the natural heat which is expelled compounds them out of the material thrown off with it."[24] Putrefaction is strongly associated with this sort of emergence: "All which arise in this manner [*by spontaneous generation*] whether on land or in the water come to be formed, as can be seen, to the accompaniment of putrefaction and admixture of rainwater."[25] (One may here recall the Hippocratic assessment of rainwater as particularly prone to spoilage, mentioned above.) Perhaps the most striking use of putrefaction—a kind of extension *ad absurdum* of the ideas behind "spontaneous generation"—is the recipe offered in the spurious, pseudo-Paracelsian *De natura rerum* (1537), where the alchemist endeavors to create a "chymicall homunculus" from putrefying horse dung and male human semen. Later in the sixteenth century, Giambattista della Porta's work *Natural Magick* of 1584 contributed to the consolidation of this idea in the early modern European imagination; by the mid-seventeenth century, this idea of artificial creation of life from putrefying matter was quite widespread. In his *Ortus medicinae* (1648 and subsequent editions), van Helmont famously offers a recipe for making mice out of the ferment of wheat imprinted onto an old shirt as well as another recipe for producing scorpions from basil left to decay between two bricks.[26] In spite of various experiments aimed at refuting it, including Francesco Redi's famous experiment disproving the hypothesis that maggots would emerge out of decaying meat,[27] the theory of spontaneous generation would survive until the nineteenth century.

It should be noted that the types of creatures believed to emerge from putrefaction were mostly small ones—insects,rodents, small reptiles. Throughout the early modern era, these lowly creatures yielded through spontaneous generation were associated with diseases like the plague, even as it was not (yet) known in which mysterious way they were linked (the actual bacteriological causation being discovered only in 1894, when Alexandre Yersin isolated the bacterium *Pasteurella pestis*). Rats, frogs, and insects did, however, already share a common point of origin with disease in the early modern imaginary: the common milieu of putrefaction. As Lucinda Cole remarks, "Rats and other small animals nonetheless played an important role in linking environmental and supernatural accounts of disease . . . because they were still widely regarded as creatures of putrefaction, born of rot and corruption, rats signified, by their very presence, an unhealthy environment from which illness may emerge."[28]

Vermin and the environmental features that typically generate miasma relate to ancient ideas about impurity and pollution whose echo was still present in early modern times, often reworked within a Christian theological framework.[29]

The Nature of the Contagium: Material or Spiritual?

IMPORTANTLY, HOWEVER, such religious underpinnings did not preclude naturalistic explanations of disease and contagion. In fact, these two aspects—the divine and the material—were not opposed but rather intertwined across history; seeing them as opposed would be anachronistic, saying more about our own perspective than about early modern views. The material part of the argument was, especially since the sixteenth century, related to the concept of "seeds of disease," following an interrupted and rather convoluted line of tradition that links ancient theories with their early modern reincarnations. In Lucretius's *De rerum natura* (around 56 BCE), which provides a substantial part of the ancient backdrop, we read, "There are seeds of many things (*multarum semina rerum*), and this earth and heaven has enough disease and malady, from which the force of measureless disease might avail to spread abroad."[30] Lucretius's work appeared in an *editio princeps* in Brescia in 1473, with the Aldine edition following in 1500; this enabled a revival of interest in his ideas in the Latin West from the late fifteenth century onward.

Vivian Nutton suggests that Galen himself may have been aware of Lucretius's philosophical poem and his use of the idea of "seeds of disease"[31]—which included the idea that disease is caused by putrefied air containing such seeds. For his own part, Galen too championed an idea of seeds of disease lingering in the body and causing relapses of fever when the patient failed to adhere to the proper regimen;[32] Galen's rather pragmatic approach may have favored the corpuscular view. Interest in Galen's works likewise peaked in the late medieval and early Renaissance periods thanks to new Latin translations of his Greek works.

So much for the material aspect.

Yet not all theories of "seeds" considered them material entities. Stoic theories, for instance, involved a spiritual dimension. As Marina Banchetti-Robino notes, while Lucretius and other Epicurean authors "conceive of semina rerum, or the 'seeds of things,' in entirely physical terms as atoms," the Stoics "conceive of semina as immaterial active and formative principles having nothing to do with atoms."[33] This Stoic tradition, continued and modified by the Neoplatonists, found its later echo in the scholastic philosophy of St. Augustine, whose influential writings amplified its reach into Renaissance theories of disease and contagion. In this way, the ancient Stoic *logoi spermatikoi* (λόγοι σπερματικοί) turned into "rationes seminales," a term referring primarily to a set of nonmaterial information—"a rational 'program' that specifies the creative power that is immanent in nature," as Banchetti-Robino puts it.[34] This sense of the term is the one prevailing in many adepts of the chemical philosophy, such as Paracelsus. Additionally, some chemists, like van Helmont, integrate this spiritual dimension of principles with a more corpuscularist vision of particles.[35]

While deeply indebted to the corpuscularian understanding of the "seeds of disease," both Fracastoro and Kircher did not remain completely untouched by theories inspired by the Augustinian strand. Rather, like many of their contemporaries, they combined both approaches. In Fracastoro's work, the material dimension of the minutiae prevailed more decisively, in opposition to more temperate coeval views. With Kircher, the minutiae are not just material entities; they are also living bodies, approaching animal nature.

Infection and Contagion in Fracastoro's *De contagione*

PHYSICIAN GIROLAMO Fracastoro starts his treatise, *De contagione* (1546), by outlining the very definition of contagion as "an infection passing from one unto another."[36] The source of infection is "imperceptible particles," a detail that may be considered Fracastoro's most important contribution: "We say that contagion is . . . a corruption that passes from one unto another, once (after) an infection has taken hold among the imperceptible particles."[37] Fracastoro's attempted definition of these particles is, unfortunately, less than enlightening: "Indeed I call tiny and imperceptible particles those, out of which the composition and mixture is made."[38] These imperceptible particles may be understood as minutiae of a material nature, able to circulate through the air on their own; they, too, typically arose from putrefaction. However, unlike in the case of insects engendered by spontaneous generation, which were living beings, the "imperceptible particles" had an undecided status between the living and the nonliving; indeed, there is no hint whether Fracastoro thought that they were or were not endowed with life. Hence while they too were correlated with putrefaction, they constituted a different type of disease-inducing minutiae. The ambiguity inherent in the Fracastorian concept of imperceptible particles reminds us of the epistemological problems surrounding microscopy: as Daniel Liu puts it in this volume, "how it was possible to recognize and know" such minutiae that would be perceivable with the aid of the microscope. In the absence of such technology, Fracastoro might have made a decision to abstain from metaphysical speculation on the exact nature of the "imperceptible particles."

A closer look at infection (which Fracastoro defines as a corruption "to a certain extent"[39]) shows that the concept, far from being a purely medical one, was related—even in Fracastoro's time—both to religious ideas and to Galenic miasma theory. This dimension was not at odds but rather coexisted with the material nature of the minutiae. As Margaret Pelling remarks, "The term 'infection' has a root meaning 'to put or dip into something,' leading to *inficere* and *infectio*, staining or dyeing. This is a further reminder that 'an infection is basically a pollution.'"[40] In fact, the Greek translation of the Latin *inficere* is *miaino*, from which the noun *miasma* is derived. These terms reveal the close association between disease and impurity, including a religious or moral impurity that would today be understood as metaphorical. For the ancients, however, and to some extent possibly also for Fracastoro and his contemporaries, the boundary between disease and religious (moral) pollution—the pollution of sin—was trickier to draw. Hence an infection may or may not lead to disease; disease was only one of multiple categories of what could be achieved by way of "infection."[41] While this spiritual aspect is rather in the background for Fracastoro, it may still be observed, perhaps especially in Fracastoro's exposition of three pathways of contagion. These are by direct contact, by "fomes," and at a distance ("contactu solo," "fomitem," and "ad distans"). The first pathway is straightforward. The second one, *fomes*, consists in such minutiae of disease ensconced in linen, sheets, clothing, and other domestic objects, which become activated when a person makes contact with them. While it appears self-explanatory for the modern reader, who thinks in terms of microorganisms, the context out of which Fracastoro's ideas developed would have been different and most likely magicotheological. As Nutton has pointed out, "A much more likely source for the metaphor of *fomes* is theology. According to many theologians, from the fourth century onwards, even after baptism or penance had wiped away sin (which is often termed a disease or contagion), there still remained some *fomes peccati* (concupiscence), which inclined man towards evil, but which could be checked by true belief and Christian living."[42] The *fomes* of disease thus parallel the *fomes peccati*: a connection that, while embracing a Christian expression, could have been inherited from ancient Greek thought about *miaino*.

Fracastoro dedicates a whole separate chapter of *De contagione* to contagion at a distance (chapter 5). This latter sort of contagion happens in the case of "pestilential fevers and phthisis, inflammations of the eyes, pustules such as are called variola, and other similar ones."[43] There is "more than a little doubt" (*dubitatio non parva*) about the nature of the agents of such contagion; also, a majority of all contagions belong to this category.[44] Contagion at a distance exemplifies the dynamics between minutiae (the imperceptible particles) and their milieu (the air), through which the minutiae travel unseen until they reach a susceptible host. Of the three pathways of contagion outlined by Fracastoro, this remains the one commanding the most wonder—to a great extent because the existing technology (the magnifying lens they had at their disposal) did not allow for a clarification of the nature of such minutiae. But what is most important is that with Fracastoro, although minutiae and milieu act in tandem to generate disease, the minutiae have been separated conceptually from the milieu; they have a sort of being of their own, even if it is unclear whether they are alive or not.

The Nature of "Insensibilia," Fracastoro's Seeds of Disease

THE ASSUMPTION of contagion at a distance meant that one had to ask questions about *what exactly* is transported through the air and thus causes said contagion—in other words, what is the nature of that "agent X," as Nutton[45] has called it. In an era before the microscope, when one could not readily verify assumptions about potential minute physical particles, the answers to these questions could be quite varied and imaginative. Most authors assumed putrid changes in the air and environment to be at least coresponsible for causing disease. Fracastoro, too, claims that there is a sort of mixture that is more apt to become putrefied and to contract contagions: it is a mixture of warmth and humidity. Heat and humidity producing putrefactions is an ancient idea, but many authors also pointed out that such environmental conditions could not be enough to cause disease, or else plague would be everywhere. By itself, putrefaction for Fracastoro—in contradistinction to his contemporary Giambattista da Monte—is not enough to create disease and contagion; there has to be an additional element, an added principle. Such a principle consists of the minute "imperceptible particles," "which shall hence be called seeds of contagions."[46] Fracastoro is the only one in his own time to propose this elegant, though eccentric solution of "seeds," which according to him was more apt to explain both the occurrence and the transmission of plague.

The novelty of the "agent X" that Fracastoro posits resides in part with the way it confers the status of being upon the minutiae thought to emerge from putrefaction. When da Monte claims that putrefaction itself engenders disease, he too assumes that putrefaction travels and impresses itself onto a new body, but the tiny bits of putrefaction are not, in their nature, distinct from the putrefying matter from which they arose.[47] These minutiae are simply, as it were, pieces of milieu, of the same nature as the milieu itself. For Fracastoro, however, these imperceptible particles are distinct forms of being that are able to multiply by generating offspring, exist in the milieu and are favored by a certain milieu (putrid, humid, warm), but are not of the same nature with the milieu.

Fracastoro explains the distinct material quality of his seeds of contagion by recourse to the manner in which contagion occurs. If contagion is, as defined, the passing of disease from a body to another, then the same disease that is in the first body should also be found in the second and the next ones after that. Such a quasi-identical transmission cannot, however, be achieved by "spiritual" means;[48] hence, the generation of disease, like the generation of all things, must come from "primary qualities"—that is, from a physical entity or entities: "We should expect these contagions to develop not through putrefaction alone, but from initial seeds, and from others who are generated and propagated, which are of a similar nature to those [initial ones]."[49]

The way these particles work is seen, for instance, in the manner in which onion or garlic affects our eyes.[50] The eyes, in fact, play a double role here.

Passively, they may be the recipients of "morbific" particles, and hence they can play the role of an entry point for these particles into the body. Actively, the eyes themselves may be the ones sending off the disease particles, from a vitiated body that initiates contagion: "From the eyes afflicted by inflammation, [the seeds of contagion] can be projected into other eyes and can bring about a similar infection"[51] (*consimilem infectionem inferre*). When he considers the eye as causative of (another individual's) disease, Fracastoro's judgment differs much from our own, showing him closer to antiquity[52] and ancient ideas about contagion transmitted from the eyes. As for the second (target) body, from the eye as entry point, the disease spreads into this second body through the veins and arteries, ultimately reaching the heart. The seeds of disease achieve this by "propagation," by producing offspring (*sobolem*) that resemble them.

Another common way for the seeds of disease to enter the body is by inhalation. Once having entered the body through the nose, the seeds cannot easily be expired back; they then, by their glutinous nature, attach to the humors and to the parts of the body, and even to animal (vital) spirits, which, as they flee a species that is contrary to themselves, manage to bring themselves and the seed enemy up unto the heart.[53]

As we will see, Kircher will argue the same manner of propagation of physical minutiae inside the body more than a century later. Between Fracastoro and Kircher, van Helmont's theory of ferments will exert additional influence, imbuing these contagion theories with a chemical dimension.

Ferments and Seeds of Disease in Van Helmont: A "Hybrid Ontology"

BY THE mid- to late seventeenth century, the chemical philosophy had gained considerable ground in Europe. With it, phenomena like putrefaction and fermentation were reinterpreted chemically and recontextualized to inform a more modern theory of disease and contagion. While Paracelsus had been the main initiator of chemical philosophy in the sixteenth century, the Flemish physician Jan Baptista van Helmont is responsible for reinterpreting ferments and fermentation as well as putrefaction, turning them into newer concepts with greater explanatory power.[54] In particular, van Helmont's ideas around ferments changed the way early modern physicians and natural philosophers thought about disease with regard to both contagion and spread of disease within the body.

In his flagship work, *Ortus medicinae*, published posthumously in 1648, van Helmont discusses at length general environmental factors such as air[55] and water. However, compared to his predecessors, he diminished the role of air and water in causing disease in order to emphasize a concurring factor—the entities of specific diseases. These entities for van Helmont were not necessarily material; that is, the disease may be occasioned by a physical thing outside the human body that infects it, but the growth of disease in the human body itself occurs in a more spiritual manner, through the mediation of ferments and of an entity called an Archeus (a term that van Helmont borrows from Paracelsus). The air's role is relegated to being the indirect medium of disease by harboring disease-causing seeds and ferments that bring about epidemics such as the plague.

Van Helmont sets up a theory of ferments and fermentation wherein he imports elements from alchemy to account for changes and "transmutations" in the human body. For him, as for Fracastoro, the minutiae that constitute the seeds of disease have indeed a physical existence, one initially external to the body to be infected. However, they are also accompanied by "ferments" that are principles ("dynamic principles," as Pagel has called them[56]) of a more abstract nature, akin—as Hiro Hirai has noted[57]—to the Stoic and later Augustinian *rationes seminales*. These have ontological priority; the ferment is "prior to the seed," being the source of the seed's generation.[58] For van Helmont, there are, moreover, two types of ferments: the perishable seminal ferments, which are closer to matter (*fermentum seminale*), and the more important "principiating" ferment (*fermentum principians*).[59] It is the principiating ferment that carries more ontological weight for van Helmont, since it constitutes the formal cause ("formall created Being");[60] all bodies, according to van Helmont, are generated from water imbued with the power of this principiating ferment, which acts as a milieu for the process of generation (including spontaneous generation but also other fundamental transformations of matter that count as new generations, such as the "fermentation" of water into vinegar). This ferment is the "original beginning of things."[61] On the other hand, the exact qualities of the generated thing are determined by the *fermentum seminale*, which is more specific and ontologically closer to physical matter and its seeds (it contains the *Archeus seminalis*[62]). The seminal ferments permanently encode a specific propensity toward one disease or another. Seminal ferments, hence, are immaterial packages of preformed information put together by God and predetermine the type of seeds that are to grow,[63] including those that are specific for particular diseases.[64] Their existence is sustained by "continuous propagation."[65] The seeds themselves, in turn, are indeed the immediate (efficient) cause of disease.

As for the mechanism of how disease is engendered, van Helmont's theory is very different from Fracastoro's.[66] Though at first glance both champion the idea of seeds, each imagines the process of contagion differently. For van Helmont, once the seeds—aided by the ferments—enter into contact with the person to be infected, they trigger a sort of spiritual mechanism: instead of entering the body from outside, the disease is re-created inside the body, as an image, thanks to the imaginative power of the Archeus that lives in the body. According to van Helmont, disease is thus engendered after the vital principle has been penetrated by a noxious force that excites the inner Archeus, which then forms the images (ideas) of disease, like seals.[67] This usually happens after being excited or triggered by external entities, including material ones.[68]

Two types of minutiae are therefore at play in van Helmont's theory of contagion and disease. One type is constituted by the physical seeds, the other by the incorporeal *seminal* ferments that precede and engender them. For both of these, the principiating ferment acts as a milieu: it is only in its presence that the seminal ferments can act and produce seeds that ultimately lead to the growth of plant or animal bodies or to a fermentive alteration of matter (such as the chyle fermented into blood). There is some difficulty with categorizing the seminal ferments as minutiae, however: their nature is not necessarily, or not clearly, material. Instead, van Helmont claims they are closely related to and acting in tandem with odor,[69] suggesting a hybrid or an intermediary nature between the material and the spiritual. Van Helmont's reference to odor in connection with the generation of disease is not novel, foul smells having been the marker for potential infection for many centuries, if not millennia. This is unsurprising, odor being the first element that is perceptible for human senses and hence being grounded in everyday experience. Van Helmont's innovation, however, consists in assigning odor the role of exciting the imagination and the inner Archeus. In this manner, what would otherwise be an environmental feature gains a different sort of agency, a type of mediation between the spiritual and the corporeal.

This mediation is a feature typical of van Helmont. As many scholars have shown, van Helmont's ontology reunites a vitalistic and a corpuscularian aspect. In the 1990s, William R. Newman coined the expression "vital corpuscularianism" to describe van Helmont's ontology.[70] More recently, Marina Banchetti-Robino has written of "Helmont's hybrid ontology, which fuses a corpuscular conception of *minima naturalia* with a non-corporeal conception of *semina rerum*." Van Helmont believed in the existence of minute physical corpuscles generative of disease, but he thought that these would merely be occasioning disease by sending its image and information into the body, not creating it by material reproduction once they had physically entered the body (as Fracastoro and, later, Kircher believed as well). In Banchetti-Robino's words, "For Helmont, chemical alterations involve the *minima* as physical units but also depend upon ferments that are contained in the *semina*, which function as formative spiritual agents."[71]

Kircher and Contagion by (In)Visible Vermin

A DECADE after *Ortus medicinae* and more than a century after Fracastoro's *De contagione*, Jesuit polymath Athanasius Kircher dedicated his 1658 work, *Scrutinium pestis*, to explaining contagion in plague via his own concept of seeds of disease. His innovation consists of positing two types of minutiae, both physical: subvisible worms (*vermes*) and the poisonous putrid corpuscles or seeds that precede and engender these worms by a process analogous to macroscopic spontaneous generation.

Kircher distances himself from van Helmont with respect to a central issue: the physicality of the germs of disease and the manner in which disease catches hold of a body. For Kircher, it is the external agent that is ingested. While van Helmont had stressed the role of imagination, since the Archeus works in a spiritual dimension, Kircher negates this, claiming that imagination by itself will *not* cause the plague—there has to be that *peregrinum virus*, that foreign venom; imagination solely creates a predisposition in the host body.[72] On the other hand, Kircher draws on Helmontian and chemical tropes to explain the manner in which contagion spreads (namely, like fermentation) and in part also to sustain an ontology of disease.

In the beginning of the work, a religious aspect is undeniable—even more overtly so for Kircher than for his precursor, Fracastoro. The *Scrutinium*'s very first sentence is "Humankind is polluted with the contagion of original sin."[73] As God became aware of the "iniquity of mortals," He threw a threefold curse upon mankind—war, famine, and the plague,[74] the "filthiest and most abominable" of diseases. Despite deferring to this religious narrative, Kircher makes it clear that "while [epidemics of plague] are sent by God, they do not lack natural causes."[75] These make up the proper subject of his treatise.

In his broad assumptions about natural causes, Kircher follows Fracastoro, along with ancient authorities. There are "three types of the disease [plague],"[76] according to Hippocrates and Galen. The pathways of contagion mirror Fracastoro's ("per contactum ... per fomitem ... ad distans,")[77] though it is not a perfect correspondence; for Kircher, it is the Fracastorian *fomes* that really are "at a distance," while the second Kircherian category is through the mediation of air. This change—concerning classifying the contagion through aerial mediation as the second one, as opposed to *ad distans*—may reflect the shifting role of the air: for Kircher, air as a milieu is more closely bound with the minutiae that constitute the seeds of disease. Hence, he conceives of contagion by air as a closer contact. This can also be due to the simultaneity (as opposed to when corpuscles remain dormant in clothes or linen, infecting people only after a certain time has passed, separating the initial trigger—the initial sick body spreading contagion—from the end target, the second body who gets contaminated from the linen).

Kircher follows Fracastoro closely in one important respect: the seeds of disease as a physical agent of contagion. For Kircher, too, contagion comes from a combination of putrefaction and an "agent X," the seeds of disease that Kircher calls "nomad seeds" (*seminaria peregrinis*). Putrefaction, meanwhile, is defined classically in Aristotelian terms as "nothing other than the corruption of innate heat of a mixed body ... realized by a heat which is external and nomad/foreign (*peregrinus*)."[78] Out of this putrefaction, plagues may emerge "in various ways,"[79] such as "from those putrid exhalations expired in places which abound in filthy and putrid slime of stagnant waters and marshes, which, receiving external heat from the sun, conceive I don't know what poison [*nescio quid virulentum concipiunt*]; from which necessarily follow the seeds [*seminaria*] of the pestiferous disease."[80]

Thus, the contagious putrefaction stems "not from the contamination of the four elements" but from the action of an additional, foreign, nomad (in the sense of transitory) element constituted by the seeds[81]—hidden corpuscles of very small size, invisible to the naked eye.[82] These multiply and propagate, generating more of their own kind.[83] Notably, each seed generates only its own kind.[84] Kircher's argument may hence be said to support an ontology of disease, a feature in which he continues the Helmontian project.

Within the human body, the seeds work to deteriorate the humors and predispose them toward the generation of the plague. Kircher follows Fracastoro when he explains how putrefaction, initially in the external air, reaches the lungs by way of inhalation of this external heat, from whence it then arrives at the heart by hitching a ride on the unsuspecting, fleeing animal spirits whose nature is contrary to the putrefaction itself.[85] At this stage, it is the human body that takes over the role of milieu for the minute seeds of disease; the latter, in turn, makes use of (supposed) existing physiological processes and features of the body, such as the animal spirits or the function of the liver, in order to establish their power and take over the whole organism—at which point, according to a common medical trope, the "natural principle" of the body would become "vitiated."[86]

The environmental dimension of contagion in Kircher roughly comprehended several areas, each of which constituted a specific milieu in which the seeds of disease could arise and thrive. Firstly, there were effluvia, which could mean the emanations from stagnant waters or exhalations from Earth (for instance, after earthquakes). All such effluvia shared the feature of air as a transporter of "seminaria contagionis," affecting humans by way of inhalation. Secondly, contaminated waters, too, could be a source of disease. Thirdly, fruit and grain could be imbued with poison or venom. Certainly, all these aspects were ultimately related, since the underlying belief referred to hypotheses about Earth's entrails. If the "bowels of the Earth" teemed with noxious effluvia, these would naturally be not only emanated by exhalations but also passed on to the waters and to plants growing out of the Earth—an idea that Kircher himself claims to have borrowed from Georgius Agricola. Yet Kircher changes the way in which the minute contagium is understood; he fuses Fracastorian views about imperceptible particles with the views concerning spontaneous generation. As mentioned above, Fracastoro's particles had an unclear status in terms of whether they were living beings or inanimate corpuscles: they seemed largely inanimate, in spite of moving of their own accord and producing offspring of their own kind.[87] Kircher's own minutiae, in turn, are twofold. First come the imperceptible corpuscles residing in poisonous effluvia (in nature) and in humors that they cause to turn putrid (once inside the body);[88] these are similar to the Fracastorian particles. These corpuscles cause disease not directly but indirectly through the mediation of tiny animal creatures—mostly vermicles—which they engender by way of spontaneous generation, as "all that is putrid generates worms [vermes] from itself and from its own nature."[89] These worms may be macroscopic, aligning with the sort of beings commonly assigned to spontaneous generation. Experience teaches us, Kircher claims, that from the seeds of weeds, which putrefy, a multitude of locusts and other noxious insects emerge, which devastate the fields and the fruits of trees, fill the air with virulent effluvia, and even contaminate dead things (et mortuæ contaminant), which are "infallible precursors of famine and plague."[90] Everything edible becomes infected, from the roots of herbs to the meat of animals that are consumed by humans and that then generate plague in the human body. This can be verified by common experience such as how water enclosed in a vessel is quickly "animated" into worms.[91] Interestingly, the vermes can also be invisible to the naked eye. As such, they commonly proliferate in vinegar, milk, and blood, especially with fevers. Kircher claims that while such worms are invisible to the naked eye ("oculo non armato insensibilibus"), their existence was confirmed recently thanks to the microscope.[92] While it is not clear what exactly Kircher could have seen under the microscope, it seems that he set up an analogy between spontaneous generation believed to happen at the macroscopic level with visible worms and insects and a presumed similar phenomenon happening at subvisible level, with these latter microworms as a type of minutiae different from the poisonous corpuscles themselves.

Importantly, the fact that the poison is ingested means that it is not produced by the natural (innate) heat of the human body[93] but comes from a different body (the ingested one). However, once these putrid elements are either inhaled or ingested, they make their way throughout the human body, corrupting the whole mass of blood, much in the manner of fermentation.[94] This trope of thought, while not originally Helmontian, had become popular as an effect of van Helmont's influence and the way in which he had made fermentation central to the process of engendering disease[95]—the difference being that for van Helmont, the process was initiated by a spiritual agent, while for Kircher, the materiality of the disease-causing entities is beyond doubt. Thus, the "putrid blood of those who suffer with fevers"[96] is full of worms, as are the bodies of the dead: "Thus I have been convinced that innumerable, though imperceptible, vermicles teem in the human body, as much in the living as in the dead one; hence here, too, that adage of Job applies: *to putrefaction I say, you are my father; to worms, my mother and my sister.*"[97] Kircher thus envisages putrefaction as linking the living and the dead, suggesting the participation of human bodies in a cycle of life that transcends individual species.

Conclusion

THROUGHOUT THE seventeenth century and beyond, putrefaction retains an important role in the etiology of disease and hence also in public health.[98] In the Galenic tradition, still upheld by some medical writers, putrefying air—emanating either from decomposing bodies or straight out of the "bowels of the earth"—constitutes a milieu that itself causes disease directly. As this Galenic model is gradually supplanted (and in some cases replaced) by the newer chemical philosophy, putrefaction turns into a milieu favorable to the causative agents of disease, which are now understood as minute corpuscles circulating through the putrid air but ontologically distinct from air itself. The nature of these minutiae, in terms of their degree of physicality as well as whether they were explicitly alive or not, varied from author to author; the common features, on the other hand, included independent travel by air and a fermentative propagation of their own kind in a target body.

Thus, the roles of putrefaction and fermentation are not fixed but rather constantly renegotiated between iatrochemistry, traditional Galenic medicine, and a more modern version of medicine that incorporates elements from both. Iatrochemists conceived putrefaction differently than Galenists thanks to, on the one hand, the practical traditions within alchemy—for which putrefaction was one of the preliminary stages of transmutation—and, on the other hand, the context provided by the ferments, which incorporate the Helmontian-Augustinian spiritual dimension alongside a corporeal one. The ferments and their action, including but not limited to processes of "fermentation," offered two things. Firstly, the fact that ferments, in the Helmontian sense of *fermentum seminale*, featured a high degree of specificity contributed significantly to the development of an ontological view of disease: diseases were specific entities because they replicated a single underlying principle. Secondly, the ensuing "fermentation" offered a blueprint of propagation—be it within the "veins of the earth" or within the human body—for the seeds of disease conceived in a corpuscularian way. In this sense, putrefaction was no longer contrary to fermentation, as had been the case with Fracastoro, who had firmly opposed vinegar (a substance considered fermented) to "putrefactio simplex," or simple decay; rather, fermentation and putrefaction worked hand in hand to produce an alteration of substance within the body of the infected. As far as their function of effecting a teleologically oriented alteration was concerned, their task was not much different from when they had been stages in transmutation in earlier alchemical texts.

Notes

1. I quote this edition: Jan Baptist van Helmont, *Ortus medicinae, id est initia physicae inaudita*, 3rd ed. (Amsterdam: apud Ludovicum Elzevirium, 1652), 92.
2. Hiro Hirai also notes this proclivity. See Hirai, "Kircher's Chymical Interpretation of the Creation and Spontaneous Generation," in *Chymists and Chymistry: Studies in the History of Alchemy and Early Modern Chemistry*, ed. Lawrence M. Principe (New York: Science History Publications, 2007), 77–87, 80.
3. I am using this term here in its modern connotation. For current discussions of virulence, see Liise-anne Pirofski and Arturo Casadevall, "What Is Infectiveness and How Is It Involved in Infection and Immunity?," *BMC Immunology* 16, no. 1 (December 2015): 13.
4. Even today, a multicausal framework is the norm, with the concept of virulence as emergent from a concurrence of factors. See Arturo Casadevall and Anne-Liise Pirofski, "What Is a Host? Attributes of Individual Susceptibility," ed. Helene L. Andrews-Polymenis, *Infection and Immunity* 86, no. 2 (October 30, 2017): e00636-17; as well as Arturo Casadevall, Ferric C. Fang, and Liise-anne Pirofski, "Microbial Virulence as an Emergent Property: Consequences and Opportunities," ed. Glenn F. Rall, *PLoS Pathogens* 7, no. 7 (July 21, 2011): e1002136.
5. See, for instance, Aristotle's framework of the four kinds of causes in *Physics*, 194b17–195a8.

6. See R. J. Hankinson, ed., *Galen on Antecedent Causes* (Cambridge University Press, 1998); R. J. Hankinson, "Galen's Theory of Causation," in *Aufstieg und Niedergang der Römischen Welt (ANRW): Geschichte und Kultur Roms im Spiegel der neueren Forschung: Medizin und Biologie: Fortsetzung*, ed. Wolfgang Haase and Hildegard Temporini (Berlin: W. de Gruyter, 1994).
7. I use this term in its etymological sense of "disease generating" and not for an anachronistic implication about "seeds" as modern pathogens.
8. Hippocrates, *Airs, Waters, Places*, in *Hippocrates*, vol. 1: *Ancient Medicine; Airs Waters Places; Epidemics I and III; The Oath; Precepts; Nutriment*, trans. W. H. S. Jones (London: Heinemann, 1923), 73.
9. Hippocrates, 87.
10. Hippocrates, 85.
11. Hippocrates, 91. A common trope in antiquity would have it that heterogeneous parts mixed together would yield a more powerfully motioned mixture, leading to fermentation and putrefaction (hence spoilage).
12. Aristotle, *Meteorologica*, trans. H. D. P. Lee (Cambridge, MA: Harvard University Press; London: William Heinemann, 1952), 299 (379 b 15–25).
13. Aristotle, 295 (379 a 10–15).
14. Armelle Debru, *Le corps respirant: La pensée physiologique chez Galien*, Studies in Ancient Medicine, vol. 13 (Leiden: E. J. Brill, 1996), 238.
15. On Galen's views on putrefaction and disease, see Vivian Nutton, "The Seeds of Disease: An Explanation of Contagion and Infection from the Greeks to the Renaissance," *Medical History* 27 (1983): 1–34, esp. 6.
16. Galen, 7.289,4–291,11 K (*On Differences between Fevers*, book 1, chap. 6). Translation by Jacques Jouanna, "Air, Miasma and Contagion," in *Greek Medicine from Hippocrates to Galen: Selected Papers*, by Jacques Jouanna (Leiden: Brill, 2012), 130–31.
17. Jouanna, "Air, Miasma and Contagion," 129.
18. On this, Robert Parker adopts a more radical stance, arguing that pollution from putrefying corpses was to be identified not with contagion in the medical sense but primarily with religious taboo (the medical aspect being more akin to a secondary side effect). See Robert Parker, *Miasma: Pollution and Purification in Early Greek Religion* (Oxford: Clarendon Press, 1983), 64.
19. On the topic of the red thread linking ancient miasma theories with their Renaissance and early modern reincarnations, see once again Nutton, "Seeds of Disease."
20. Parker, *Miasma*, 3.
21. Parker, 3.
22. Parker, 3.
23. In book 4 of his *Georgics* (29 BCE), Virgil famously describes how to produce bees from the body of a dead bullock. Virgil, *Eclogues: Georgics. Aeneid Books 1–6*, trans. H. Rushton Fairclough (London: Heinemann, 1916), 217–19.
24. Aristotle, *Meteorologica*, 297 (379 b 5).
25. Aristotle, *Generation of Animals*, trans. A. L. Peck (London: William Heinemann, 1943), 355 (762 a 10).

26. Van Helmont, *Ortus*, 92.
27. See Francesco Redi, *Esperienze Intorno alla Generazione degli Insetti* [*Experiments on the Generation of Insects*], 1668, https://bibdig.museogalileo.it/Teca/Viewer?an=323861.
28. Lucinda Cole, *Imperfect Creatures: Vermin, Literature, and the Sciences of Life, 1600–1740* (Ann Arbor: University of Michigan Press, 2016), 18.
29. For instance, Margaret Pelling links "the themes of shame, guilt and pollution" in Christianity with "the idea of disease as having followed the Fall." Pelling, "The Meaning of Contagion," 21. A case in point for this Christian reworking is Athanasius Kircher's *Scrutinium pestis* of 1658, discussed later, whose very first sentence alludes to original sin.
30. Lucretius, *De natura rerum*, VI, 655–64, in *Lucretius on the Nature of Things*, trans. Cyril Bailey (Oxford University Press, 1929), 257.
31. Nutton, "Seeds of Disease," 9–10.
32. Galen, *Commentary on Epidemics* I, book 3.7: CMG V 10, 1, 1934, I 19f. = XVIIA 239 K. However, Nutton suggests Galen might have used "seeds" figuratively or loosely.
33. Marina Banchetti-Robino, *The Chemical Philosophy of Robert Boyle* (Oxford: Oxford University Press, 2020), 20.
34. Banchetti-Robino, 20.
35. Banchetti-Robino notes that "as late as the early 17th century, many of the most significant chemical philosophies were those that reconciled a corpuscularian theory of matter with a vitalistic ontology." Banchetti-Robino, 10.
36. Girolamo Fracastoro, *De sympathia et antipathia rerum Liber unus; De contagione et contagiosis morbis et curatione Libri III*, 1st ed. (Venice, 1546), 28 (liber I, cap. 1 in *De contagione*). Translations from Fracastoro's *De sympathia* are all my own, unless otherwise noted.
37. "Infectione in particulis insensibilibus primo facta"; Fracastoro, 29.
38. Fracastoro, 28.
39. Fracastoro, 28. Corruption "in totality," meaning of the whole body, which would translate into imminent death.
40. Pelling, "Meaning of Contagion," 20.
41. Putrefaction, on the other hand (designated as "corruption"), is a constant companion of infection and contagion when these are linked to disease: one might say that while infection is the procedure, putrefaction is the manifestation of it—i.e., what the body is infected with, the concrete alteration of matter that occurs and engenders disease.
42. Nutton, "Seeds of Disease," 34. Nutton also points out that Fracastoro was a theologian himself and that "the doctrine of *fomes peccati* was formally proclaimed on 17 June 1546 by the Council of Trent, which Fracastoro attended as doctor to the Council."
43. Fracastoro, *De sympathia*, 28.
44. Fracastoro, 30.

45. Vivian Nutton, "The Reception of Fracastoro's Theory of Contagion: The Seed That Fell among Thorns?," in *Osiris*, vol. 6: *Renaissance Medical Learning: Evolution of a Tradition* (1990), 196–234, 232.
46. "Principium autem sunt particulæ illæ insensibiles . . . quæ deinceps seminaria contagionum dicantur"; Fracastoro, *De sympathia*, 29.
47. With regard to the transmission of the French disease, for instance, he claims that it originates in a distemper of the liver, which is brought about by a contagious poison contracted from outside, containing a bad and venomous quality: "Aliquod virus, in quo existit illa mala & venenosa qualitas.'" "Virus" is the Latin contemporary word for "poison" or "venom"; there is no mention of any particles, corpuscles, or vermicles that might be identified as minutiae. Giovanni Battista Montanus, "De morbo gallico tractatus," in *De excrementis libri II* (Venice: apud Balthasarem Constantinum, 1554).
48. "Quod spiritualium nullum efficere e per se potest"; Fracastoro, *De sympathia*, 32.
49. Fracastoro, 32.
50. Fracastoro, 32.
51. Fracastoro, 33.
52. See Shadi Bartsch's account of the two theories of the evil eye in the ancient world: Shadi Bartsch, *The Mirror of the Self: Sexuality, Self-Knowledge, and the Gaze in the Early Roman Empire* (Chicago: University of Chicago Press, 2006).
53. Fracastoro, *De sympathia*, 33.
54. For an overview of the influence of Helmontian iatrochemistry in England in the seventeenth century, see Antonio Clericuzio, *Elements, Principles and Corpuscles* (Dordrecht: Springer Netherlands, 2000), esp. chaps. 3 and 5.
55. Van Helmont, *Ortus medicinae*, 49ff., chap. "Aër."
56. See Walter Pagel, *Jan Baptista van Helmont* (Berlin: Springer Berlin Heidelberg, 1930).
57. Hiro Hirai, "Les *logoi spermatikoi* et le concept de semence dans la minéralogie et la cosmogonie de Paracelse," *Revue d'histoire des sciences* 61, no. 2 (2008): 245, 456.
58. "Fermentum vero, semine saepe prius, & hoc de se generat"; van Helmont, *Ortus*, 30.
59. This discussion is found in van Helmont, 91. On the distinction between principiating and seminal ferments, see also Georgiana D. Hedesan, *An Alchemical Quest for Universal Knowledge: The "Christian Philosophy" of Jan Baptist van Helmont (1579–1644)* (Routledge, 2016), 106ff.
60. As rendered in the English translation: *Oriatrike or Physick Refined*, 1662, 31. Latin original: "Est autem fermentum, ens creatum formale"; van Helmont, *Ortus*, 29.
61. Van Helmont, *Ortus*, 30.
62. Van Helmont, 91.
63. "Ut semina praeparet, excitet, et praecedat"; van Helmont, "Causae et initia naturalium," §24, in *Ortus*, 29.

64. Ferments are "individualiter per species distincta"; van Helmont, *Ortus*, 29.
65. Van Helmont, 29.
66. For van Helmont's theory of disease, see "De ideis morbosis" and "De morbis archealibus," in *Ortus*, 431–43, esp. 432.
67. "Phantasia est virtus sigillifera"; van Helmont, *Ortus*, 432.
68. The external, material triggers are in fact neither sufficient nor necessary in causing disease: what is decisive is the power of imagination. As in one of his examples, the fear of the plague will bring about the plague (van Helmont, 442: "Pavor Pestis, Pestem creat"). Andrew Wear has shown that this view, which "reversed the arrow of causality from the physical to the psychological," also had a strong religious (Christian) dimension: "Although God did not create diseases, within the fruit that Adam ate lay the power to disorder the soul, to produce lust and then 'Irregular Imagination'"; Andrew Wear, *Knowledge and Practice in English Medicine, 1550–1680*, 1st ed. (Cambridge University Press, 2000), 369–70. On van Helmont and the role of imagination in disease causation, see Guido Giglioni, *Immaginazione e malattia: Saggio su Jean Baptiste van Helmont* (Franco Angeli, 2000).
69. Van Helmont, *Ortus*, 92.
70. William R. Newman, "The Corpuscular Theory of J. B. van Helmont and Its Medieval Sources," *Vivarium* 31 (1993): 174–75.
71. Marina Banchetti-Robino, "Van Helmont's Hybrid Ontology and Its Influence on the Chemical Interpretation of Spirit and Ferment," *Foundations of Chemistry* 18, no. 2 (2016): 103–12, esp. 103.
72. Kircher, *Scrutinium*, 116. Kircher discusses van Helmont on 110–17.
73. "Humanum genus originalis peccati contagione pollutum"; Athanasius Kircher, *Scrutinium physico medicum de contagiosae luis quae pestis dicitur*, 1st ed. (Rome: Mascardi, 1658), 1. Translations from Kircher's *Scrutinium* are all my own, unless otherwise noted.
74. Kircher, *Scrutinium*, 2.
75. "A Deo immissæ, naturalibus tamen causis non carent"; Kircher, 4.
76. Kircher, 5.
77. Kircher, 7.
78. Kircher, 18.
79. Kircher, 9.
80. Kircher, 10.
81. Kircher, 21.
82. Kircher, 29: "Vere & proprie corpuscula minima, nulla visus potentia attingibilia."
83. Kircher, 32.
84. Kircher, 46.
85. Kircher, 32.
86. See Giambattista da Monte's remark to this effect: "Extat una propositio universalis in medicina, quòd quando aliquis affectus æqualiter in toto corpore est impressus tunc principium naturale est vitiatum." Montanus, "De morbo gallico," 3.

87. While today this is commonly seen as a marker of biological life, in Fracastoro's time, this connection was not mandatory.
88. Kircher, *Scrutinium*, 50: "Evaporatio humoris putridi . . . ex innumerabilibus & insensibilibus corpusculis composita."
89. Kircher, 38.
90. Kircher, 27.
91. Kircher, 39.
92. Kircher, 40.
93. Kircher, 127.
94. Van Helmont, *Ortus*, 91.
95. See van Helmont, 92–95.
96. Kircher, *Scrutinium*, 141.
97. Kircher, 142.
98. For instance, Andrew Wear argues that while "putrefaction and corruption are pivotal to early modern medicine" (Wear, *Knowledge and Practice in English Medicine*, 5), "the presence of putrefaction motivated nineteenth-century public health measures" (Wear, 471). See also Michael Worboys, "Public and Environmental Health," in, *The Cambridge History of Science*, vol. 6: *The Modern Biological and Earth Sciences*, ed. Peter J. Bowler and John V. Pickstone (Cambridge: Cambridge University Press, 2009), 145.

Bibliography

Aristotle. *Generation of Animals*. Translated by A. L. Peck. Loeb Classical Library 366. London: William Heinemann, 1943.
———. *Meteorologica*. Translated by H. D. P. Lee. Loeb Classical Library 397. Cambridge, MA: Harvard University Press, 1952.
———. *Physics: Books I and II: Transl. with Introd., Commentary, Note on Recent Work, and Rev. Bibliogr. by William Charlton*. Reprinted. Clarendon Aristotle Series. Oxford: Clarendon Press, 2006.
Banchetti-Robino, Marina Paola. *The Chemical Philosophy of Robert Boyle: Mechanicism, Chymical Atoms, and Emergence*. New York: Oxford University Press, 2020.
———. "Van Helmont's Hybrid Ontology and Its Influence on the Chemical Interpretation of Spirit and Ferment." *Foundations of Chemistry* 18, no. 2 (2016): 103–12.
Bartsch, Shadi. *The Mirror of the Self: Sexuality, Self-Knowledge, and the Gaze in the Early Roman Empire*. Chicago: University of Chicago Press, 2006.
Casadevall, Arturo, and Liise-anne Pirofski. "What Is a Host? Attributes of Individual Susceptibility." *Infection and Immunity* 86, no. 2 (October 30, 2017): e00636-17.
Chalmers, Gordon Keith. "Effluvia, the History of a Metaphor." *PMLA* 52, no. 4 (December 1937): 1031–50.
Clericuzio, Antonio. *Elements, Principles and Corpuscles*. Dordrecht: Springer Netherlands, 2000.

Cole, Lucinda. *Imperfect Creatures: Vermin, Literature, and the Sciences of Life, 1600–1740*. Ann Arbor: University of Michigan Press, 2016.

———. "Of Mice and Moisture: Rats, Witches, Miasma, and Early Modern Theories of Contagion." *Journal for Early Modern Cultural Studies* 10, no. 2 (2010): 65–84.

Debru, Armelle. *Le corps respirant: La pensée physiologique chez Galien*. Studies in Ancient Medicine, vol. 13. Leiden: E. J. Brill, 1996.

Durkheim, Émile. *The Elementary Forms of Religious Life*. New York: Free Press, 1995.

Fracastoro, Girolamo. *De sympathia et antipathia rerum Liber unus; De contagione et contagiosis morbis et curatione Libri III*. 1st ed. Venetiis, 1546.

———. *Hieronymus Fracastor's Syphilis from the Original Latin*. Translation by the Philmar Company. Saint Louis: Philmar, 1911.

Giglioni, Guido. *Immaginazione e malattia: Saggio su Jean Baptiste van Helmont*. Franco Angeli, 2000.

Hankinson, R. J., ed. *Galen on Antecedent Causes*. Cambridge, 1998.

———. "Galen's Theory of Causation." In *Aufstieg und Niedergang der Römischen Welt (ANRW): Geschichte und Kultur Roms im Spiegel der neueren Forschung: Medizin und Biologie: Fortsetzung*, edited by Wolfgang Haase and Hildegard Temporini. Berlin: W. de Gruyter, 1994.

Hedesan, Georgiana D. *An Alchemical Quest for Universal Knowledge: The "Christian Philosophy" of Jan Baptist van Helmont (1579–1644)*. Routledge, 2016.

Helmont, Jan Baptist van. *Ortus medicinae, id est initia physicae inaudita*. Amsterdam: apud Ludovicum Elzevirium, 1652 (first edition 1648).

Hippocrates. *Airs, Waters, Places*. In *Works: Translated by W. H. S. Jones*. Vol. 1. Loeb Classical Library. London: Heinemann, 1923.

Hirai, Hiro. "Kircher's Chymical Interpretation of the Creation and Spontaneous Generation." In *Chymists and Chymistry: Studies in the History of Alchemy and Early Modern Chemistry*, edited by Lawrence M. Principe, 77–87. New York: Science History Publications, 2007.

———. "Les *logoi spermatikoi* et le concept de semence dans la minéralogie et la cosmogonie de Paracelse." *Revue d'histoire des sciences* 61, no. 2 (2008): 245–64.

Jouanna, Jacques. *Greek Medicine from Hippocrates to Galen: Selected Papers*. Brill, 2012.

Kircher, Athanasius. *Scrutinium physico medicum de contagiosae luis quae pestis dicitur*. 1st ed. Rome: Mascardi, 1658.

Lucretius, *On the Nature of Things*. Translated by Cyril Bailey. Oxford, 1929.

Montanus, Giovanni Battista. "De morbo gallico tractatus." In *De excrementis libri II*. Venice: apud Balthasarem Constantinum, 1554.

Newman, William R. "The Corpuscular Theory of J. B. van Helmont and Its Medieval Sources." *Vivarium* 31, no. 1 (1993): 161–91.

Nutton, Vivian. "The Reception of Fracastoro's Theory of Contagion: The Seed That Fell among Thorns?" *Osiris* 6 (1990): 196–234.

———. "The Seeds of Disease: An Explanation of Contagion and Infection from the Greeks to the Renaissance." *Medical History* 27, no. 1 (January 1983): 1–34.

Parker, Robert. *Miasma: Pollution and Purification in Early Greek Religion.* Oxford: Clarendon Press, 1983.
Pelling, Margaret. "The Meaning of Contagion: Reproduction, Medicine and Metaphor." In *Contagion: Historical and Cultural Studies,* edited by Alison Bashford and Claire Hooker. Routledge Studies in the Social History of Medicine 15. London: Routledge, 2001, 15–38.
Pirofski, Liise-anne, and Arturo Casadevall. "What Is Infectiveness and How Is It Involved in Infection and Immunity?" *BMC Immunology* 16, no. 1 (December 2015): 13.
Principe, Lawrence. *The Secrets of Alchemy.* Synthesis. Chicago: University of Chicago Press, 2013.
Redi, Francesco. *Esperienze Intorno alla Generazione degli Insetti* [Experiments on the generation of insects]. 1668. Accessed April 30, 2018. https://bibdig.museogalileo.it/Teca/Viewer?an=323861.
Roger, Jacques. *The Life Sciences in Eighteenth-Century French Thought.* Edited by Keith R. Benson. Translated by Robert Ellrich Stanford. Stanford, CA: Stanford University Press, 1997.
Wear, Andrew. *Knowledge and Practice in English Medicine, 1550–1680.* 1st ed. Cambridge: Cambridge University Press, 2000.
Worboys, Michael. "Public and Environmental Health." In *The Cambridge History of Science. Vol. 6: The Modern Biological and Earth Sciences,* edited by Peter J. Bowler and John V. Pickstone. Cambridge: Cambridge University Press, 2009.

Confronting the Limits of Optical Interpretation

Some Philosophical Considerations of Microscopic Details

DANIEL LIU

✦ ✦ ✦

Prologue: Digital versus Analog Imaging

THIS IS an essay on optics, the branch of physics that deals with light, its interactions with matter, and the instruments that use and detect it. We usually think of "optics" as being limited to lenses or, by analogy, to notions of perception in public-relations speak. Here we'll deal more holistically with optics as the science of images (n.) and imaging (v.), broadly construed. (Hopefully a more philosophically rigorous definition of *image* is not needed here, if one is even possible.) A comparison will be helpful between digital imaging and imaging in the analog age. Take your basic camera phone. When you tell your phone to take a picture, light is arriving at the front of the camera's objective lens; the lens refracts (bends) this light and projects an image onto a digital sensor. In the case of, say, an iPhone 6 from 2014, that focused image is being projected on a small rectangular sensor, about 17 mm². The sensor is divided into about 8 million pixels—2 million of which are red, 2 million blue, and 4 million green—each of which is 1.5 μm across. When you take a picture, each pixel measures the tiny bit of light that hits it and gives that bit of light a numerical value from 0 to 255, with 0 being 100 percent dark and 255 being 100 percent bright. Because each pixel is one of three colors (red, green, or blue), a signal processor takes each pixel's 0–255 value, compares it to the values of the neighboring pixels, and creates an 8-million-pixel digital image, where each pixel is given three values of 0–255 for each of red, green, and blue, for a possible 16,777,216 (256^3) discrete color values for each pixel.[1]

If you now view this eight-megapixel digital image at "100 percent zoom," where one pixel on your screen corresponds to one pixel recorded in the digital image file, you are viewing the image at its *maximum* detail. You can consider each pixel as a "unit of detail" that has numerical value attached to it, and each one is clearly and distinctly marked off from the next pixelated unit of detail. In our iPhone 6 picture, there are quite a lot of these detail units, and because we are dealing with a digital image, each of these units has the same dimensions and basic properties as every other unit of detail. There is also no such thing as *more* detail than this: if we were to "zoom and enhance," as they liked to say in TV crime dramas, we would be seeing exactly the *same* detail but blotchier. Even though our eight-megapixel camera sensor captures eight million units of detail, this is a very *finite* amount of detail, and there is not *more* available.[2]

Now blink a couple of times. Look out your window or across the room at the middle distance, and think about how it is you're able to distinguish one faraway thing from another. (If you're visually impaired, as many of us are, this exercise also works for any of your other senses as well.) Our natural senses—sight, hearing, touch, smell, and taste—don't have such clearly defined "units" of perception as we do with a digital image, and most of us do not perceive the world in precise 0–255 digitally graded steps. Instead, we are dependent on two properties of the world we perceive: the actual *sharpness* of an object's boundary and the relative *contrast* in color and luminosity (or, by analogy, aroma, pitch, texture, etc.) between two objects. We have to strain ourselves to detect fine distinctions, and our ability to determine these distinctions is often based on our familiarity with our surroundings and the day-to-day training of our perceptions. It's not quite right to call these our "natural" senses, and it's probably even less accurate to call our senses "analog"—but in the twenty-first century, we are so accustomed to digital conventions that this comparison is useful, nevertheless.

If you're a scientist, then the difference between analog and digital detail is even more dramatic. In scientific imaging, where the goal is to *detect* and *measure* as much fine detail at the finest degree possible, the tiniest differences matter. In a digital image, the differences are obvious: if you have two adjacent pixels, and one measures a value of 100, and the neighboring pixel measures a value of 102, the difference is 2, and your digital image sensor has measured it. The numerical value of each individual pixel matters a great deal, especially in demanding fields like astronomy, where a 0 is the vast nothingness of space and a 1 could be a faraway galaxy, star, or planet.[3]

But in the more mundane, analog world, this would be like trying to specify the difference between the "puce" and "rose gold": it can be done, but it takes both mental and infrastructural effort. As the physicist and optical engineer August Köhler (1866–1948) laid out the problem in 1926, "One must ask: where is the limit up to which one can still interpret the images of optical instruments in the old and familiar way? Is there a sharp boundary or a gradual transition between the area where, on the one hand, a certain differentiation of shapes and sizes is possible, and on the other, where the picture supplies only more or less questionable signs of a finer structure? What else can you conclude from such evidence with security? On what circumstances does the location of this limit depend?"[4] That is to say that rather than assume that the microscope self-evidently reveals fine structures, Köhler here argued that modern, scientific microscopy had become a science of the scrutiny of the limit of optical interpretation. To use Hans-Jörg Rheinberger's terminology, we might say that scientific microscopy is a kind of microperformance dedicated to recovering an experimental trace as it threatens to disappear. In the rest of this essay, I endeavor to provide a history of Köhler's rather profound statement about distinctions, differences, and the limits of optical interpretation. I will show that the history of microscopy actually furnishes us with a set of scientists who grappled with the metaphysical problem of detail—that is to say, the problem of what a detail *is* and the epistemological problem of how it was possible to recognize and *know* it—and I will make some tentative suggestions of what all of this detail actually amounts to.

Magnification and the Microworld

IN THE history of science, we might be able to make a fundamental distinction between the study of small things and the study of the microworld (see table 1).[5] In this essay, I will make the corresponding shorthand distinction in procedures between *magnification* and *resolution*. It should first be said that the small and the microworld, magnification and resolution, are overlapping reductionist epistemological categories and are not categorically separate. Nevertheless, this distinction can be a useful one, and I've derived it by examining the history of observational practice rather than relying only on a priori categorizations. The first category, magnification to study things that are very small, requires both practical and conceptual isolation and purification. The histories of modern biochemistry, molecular biology, and pharmacology are ground zero for the focus on small things today, at least in the history of the life sciences: a molecule must be isolated by centrifugation, titrations, electrophoresis, or any number of steps to ensure that the object or material under study is pure.[6] The art and science of this isolation are merely the prerequisite to a further understanding of the structure and function of the object. The discovery of the double-helix structure of DNA is one of the most famous exemplars of this general principle: before any X-ray crystallographic imaging could be done, the DNA in its two different crystalline forms had to be extracted, purified, and crystallized in order to obtain a "clean" X-ray diffraction pattern, and this, in turn, was the essential empirical basis from which the molecular model of interlocking nucleic acid pairs could be derived.[7] In these sciences, isolation and purification are thus prerequisites for magnifying and understanding the object in question. For many philosophers and critics in the twentieth century, this isolation has been essentially hell: the reduction of life, of experience, of emotion, of meaning, down to the activity of a single object or substance. Its boogeymen are the genetic essentialism of Jim Watson and Richard Dawkins, its prophets the magnetic personalities of René Descartes—or, more proximately in nineteenth-century German philosophy, Vogt, Pettenkofer, and Moleschott, each of whom was famous for equating intellect and cognition with the digestion of phosphorous.[8] Reductionism qua distillation down to an essence has always been the core practice of alchemists and chemists, and rhetorical power of doing the same has been recognized since the beginnings of Western philosophy itself.

A second form of reductionist epistemology is the resolution of detail and the construction of the microworld. The resolution of detail is rarely praised, condemned, or even much noticed. The important exception to this is when there is "too much detail," where one "can't see the forest for the trees" or one "misses the big picture": this is the point at which "richly detailed" crosses over into being too detailed, such that one's powers of analysis and synthesis start to break down. When most of us beginners look into a microscope for the first, second, or even third time, we experience peering into a different world with different rules. True, one can potentially see lots of detail, but in a basic microscope, left is literally right, up is down, your field of view is restricted to a tunnel, and the world in focus is a flat plane rather than a three-dimensional world of whole bodies.[9] More importantly, however, when looking at the physical world so closely, the contexts and milieus that make an object familiar are left behind, either because the context itself is so magnified that it is unrecognizable or because it is out of the microscope's field of view entirely. Just as a reader can only understand a new literary form through repetition and contextualization, the world revealed by the microscope demands routes toward familiarity and trust in novel observations. Thus, the histories of the microscope and microscopy illustrate why the study of small

TABLE 1. An extremely schematic suggestion for the
differences between the small and the microworld

Small	Microworld
Magnification/detection (of an object)	Resolution (of fine structures)
Isolation/purification	Contrast/context
Monocausal reductionism	Reductionism to parts
Natural philosophy of causes	Natural history and taxonomy of structures

things and the resolution of small details are two distinct but not separated epistemological regimes.

From its invention during the Scientific Revolution in Europe up to the mid-nineteenth century, the microscope presented a significant problem of validation. Early modern microscopy began as an art of magnification of small things, and early microscopy books were filled with illustrations of magnified objects in glorious and gory detail: Hooke's famous illustrations of the cellular structure of cork, the hairy scales of fleas, or the teeming animalcules in pond water. This magnifying operation centered on individual objects. Historian of science Catherine Wilson has argued that, in its infancy, microscopy was plagued with problems of "overseeing and overrepresenting" a plethora of individual objects and lacking much connective logic among them.[10] For Hooke and the learned ladies and gentlemen of his era, the microscope may have been the instrument par excellence for an "extension of the empirical horizon," but this empiricism was fraught with both technical and epistemological difficulties. It was not well understood how lenses worked differently in microscopes versus telescopes, a significant problem given that telescope lenses are large and curved relatively gently, while microscope lenses are small and had extreme curvatures to match. Spherical and chromatic aberrations of microscope lenses rendered the edges blurry and fringed with unnatural colors, while the lens's distortions rendered straight lines into curves (a feature familiar in peephole and ultrawide lenses). Until the nineteenth century, these aberrations and distortions had to be corrected in the mind's eye, more so than in the instrument itself (see fig. 1).[11] In microscopy's early years, instrument makers took design instructions from savants like Hooke and made them work to the best of their ability, drawing on their experience making spectacles, magnifying glasses, or telescopes: many optical designs in early microscopes were thus unique to the instrument makers or even unique to individual instruments.[12] For a long time, serious scientists and natural philosophers preferred to leave microscopes to the amateurs, for

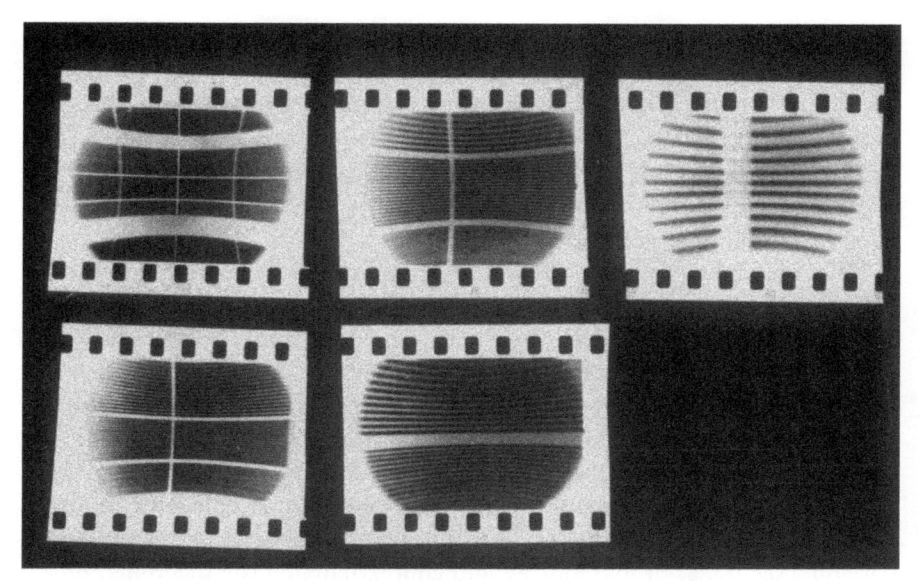

FIG. 1. Twentieth-century examination of late seventeenth-century microscopes (1660s–90s), using an Abbe test plate that consists of engraved parallel lines. Note the high degree of pincushion distortion, which causes straight, parallel lines to appear curved, and the blurred fringes of what should be clear lines, caused by chromatic and spherical aberration (Brian Bracegirdle, "The Performance of Seventeenth- and Eighteenth-Century Microscopes," *Medical History* 22, no. 2 [April 1978]: 187–95, https://doi.org/10.1017/S0025727300032312, courtesy of Cambridge University Press; cf. Savile Bradbury, "The Quality of the Image Produced by the Compound Microscope: 1700–1840," in *Historical Aspects of Microscopy*, ed. Savile Bradbury and G. L'E. Turner [Cambridge: W. Heffer & Sons, 1967]).

whom the novelty of microscopic vision was more important than its absolute trustworthiness.[13]

It was only when microscopy stabilized as a routine practice in the eyes of its users *and* audiences that the microscope started to do something more interesting than magnify things. Until recently historians have regarded microscopy in the eighteenth century as being in a fallow or moribund state, with the microscope itself being cast aside as an amusing toy rather than an instrument that could reshape science or philosophy. However, Marc Ratcliff's pathbreaking survey of eighteenth-century microscopy literature has shown how the microscope was *used* in addition to how natural philosophers thought about it. If seventeenth-century microscopy was notable for what practitioners could wring out of a single microscope, then the headline trend in eighteenth-century microscopy was the simultaneous use of many different microscopes by individual scientists seeking to make reproducible observations. Whereas Leeuwenhoek's peerless microscopes in the seventeenth century mounted unique specimens semipermanently to a unique lens, in the eighteenth century, an individual microscopist would observe a mobile specimen using many different microscopes at their disposal.[14] In other words, microscopists in the seventeenth century freed the specimen somewhat, making it observable multiple times under different microscopes from many different perspectives.

The specimen was not the only thing that became more mobile in seventeenth-century microscopic practice. From the 1740s, the number of published microscopic *images* began to increase rapidly. Scientists in the eighteenth century collected and systematized individual observations and images into detailed, taxonomic descriptions of the anatomy of animals, plants, and minerals. From the middle of the eighteenth century, microscopists added standardized measurements and degrees of magnification to accompany their engravings; these were meant to supplement an older, iconographic convention, what Ratcliff calls the method of "natural comparison" (see fig. 2) of juxtaposing and contextualizing the magnified object with its true-to-nature size.[15] It was through these kinds of communicable and replicable conventions, Ratcliff argues, that the creation of regular, repeated, and communicable observations in image and print had a truly profound impact on what it meant to look through the microscope in the first place. To wit, "scholars were dealing now with species and no longer with specimens," with generalized ideas and theories rather than with this particular fly or that particular rock.[16] Microscope observations could now fit into an existing matrix of knowledge, brought about by scientists' social practices and milieus. And though microscope optics did not improve much in the eighteenth century, the instrument did undergo important changes. Whereas seventeenth-century microscopes were made of wood or paper and decorated with exotic luxury materials (snakeskin, gold, ivory, etc.), after the 1730s, microscope makers switched to more practical brass and other alloys that held lenses in more consistent alignment. As microscope makers began standardizing their designs, the microscope as a sumptuous ornament transformed into a much plainer tool that projected an aesthetics of instrumental reliability.[17]

Thus, even though microscopes before the mid-nineteenth century were plagued by severe optical problems, the accumulation of observations, images, and accompanying descriptive and philosophical texts stabilized microscopy as a useful scientific practice. The initial natural philosophical excitement around the microscope in the seventeenth century, with Hooke's outlandish claim that the microscope overcame the fall of man by letting him

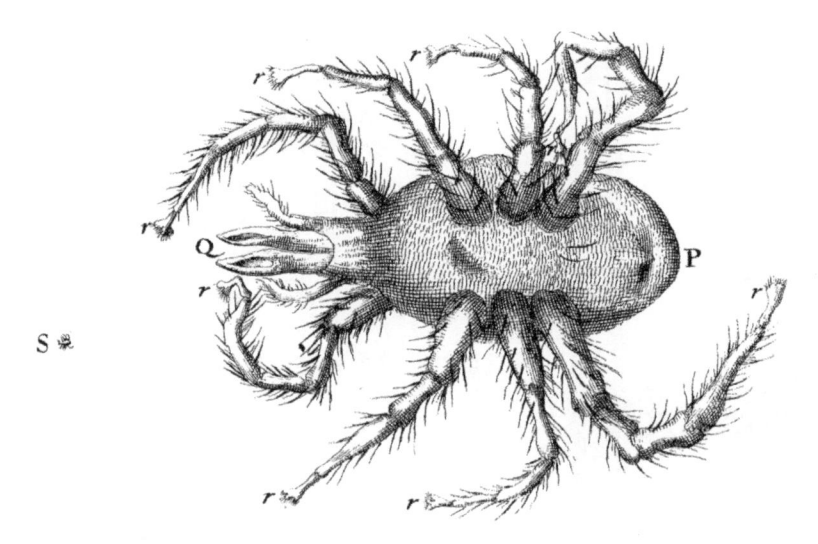

FIG. 2. The method of "natural comparison" of a flea, unmagnified (*left*) and magnified (*right*) (James Wilson, "The description and manner of using a late invented set of small pocket-microscopes, made by James Wilson; which with great ease are apply'd in viewing opake, transparent and liquid objects: as the farina of the flowers of plants, etc. the circulation of the blood in living creatures, etc. the animalcula in semine, etc.," *Philosophical Transactions* 23 [1702]: 1241–47).

"[taste] too those fruits of Natural knowledge, that were never yet forbidden," thus gave way in the eighteenth century to more mundane efforts to compile, compare, and communicate microscopic observations—an effort that was as much about the variability of microscopes as it was the diversity of European microscopists and their agendas.[18]

The Revolution in the Theory of the Microscope and the Milieu of the Microscope's Physical Limit

BECAUSE WE tend to emphasize the microscope as a tool for image making or "seeing the invisible," our attention is usually turned to the primary optical system of the microscope—that is, the objective or the primary imaging lens and the ocular or eyepiece. If you step into a modern microscopy lab, you might notice the imposing physical size of today's most advanced microscopes, in comparison to either school microscopes or most microscopes in the eighteenth or nineteenth century (see fig. 3). Notice, also, the accumulation of other accessories, including the computer, digital camera equipment,

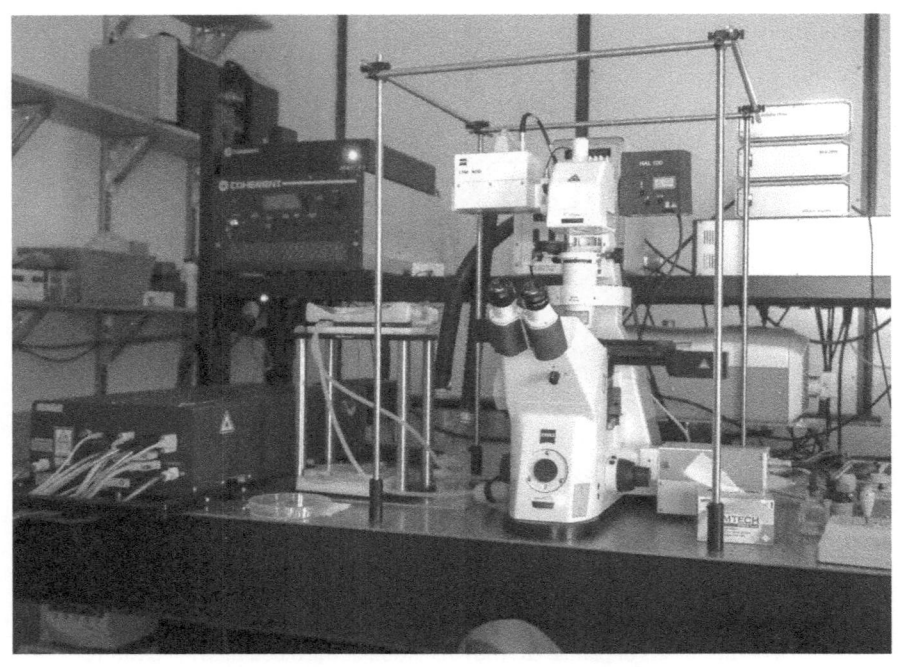

FIG. 3. A modern, high-end microscope setup. Note the overall size and scale of the microscope accessories in comparison to the eyepiece lens, which itself is much larger than the main objective lens, not visible in this picture. Most of the equipment here is sitting on a heavy, pneumatically stabilized table to prevent vibration (photo by the author, 2017).

amplifiers, lasers, and so on, not to mention the pneumatically stabilized, antivibration table on which the microscope sits.

All of these accessories in the microscope's milieu are designed to help microscopists confront the absolute physical limit of the optical microscope: the smallest detail an ordinary light microscope can resolve is 250 nm. For scale, we know today that a single *E. coli* bacterium is about 600 nm long, an influenza virus is 80–120 nm in diameter, and the cell membrane is 7–10 nm across; the latter two were invisible to direct observation until the invention of the electron microscope. To be even more specific, any object of any size that emits light could be *detected* against a black background—telescopes can detect individual luminous points—but the microscope's ability to distinguish *two* points is limited, and these two points must be 250 nm or more apart. This is what is known as the *resolution* limit, and it is this limit that is relevant to the question of how much detail we can see. Moreover, the 250 nm limit is one that is specified for ideal conditions only, and observational conditions are usually far from ideal. All of the microscope's accessories, down to the table the microscope sits on, are designed to make these conditions as ideal as possible and to help the microscopist confront the microscope's absolute physical limits. Many of these accessories only started to appear in their modern form in 1873, when the fact of the physical limit of optical interpretation became theoretically clear.

It goes without saying that in the last three hundred years the optical systems in telescopes and microscopes have improved dramatically. Yet even looking at them superficially, telescopes and microscopes have improved in very different ways. The basic optical layout (an objective next to the specimen and an ocular near the eye) and physical size of microscope lenses did not change much from the mid-seventeenth to the mid-nineteenth centuries, and while the glass lens elements in microscopes today are far more advanced than they were a century ago, this progress is not superficially evident by the mere size of the lenses in question. By contrast, the optical systems of telescopes have grown in size continuously: the objective lenses of Galileo's telescopes circa 1610 were about 37 mm in diameter, and by the end of the nineteenth century, telescopes were being made with objective lenses reaching over one meter; the Keck telescopes, built on Mauna Kea in the mid-1990s, have primary mirrors that are ten meters in diameter, and in the 2020s, the largest telescopes will reach twenty to forty meters in diameter. The microscope and the telescope do not occupy opposite ends of the same spectrum but rather are fundamentally different from each other: seeing ever farther away objects requires bigger lenses, but seeing ever-smaller objects does not.

Despite this obvious fact, until the 1870s, microscope optics were designed using geometrical methods and optical theories borrowed from telescope optics. It was long assumed that the basic principles that lent the telescope its magnifying power also applied to the magnifying power of the microscope, and many early microscope lens designs were merely smaller versions of telescope lens designs.[19] In 1830, the British wine merchant and amateur optical engineer Joseph Jackson Lister (1786–1869) became the first to show on a theoretical level why an achromatic microscope could not be designed like an achromatic telescope; he proposed a very un-telescope-like arrangement of lenses and a way of calculating how best to arrange them to produce an image that was both clearer and flatter.[20] Yet even with Lister's dramatic improvement, microscope designers continued to have difficulty improving the magnifying power of the microscope without further losing clarity and brightness. This was because early theories of telescope *and* microscope optics still depended on following the geometrical path of light as it was *refracted* or bent by each glass element in the lens. Improving a telescope in this period meant increasing the diameter of the front objective lens to collect more light and increasing the power of the rear eyepiece to magnify the image projected by the objective.

When applied to the microscope, this older theory demanded an operation akin to zooming in in photography or cinematography, for which the locus of improvement was in an increase in magnification and overall brightness. That is to say, by narrowing the field of view, one does something analogous to magnifying and *bringing an object closer* to reveal its details. By the mid-nineteenth century, some British microscope designers were attempting to do exactly this, pursuing higher magnification and using larger objective lenses to increase the overall brightness of the image.[21] In the schematic argument of this paper, the old theory sought to look at an object more and more closely by isolating and magnifying it. Had magnification and image brightness alone guided microscope development, we would have had ever larger microscope lenses rather than what we see in figure 3. Yet these attempts at increasing magnification led only to a decade-long "divisive and embarrassing" public debate in Britain in the 1860s rather than any real improvement to the microscope.[22]

In 1873 Ernst Abbe (1840–1905) and Hermann Helmholtz (1821–94) independently "solved" the physics of the microscope's optical system and ushered in the era of truly scientific microscope design.[23] (The philosopher of science Ian Hacking once wrote, "I suppose no one understood how a microscope works before Ernst Abbe."[24]) Abbe and Helmholtz discovered in 1873 that the microscope's core problem was how light *diffracts* as it moves through the *specimen* and into the objective lens—that is, that the fine details of the specimen itself affect light in complex ways before the objective lens captures it. By contrast, the older theories of the microscope and the telescope were primarily concerned with how light *refracts* through the objective lens—that is, how the curved glass lens bends and focuses light to create an image. In other words, rather than analyze the geometry of the rays of light as they were refracted by the lens, Abbe and Helmholtz examined the physics of light itself, especially how light *waves* bend and interfere with stationary obstacles, such as the microscopic structures of cells. Abbe's formulation of the mathematical problem in 1873 has become the more famous of the two:

$$d = \frac{\lambda}{2n \sin\theta}$$

where d is the smallest object that could be resolved, lambda (λ) is the wavelength of light, n is the refractive index of the imaging medium, and $2\sin\theta$

is the angle of the cone of light that can enter the lens at a given distance to the object. In this equation, lambda, the wavelength of light, becomes the absolute limiting factor to the microscope's ability to resolve detail: the eye perceives blue light at a wavelength of about 450–500 nm, and therefore the smallest detail the microscope can theoretically resolve is half that, or about 250 nm. The rest of the equation, $n \sin\theta$, is the space in which the microscope designer can make improvements: this is known as the numerical aperture (NA) of a microscope lens and is important enough that objective lenses have this number engraved on the barrel along with the magnification.

Thus, if the discovery of the microscope in the seventeenth century had created a whole world of the potentially visible, then the discovery of the 250 nm "Abbe diffraction limit" foreclosed this particular sense of limitless knowledge of the small: the laws of physics stated that the realm of the microscopically visible had a hard end.[25] Abbe rather pessimistically wrote in 1873 that "There is no microscope that can objectively see—or will ever be able to see—the true nature of an object that cannot already be recognized by a normal eye with a sharp, 800× immersion objective."[26] One could make and use a microscope with higher magnification, but it would not actually reveal any *more* or *finer* detail than a high-quality lens with lower magnification. In fact, even if you had two lenses of different magnifications that showed the same degree of detail, working at a higher magnification would be more difficult because the larger image is dimmer, shows less context, and is prone to errors such as vibration of the table or the eye. From 1873 onward, it would never be possible to magnify one's way down to the infinitesimal world of molecules, atoms, and beyond, and since the end of the nineteenth century, maximum *useful* magnification has not increased beyond about 1,500×.[27]

Abbe and Helmholtz's new theory of the microscope led to a dramatic change in how microscopes were designed and used. The end to naive microscopic reductionism led microscope designers like Abbe to find ways of making the microscope easier to use and making the limits of optical interpretation easier to reach. From the 1870s onward, most of the greatest technical advances in microscopy would be found in the microscope's accessories and the microscope's milieu; objective lenses continue to improve, but this is either incremental or in response to other technical changes, such as the use of laser light sources or digital image capture. The equation above by Abbe immediately suggested several routes of improvement. Abbe published his version of the new theory in the biological journal *Archives of Microscopic Anatomy*, and in the very next article, he revealed its first technological fruit: not a new microscope objective lens but rather a new *condenser*, a set of lenses mounted below the microscope stage that adjusts the light *before* it reaches the specimen (see fig. 4).[28] This was certainly not the first microscope condenser—Hooke used a glass sphere filled with water to focus light on the specimen—but Abbe's condenser was designed to make it easier to adjust the angles at which rays light approached the specimen and the objective lens, maximizing the term $2\sin\theta$ in the above equation. A next step was controlling the term n, the somewhat vaguely conceptualized "refractive index of the imaging medium": pragmatically speaking, this meant standardizing the thickness and the optical quality of the glass slide and the cover slip used to mount the specimen in the microscope, materials that hitherto had been thought of only as specimen holders rather than integral parts of the microscope's optical system. Making these absolutely mundane working materials uniform became a top priority, and microscope makers like Carl Zeiss (who by this point had hired Abbe to be its scientific director) created partnerships with optical glass manufacturers to make high-quality and disposable microscope supplies.[29] And in order to control lambda (λ), the

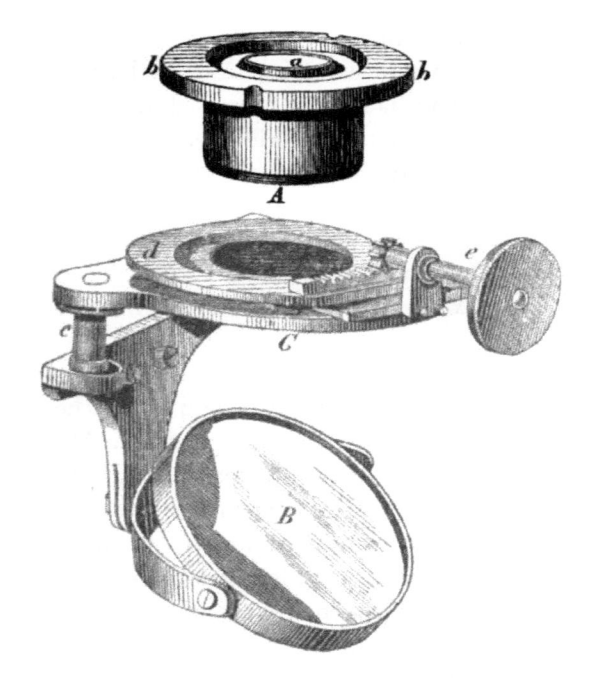

FIG. 4. Abbe's condenser, mounted between the specimen stage and the light source, in this case a mirror (Abbe, "Ueber einen neuen Beleuchtungsapparat am Mikroskop," 475).

wavelength of light, microscope makers began making electric lightbulbs that could create consistent light; this had the side effect of liberating microscopists from the need to work at large windows. Electric lightbulbs were even developed to create monochromatic light, again in the hopes of controlling the terms of the equation, and this monochromatic light in turn demanded the development of microphotography in order to actually *see* the image created. This, in turn, led to the development of the unusual ultraviolet light microscope in 1904 (see fig. 5), which used wavelengths of light so short that the eye could not perceive them—solely with the goal of seeking smaller details by reducing λ in the equation above.[30] From this point of view, microphotography was not developed in the service of making a more "objective" and thus less "subjective" image; instead, it was a technical achievement of illumination and translation by image capture that might resolve details the eye could not.[31] Note too that the microscope "itself" is being augmented by a growing number of accessories for image making, each of which represents a different facet of optical technology (e.g., photography, electrical illumination).

The discovery of the microscope's absolute physical limit also freed optical scientists to design microscope lenses and accessories to do things other than achieve high magnification: after the 1880s, microscope lens designers began to prioritize optical clarity, vividness, and contrast rather than pursue high magnification at any cost. The microscopes Abbe designed for Zeiss in the 1880s prioritized giving a clear and sharp picture across the whole field of vision, such that an object in the middle of the image was as sharply defined as at the image's periphery. Such highly "corrected" optics—compound lenses designed to eliminate spherical and chromatic aberration—made an image at a given magnification more usable and had the added benefit of presenting truer-to-nature colors than previous microscope optics did.[32] "Clarity" here not only means rendering objects with sharp edges; it means making them distinguishable by projecting images with high contrast. Relatedly, one of the simpler developments in late nineteenth-century microscopy was the commercialization and improvement of the technique of dark-field illumination: by inserting a metal disk into the condenser and blocking out *direct* illumination, light could be diffracted through the sides of the specimen, giving its features an otherworldly glow against a dark background (see fig. 6).[33] In fact, such an image has *less* absolute resolution, and its finest details are far coarser than the 250 nm limit that direct bright-field microscopy can achieve; nevertheless, it gives the impression of showing more detail by considerably increasing the *contrast* of adjacent structures.

Dark-field microscopy was a relatively simple technology compared to other contemporary developments in optics and microscopic technique: it

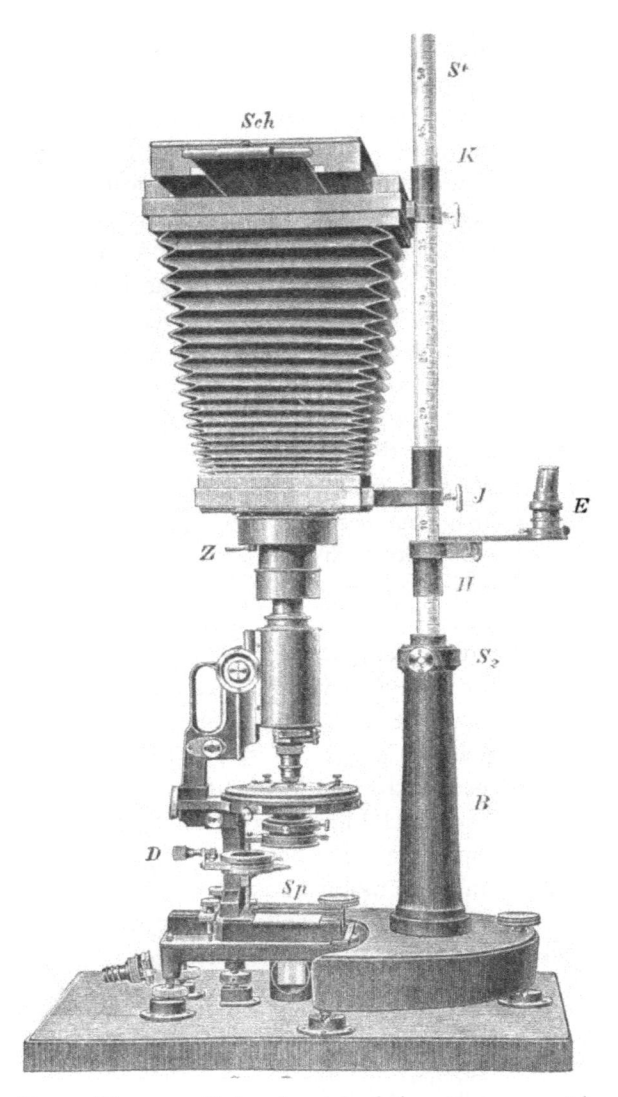

FIG. 5. The 1904 Zeiss ultraviolet light microscope with photographic apparatus—a necessary feature of this microscope, since ultraviolet light is not visible to the human eye. Not shown here is the essential ultraviolet lamp, which, in 1904, would have been about as large as the camera bellows pictured here (Köhler, "Mikrophotographische Untersuchungen mit ultraviolettem Licht," 276).

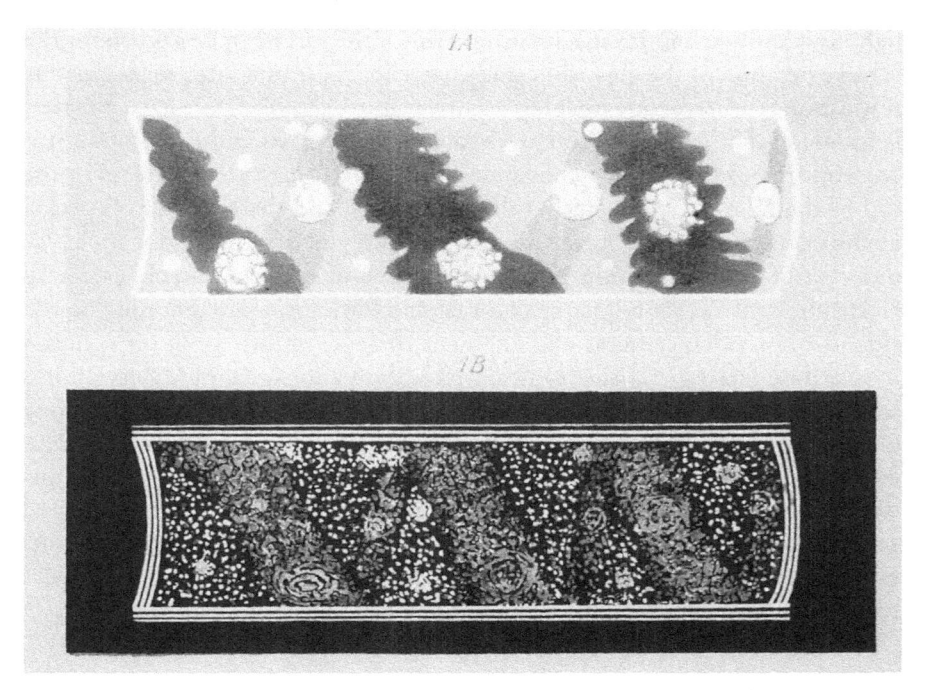

CONFRONTING THE LIMITS OF OPTICAL INTERPRETATION 137

FIG. 6. Ordinary or "bright-field" microscope image (*above*) versus dark-field microscope image (*below*) of a cell of *Spirogyra*, a common pond scum algae. The dark-field image has less absolute resolution but reveals detail due to its ability to impart high contrast (from Nikolai Gaidukov, *Dunkelfeldbeleuchtung und Ultramikroskopie in der Biologie und in der Medizin* [Jena: Gustav Fischer, 1910], plate 5).

was easy to commercialize and popularize, producing dramatic results by, at its most basic, merely adding a small metal disc to an existing microscope accessory. Despite this simplicity, I would argue that it is emblematic of what it has meant to improve microscopic vision after the discovery of the microscope's diffraction limit: it is a technique that serves to highlight details in context by showing them clearly, sacrificing magnification or the ability to resolve the absolute smallest details in favor of improving contrast across the whole image. The details in figure 6 that dark-field illumination makes possible are not close to the microscope's theoretical limits but rather are examples of techniques that enhance the microscopist's powers of optical interpretation.

In the years immediately after the discovery of the diffraction limit, much of this kind of improvement happened below the objective lens: the Abbe condenser and dark-field illumination are both techniques for controlling light as it approaches the specimen, and much the same can be said about the standardization of the glass slide and coverslip. A century later, an analogous revolution in microscope imaging occurred above the objective lens, one that is more familiar to us digital natives today. In the 1970s and early 1980s, microscopists began to attach consumer-grade electronic video cameras to the top of the microscope, replacing the ocular lens and the human eye's retina as the locus of *imaging*.[34] Live video cameras are not only useful for the obvious task of making motion pictures or videos of microscopic objects: easy electronic controls for adjusting contrast and luminosity are mounted on the side of most video cameras and augment the eye's ability to detect contrast in real time (see fig. 7). Electronic amplification and signal modification has thus become a way to help the eye detect contrast and differences among objects that were previously limited by the nature of our biological tissues: too much light can fry the specimen and overwhelm or even damage the microscopist's own eyes. Properly calibrated, vidicon tubes and digital imaging sensors are even able to quantify these minute differences (recall our previous example of the adjacent digital pixels measuring values of 100 and 102). Electronic amplification does not show us what was not there before but merely

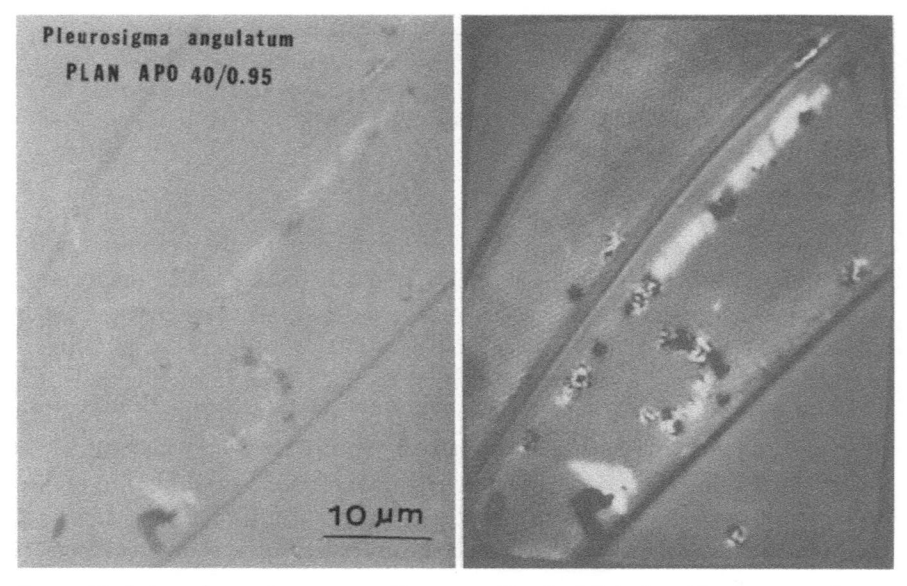

FIG. 7. Real-time electronic contrast adjustment. On the left is the video image as the eye might see it; on the right is the image after the contrast dial on the video camera is given a twist (Inoué, "Video Image Processing Greatly Enhances Contrast," 351; permission conveyed through Copyright Clearance Center).

amplifies differences to make them more apparent, giving us a different way to see or read the details by highlighting their differences—all without needing to interfere with the specimen sitting under the microscope itself.

"Mapping the Interactions between the Specimen and the Imaging Radiation" at the Limits of Microscopic Interpretation

IN THE early twentieth century, the microscope was instrumental in ending two major scientific debates about the nature of the physical world below the microscope's 250 nm limit—the world of "submicroscopic" dimensions. The first of these, the question of the atomicity of matter, had vexed scientists and natural philosophers since Democritus. In 1909, the French physicist and physical chemist Jean Perrin (1870–1942) observed a colloidal preparation of gamboge (a yellow plant pigment) using the newly invented ultramicroscope (see fig. 8). These measurements of the gamboge particles' Brownian motion or "random walk" became the empirical proof of the atomicity and discontinuity of matter itself (see fig. 9).[35] The ultramicroscope Perrin used to make these measurements was based on an elaboration of the principles used in dark-field microscopy: by illuminating a specimen from the side instead of reflecting it from above or transmitting it from below, the ultramicroscope could indirectly *detect* particles hundreds of times smaller than the smallest objects that could be *seen* under an ordinary microscope. However, the ultramicroscope could not *resolve* those same particles, and as a result, Perrin could not directly see them "in the old and familiar way": their presence could be discerned, and their movements measured and quantified, but they appeared to the observer merely as fuzzy spots of light against a dark background. Perrin's exercise thus exemplified Köhler's statement that there were advantages in *not* using the microscope in the old and familiar way. In order to press beyond the 250 nm limit that Abbe and Helmholtz discovered in 1873, the microscopist could not see but had to infer what physical interactions led to the creation of the visible image, with the aid of microscope accessories that helped the microscopist control and understand the path of the light moving through the whole microscope system. As the microscopist Elizabeth M. Slayter put it, the microscopist could "map the interactions between the specimen and the imaging radiation" in order to gain access to the *sub*microscopic scale.[36]

Slayter's use of the word "map" is not metaphorical: from the beginning of the twentieth century, there were many scientific contexts in which microscope images themselves could be less important than the diagrammatic

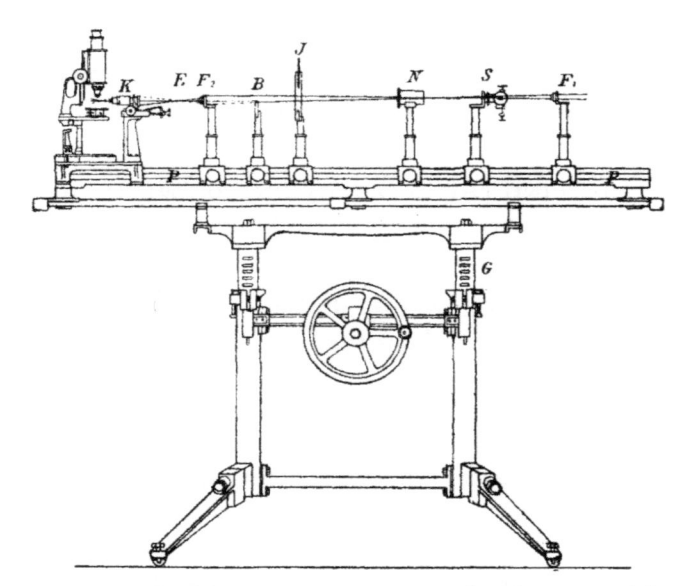

FIG. 8. The slit ultramicroscope. Note again the relative size of the accessories used in addition to the standard microscope (*upper left*), including the heavy adjustable table. Not shown in this image is the huge, 5,000–10,000-watt arc lamp placed on a separate table to the right, providing the necessary light for the instrument (Henry Siedentopf and Richard Zsigmondy, "Über Sichtbarmachung und Größenbestimmung ultramikoskopischer Teilchen, mit besonderer Anwendung auf Goldrubingläser," *Annalen der Physik* 315, no. 1 [January 1903]: 1–39).

maps that were drawn to explain those images. Perrin was a case in point: historian of science Charlotte Bigg has shown that earlier microphotographs of gamboge and Brownian motion provided "objective" evidence in the naive sense, but Perrin ultimately used a diagram to portray the *reality* of the random walk of atoms and molecules.[37] Over the course of the twentieth century, microphotographs replaced engravings and hand-drawn illustrations as the former became easier in routine microscopy. But at the microscope's interpretive limits, a diagram was more useful for showing how invisible structures could cause visible images to form. The second major scientific debate that the microscope ended was over the molecular constitution of the cell and other biological materials, and this was solved with the aid of diagrams and maps, which were necessary to piece together what the microscope's multiplying accessories were doing in concert.

In its most basic form, the microscopic technique that revealed the molecular structure of cells and tissues, *polarized light microscopy* or *polarization*

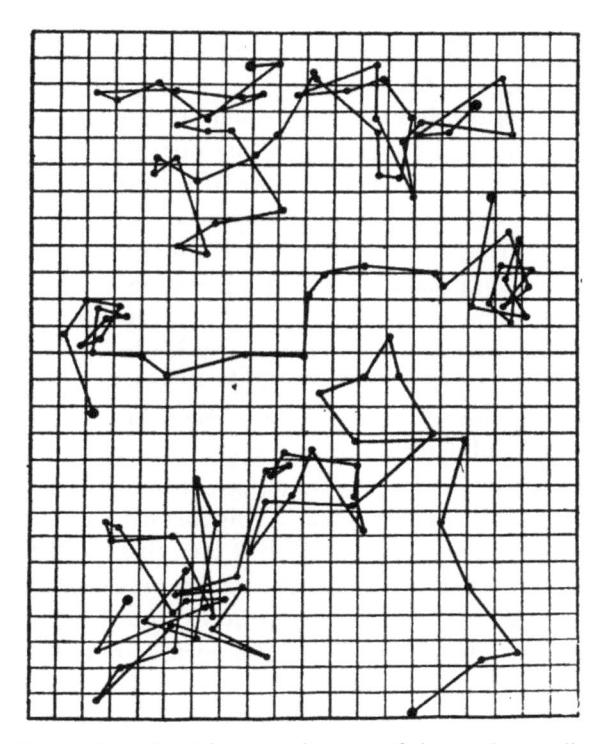

FIG. 9. Jean Perrin's iconic diagram of the random walk represents the average displacement of three individual particles: each point was marked every thirty seconds for twenty-five minutes after the last, then connected, and each connection measured. The value of the average of the square of each segment confirmed Einstein's hypothesis that the movement of molecules corresponds to the temperature and viscosity of the system (Perrin, "Mouvement brownien et réalité moléculaire," 81).

microscopy, was just another technique to improve image contrast: polarized light microscopy relies on the same principles, even the same materials that make polarizing sunglasses boost contrast and reduce road glare when driving. After inserting one polarizer accessory between the condenser and the specimen stage and a second polarizer accessory between the objective lens and the eyepiece, the appearance of the specimen changes, rendering even a humdrum hair or starch grain into a dazzling array of colors.[38] Usually these colors are useful for simple identification work and taxonomy: different minerals and shells display distinct colors, and when polarization microscopy was occasionally used by biologists in the nineteenth century, it was done in a straightforward manner. Mineralogists took to polarization microscopy more quickly in the nineteenth century, using it to show how materials bend light in different ways (see fig. 10).

But starting in 1916, Hermann Ambronn (1856–1927)—a former botanist who had been hired by the Carl Zeiss optical firm as a manager in its microscopy division—developed a novel if somewhat clumsy theory of polarization microscopy and the colors it produces. Whereas many of the microscope accessories developed in the late nineteenth century sought to strengthen contrast and make optical interpretation easier, polarized light microscopy sought to increase the *range* of contrast possible—and thus the range of possible distinctions and details—by rendering color and changing color patterns into evidence for invisible structures. Ambronn's theory showed that the pattern of colors visible in the polarized light microscope could be quantified and that these measurements could be used to infer even a soft biological specimen's molecule-scale structure. Technically, the formal mathematics of Ambronn's theory could only furnish the microscopist with two basic shapes, "rodlets" or "platelets." But even this crude theory usefully suggested that there were scales of organization between the visible realm and the molecular one—the latter being worked out by theoretical physicists and X-ray crystallographers—and the polarization microscope would provide the instrumental and conceptual bridge between them.[39]

Biologists committed to Ambronn's methodology could piece together the molecular details by mapping out how different parts of a specimen bend light and produce a characteristic pattern of brilliant colors under the

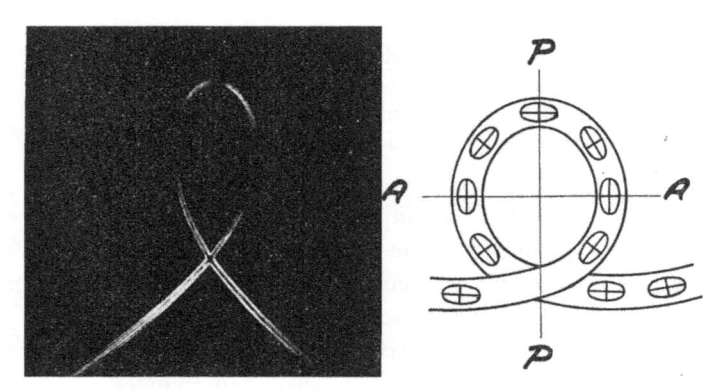

FIG. 10. A human hair bent under the polarized light microscope. The colors and patterns of light indicate the directionality or *anisotropy* of the specimen, mapped on the right with a series of ovals or "optical indicatrix" (W. J. Schmidt, *Die Bausteine des Tierkörpers in polarisiertem Lichte* [Bonn: Friedrich Cohen, 1924], 57).

polarization microscope. While Ambronn never actually applied his own theory, his student Albert Frey-Wyssling (1900–1988) was arguably one of the first "molecular biologists" before the term ever came into wide use. Painstakingly observing plant cells from different angles under polarized light, by the late 1920s, Frey-Wyssling showed that the plant cell wall was not a simple aggregate of cellulose but that each type of plant cell had many layers of different kinds of cellulose fibers running in different directions in a three-dimensional space (fig. 11).[40]

These early molecular diagrams of biological structures were the result of many microscopic observations made by dynamically rotating a specimen and examining it from different angles under polarized light. They were rarely accompanied by microphotographs, since only one would not actually present enough visual evidence of submicroscopic structure: instead, these diagrams or maps synthesized and represented many days', weeks', and months' worth of observations. Of all of the techniques used for detecting minute differences at the limits of perception, polarized light microscopy was possibly the most difficult to master: before the age of electron microscopy, only a few achieved sufficient skill with the technique to be able to reach down to molecular dimensions and create these diagrams. The polarization accessories that defined the technique drastically limited the magnification, brightness, and *direct* resolving power of the microscope. Before the advent of usable electron microscopy in the 1950s, the submicroscopic, molecular details were mapped out by careful inference, aided by tools to manipulate light rather tools that permitted direct observation.

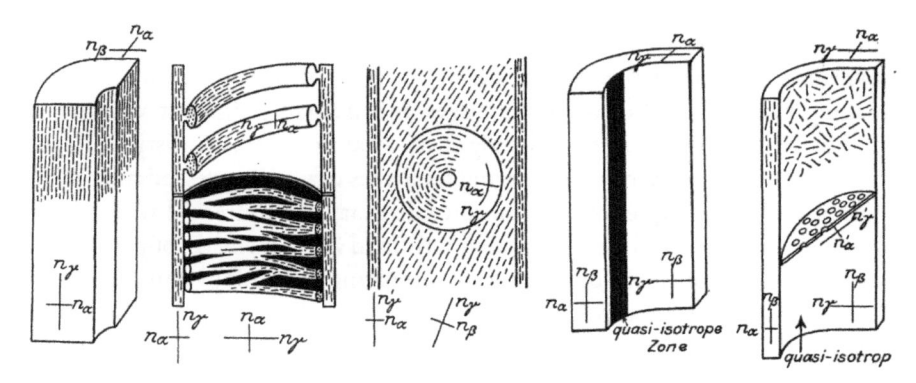

FIG. 11. Five different molecular morphologies of five different plant cell types, each hash representing a rodlet of cellulose. The labels *n* along with the marked axes indicate the way each cell wall type bends light under the polarized light microscope (Frey, "Der submikroskopische Feinbau der Zellmembranen," 765).

Conclusion

I WANT to return to my suggestion that there are really two opposite yet complementary ways of thinking about the "minute." On the one hand is a magnification or enlargement of a single object, like we see in the iconic photograph of James Watson and Francis Crick standing near their two-and-a-half-meter-tall model of DNA.[41] On the other hand is the resolving of detail showing the relationships between the whole and its parts, as we see with Frey-Wyssling's diagrams of the molecular structure of the plant cell wall. In this essay, I have sought to emphasize resolution of detail, in part because it is an activity that does not come intuitively or easily. In fact, as I hope I have shown, resolving detail in the history of microscopy required finding ways of using and improving the microscope that ran counter to our first impulse to look at one small thing more closely as quickly as possible: that exhilaration of seeing tiny things is what can make the microscope so wonderful and exciting to use for a novice. As many of these examples show, however, resolving detail requires looking at the world repeatedly, developing ways of making contrasts and distinctions between perceptions—and ideas—that might otherwise go unnoticed.

In the twenty-first century, how these contrasts, distinctions, and details are noticed is rapidly changing. To push the metaphor of something being "richly" detailed, we are now in an era where the amount of knowable detail in the world has become *too* rich for a mere human to metabolize. As I mentioned in the prologue, for most of the history of microscopy, distinguishing differences required straining one's flesh-and-blood eyes and developing ingenious ways to make the microscope easier to use. Microscopy in the digital age is considerably different, since every resolvable detail in a given image is quantified and calculable. There is now a school of thought in microscopic imaging that, in the future, human eyes and human judgment will not be the initial or intermediate-level judges of what counts as a distinct detail: with digital imaging sensors capturing terabytes of uncompressed microphotographs in rapid sequence, it is conceivable that a computer AI will identify meaningful details and that only such a trained AI can sift through the millions of discrete observations that digital imaging technology can capture.[42]

Even if such automation might be aesthetically distasteful, we should appreciate that detail represents an accumulation of knowledge, and we can be impressed by this accumulation for its own sake. A detailed picture of the world integrates and synthesizes what could otherwise be a mere heap of knowledge and has the potential to replace a crude icon or figure with a more systematic understanding of how the world works. Since March 2020, for example, icons of the SARS-CoV-2 coronavirus are ubiquitous and unavoidable. We should perhaps take some comfort in the fact that the biology and the structure of the virus are quite well known and knowable because the natural sciences have found a way to deal with key aspects of the biological microworld with incredible detail and organization. The rest of us will have to catch up in order to gain a clearer, integrated, and detailed sense of the personal and social suffering this disease has wrought.

Notes

1. A more thorough introduction to demosaicing with visual aids can be found at https://en.wikipedia.org/wiki/Bayer_filter.
2. This is a rough description of twenty-four-bit color, which is itself a conventionalized description of what can be a far more complicated mathematical description of digital color.
3. An excellent exploration is Janet Vertesi, *Seeing Like a Rover: How Robots, Teams, and Images Craft Knowledge of Mars* (Chicago: University of Chicago Press, 2015). For a description of predigital observational practice in astronomy, see Omar W. Nasim, *Observing by Hand: Sketching the Nebulae in the Nineteenth Century* (Chicago: University of Chicago Press, 2013).
4. August Köhler, "Hermann Ambronn und die wissenschaftliche Mikroskopie," *Naturwissenschaften* 14, no. 33 (August 13, 1926): 765–67, on 766.
5. In the sciences, the term *microworld* was originally a casual term used by physicists in the early twentieth century who wanted to demarcate the unusual quantum physics of the atomic and subatomic realms—the microworld—from the familiar physics of the *macroworld*. The term was quickly picked up by biologists and microscopists, possibly as a contraction of the older phrase "microscopic world." The *OED* points to an early use of the term "microworld" in English in Ralph W. G. Wyckoff, "Some Recent Developments in the Field of Electron Microscopy," *Science* 104, no. 2689 (July 12, 1946): 21–26, https://doi.org/10.1126/science.104.2689.21.
6. Many examples are provided in Angela N. H. Creager, *The Life of a Virus: Tobacco Mosaic Virus as an Experimental Model, 1930–1965* (Chicago: University of Chicago Press, 2002).
7. Rosalind E. Franklin and R. G. Gosling, "Molecular Configuration in Sodium Thymonucleate," *Nature* 171, no. 4356 (April 1953): 740–41, https://doi.org/10.1038/171740a0; Natasha Myers, *Rendering Life Molecular* (Durham: Duke University Press, 2015).
8. Frederick Gregory, *Scientific Materialism in Nineteenth Century Germany*, Studies in the History of Modern Science, vol. 1 (Dordrecht: D. Reidel, 1977); Dorothy Nelkin and M. Susan Lindee, *The DNA Mystique: The Gene as a Cultural Icon*, 2nd ed. (Ann Arbor: University of Michigan Press, 2004).

9. Ilana Löwy, "Microscope Slides in the Life Sciences: Material, Epistemic and Symbolic Objects: Introduction," *History and Philosophy of the Life Sciences* 35, no. 3 (2013): 309–18; Hans-Jörg Rheinberger, "Präparate—'Bilder' ihrer selbst: Eine bildtheoretische Glosse," *Bildwelten des Wissens* 1, no. 2 (2003): 9–19.
10. Catherine Wilson, *The Invisible World: Early Modern Philosophy and the Invention of the Microscope*, Studies in Intellectual History and the History of Philosophy (Princeton: Princeton University Press, 1995), chap. 7.
11. G. L'E. Turner has shown that microscope optics remained largely the same from the mid-seventeenth up to the early to mid-nineteenth centuries. G. L'E. Turner, "The Microscope as a Technical Frontier in Science," in *Historical Aspects of Microscopy: Papers Read at a One-Day Conference Held by the Royal Microscope Society at Oxford, 18 March, 1966*, ed. Savile Bradbury and G. L'E. Turner (Cambridge: W. Heffer & Sons, 1967).
12. Savile Bradbury, *The Evolution of the Microscope* (Oxford: Pergamon, 1967), chap. 1.
13. Wilson, *Invisible World*.
14. Marc Ratcliff, *The Quest for the Invisible: Microscopy in the Enlightenment* (Farnham: Ashgate, 2009), 6–7, 78–79.
15. Ibid., 151–67.
16. Ibid., 127.
17. Ibid., 82–90.
18. Robert Hooke, *Micrographia; Or, Some Physiological Descriptions of Minute Bodies Made by Magnifying Glasses With Observations and Inquiries Thereupon* (London: Jo. Martyn and Ja. Allestry, 1665), preface, Linda Hall Library Digital Collection, https://catalog.lindahall.org/permalink/01LINDAHALL_INST/1nrd31s/alma991863803405961.
19. Stuart M. Feffer, "Ernst Abbe, Carl Zeiss, and the Transformation of Microscopical Optics," in *Scientific Credibility and Technical Standards in 19th and Early 20th Century Germany and Britain*, ed. Jed Z. Buchwald (Dordrecht: Kluwer, 1996), 32–33; see also Bradbury, *Evolution of the Microscope*.
20. Joseph Jackson Lister, "On Some Properties in Achromatic Object-Glasses Applicable to the Improvement of the Microscope," *Philosophical Transactions of the Royal Society of London* 120 (1830): 187–200, https://doi.org/10.1098/rstl.1830.0015. On Lister, see Bradbury, *Evolution of the Microscope*, chap. 5; and Jutta Schickore, *The Microscope and the Eye: A History of Reflections, 1740–1870* (Chicago: University of Chicago Press, 2007), chap. 5.
21. This is a simplification of the convoluted debate over the relationship between angular aperture and immersion optics in the mid-nineteenth century; see Stuart M. Feffer, "Microscopes to Munitions: Ernst Abbe, Carl Zeiss, and the Transformation of Technical Optics, 1850–1914" (PhD diss., University of California, Berkeley, 1994), chap. 3.
22. Ibid., 120.

23. Ernst Abbe, "Beiträge zur Theorie des Mikroskops und der mikroskopischen Wahrnehmung," *Archiv für mikroskopische Anatomie* 9 (1873): 413–68; Hermann Helmholtz, "Über die Grenzen der Leistungsfähigkeit der Mikroskope," *Monatsberichte der königlichen preussische Akademie der Wissenschaften zu Berlin*, October 1873, 625–26.
24. Ian Hacking, *Representing and Intervening: Introductory Topics in the Philosophy of Natural Science* (Cambridge: Cambridge University Press, 1983), 187.
25. Christiane Frey, "The Art of Observing the Small: On the Borders of the 'Subvisibilia' (from Hooke to Brockes)," *Monatshefte* 105, no. 3 (Fall 2013): 376–88, https://doi.org/10.1353/mon.2013.0078; Christoph Meinel, "Adams neue Augen: Mikroskopische Imaginationen," in *Bilder sehen: Perspektiven der Bildwissenschaft*, ed. Mark Greenlee et al., Regensburger Studien zur Kunstgeschichte, Bd. 10 (Regensburg: Schnell & Steiner, 2013), 139–54.
26. Abbe, "Beiträge zur Theorie," 468 (emphasis in original).
27. Abbe and Helmholtz's theory was quickly accepted, but as late as the 1920s, there were microscope makers, such as the American engineer Royal Rife (1888–1971), who claimed to have made microscopes with magnifications of 5,000× or more. A lack of understanding of this admittedly difficult optical principle has led to some embarrassing cases in the history of pseudoscience; cf. James E. Strick, *Wilhelm Reich, Biologist* (Cambridge: Harvard University Press, 2015), chap. 3.
28. Ernst Abbe, "Ueber einen neuen Beleuchtungsapparat am Mikroskop," *Archiv für mikroskopische Anatomie* 9 (1873): 469–80.
29. David Cahan, "The Zeiss Werke and the Ultramicroscope: The Creation of a Scientific Instrument in Context," in *Scientific Credibility and Technical Standards in 19th and Early 20th Century Germany and Britain*, ed. Jed Z. Buchwald, Archimedes 1 (Dordrecht: Kluwer, 1996), 67–115, https://doi.org/10.1007/978-94-009-1784-2_3; Feffer, "Microscopes to Munitions," chap. 4.
30. August Köhler, "Mikrophotographische Untersuchungen mit ultraviolettem Licht," *Zeitschrift für wissenschaftliche Mikroskopie und mikroskopische Technik* 21 (1904): 129–65, 273–304.
31. Siegfried Czapski, "Die voraussichtlichen Grenzen der Leistungsfähigkeit des Mikroskops," *Biologisches Centrallblatt* 11, no. 20 (November 1, 1891): 609–19.
32. Feffer, "Microscopes to Munitions," 247–48; Ernst Abbe, "Ueber neue Mikroskope," *Sitzungsberichte der jenaischen Gesellschaft für Medicin und Naturwissenschaft* 20, no. 2 (1887): 107–28.
33. Dieter Gerlach, *Geschichte der Mikroskopie* (Frankfurt am Main: Harri Deutsch, 2009), 664–65.
34. Shinya Inoué, "Video Image Processing Greatly Enhances Contrast, Quality, and Speed in Polarization-Based Microscopy," *Journal of Cell Biology* 89, no. 2 (May 1, 1981): 346–56; Robert Day Allen, Nina Strömgren Allen, and Jeffrey L. Travis, "Video-Enhanced Contrast, Differential Interference Contrast (AVEC-DIC) Microscopy: A New Method Capable of Analyzing Microtubule-Related

Motility in the Reticulopodial Network of Allogromia Laticollaris," *Cell Motility* 1, no. 3 (1981): 291–302; see Rudolf Oldenbourg, "Observing the Living Cell: Shinya Inoué and the Reemergence of Light Microscopy," in *Visions of Cell Biology: Reflections Inspired by Cowdry's General Cytology*, ed. Karl S. Matlin, Jane Maienschein, and Manfred Laubichler (Chicago: University of Chicago Press, 2018), 280–300.

35. Jean Perrin, "Mouvement brownien et réalité moléculaire," *Annales de chimie et de physique* 8, no. 18 (September 1909): 5–119. For an in-depth analysis, see Mary Jo Nye, *Molecular Reality: A Perspective on the Scientific Work of Jean Perrin* (London: Macdonald, 1972); Charlotte Bigg, "Evident Atoms: Visuality in Jean Perrin's Brownian Motion Research," *Studies in History and Philosophy of Science Part A* 39, no. 3 (September 2008): 312–22, https://doi.org/10.1016/j.shpsa.2008.06.003. On the slit ultramicroscope, see Cahan, "Zeiss Werke and the Ultramicroscope."

36. Elizabeth M. Slayter, *Optical Methods in Biology*, 1st ed. (New York: Wiley-Interscience, 1970), 263; quoted in Hacking, *Representing and Intervening*, 190.

37. Charlotte Bigg, "A Visual History of Jean Perrin's Brownian Motion Curves," in *Histories of Scientific Observation*, ed. Lorraine Daston and Elizabeth Lunbeck (Chicago: University of Chicago Press, 2011), 156–79.

38. Some examples are available at "Polarizing Microscope Image Gallery," Leica Microsystems, accessed March 19, 2020, https://www.leica-microsystems.com/science-lab/galleries/polarizing-microscope-image-gallery/.

39. Hermann Ambronn, "Ueber das Zusammenwirken von Stäbchendoppelbrechung und Eigendoppelbrechung, I," *Kolloid-Zeitschrift* 18, no. 3 (April 1916): 90–97, https://doi.org/10.1007/BF01432287; Hermann Ambronn and Albert Frey, *Das Polarisationsmikroskop: Seine Anwendung in der Kolloidforschung und in der Färberei*, Kolliodforschung in Einzeldarstellungen, Bd. 5 (Leipzig: Akademische Verlagsgesellschaft, 1926).

40. Albert Frey, "Der submikroskopische Feinbau der Zellmembranen," *Naturwissenschaften* 15, no. 37 (September 16, 1927): 760–65.

41. Soraya de Chadarevian, *Designs for Life: Molecular Biology After World War II* (Cambridge: Cambridge University Press, 2002), chap. 5.

42. Yicong Wu et al., "Simultaneous Multiview Capture and Fusion Improves Spatial Resolution in Wide-Field and Light-Sheet Microscopy," *Optica* 3, no. 8 (August 20, 2016): 897, https://doi.org/10.1364/OPTICA.3.000897.

Experimental Environments
Micrologies of Knowledge

HANS-JÖRG RHEINBERGER

❖❖❖

It is characteristic for scientific experiments that minutiae play a major role for their outcome. A history of such "major" minutiae remains to be written. After all, finding the right twist is a matter of intimate acquaintance with the material at hand and of serendipity. Nobody has expressed this combination in a more vivid fashion than Claude Bernard, the great French physiologist of the nineteenth century. In his *Introduction to the Study of Medicine*, he once put it as follows: "In scientific investigation, minutiae of method are of the highest importance. The happy choice of an animal, an instrument constructed in some special way, one reagent used instead of another, may often suffice to solve the most abstract and lofty question." And, he continued, "one must have been brought up in laboratories and lived in them, to appreciate the full importance of all the details of procedure in investigation, which are so often neglected or despised."[1]

Experimenting in Vitro

In what follows, I will restrict myself to exposing and discussing a form of experimentation in the life sciences that is literally *microperformative* in its very constitution. It is what we call in vitro experimentation.[2] It operates in the dimension of microliters, and it uses micro-, nano-, or even picoamounts of substances. Alternatively, it manipulates single cells of micrometer dimensions in liquid or gel-like nutrients. In vitro means in the glass—that is, in laboratory test tubes or dishes. Test-tube experimentation became a characteristic feature of the life sciences only in the late nineteenth century. From that time onward, it has played a decisive role in biological and biomedical laboratory work around the world.

An early form of in vitro experimentation was based on so-called *homogenates* of cells. Here, cells are simply squeezed out or broken up with appropriate instruments, and the molecules contained in the cell sap are subjected to experimentation. In parallel, a different form of in vitro experiment—namely, efforts to develop *single-cell cultures* of higher organisms in the test tube, including human cells—gained prominence. The background for this form of dealing with cells of higher organisms was, of course, microbiology: work with pure cultures of bacteria had already been established in the last third of the nineteenth century.

From the 1930s on, with the availability of ultracentrifuges, *fractionation* of cellular contents via high-speed centrifugation became predominant. With it, the tiny particles of the cell sap, some of them soon duly addressed as microsomes, could become separated from one another into different layers according to their molecular weight. In the wake of World War II, *radioactive tracing* added another microdimension to that kind of test-tube assessment of metabolic processes. Radioactive tracers enhance the sensitivity of physicochemical measurements from the micro- down to the nano- and even picorange. In the in vitro line from homogenates to fraction to metabolic tracing, it was not simply the manipulation of organic substances *extra corpore*—outside the body—that defined the new test-tube cultures in their various forms.

Following physiologist and historian of science Herbert Friedmann, what was really new about this experimental regime was the claim and in-principle demonstration that tissue extracts could provide and represent not just what have been called natural products—that is, organic *compounds* synthesized in cells and tissues—but, first and foremost, natural *processes*. Friedmann claims, "From now on, extract repeats or mirrors process."[3] What is thus at stake in the transition from experimenting with living systems to (partial) test-tube systems is not just the transition from biology to chemistry—or, to put it differently, from physiological processes to organic substances. What is at stake is rather something akin to a *reduplication* of life: a life, so to speak, under artificial conditions. What is called for is a sort of minimal life: from the myriad of reactions simultaneously going on in the cell, ideally one is retained and followed in its steps. In order to do so, the milieu of the cell or the "milieu interne" of the organism, to use Claude Bernard's fitting expression,[4] has to be replaced by a milieu composed in the test tube.

In this process of repeating or of mirroring life, however, what is implied is a permanent threat of aberration that confronts its practitioners with the equally permanent question of how far we are still able to see "nature" in this mirror glass and how one can finally decide whether one still does so or has created a context in which things happen differently. And it is exactly here that the "major minutiae" mentioned in the introductory paragraph come to play out their full importance.

Micropreparations 1: Phage

WHAT I would like to do now is to concentrate in an exemplary fashion on a form of micropreparation that played a crucial role in the transition from classical to molecular genetics around the middle of the twentieth century.[5] It will allow me to make a number of points that are connected to working effectively with minute biological entities that cannot be seen, not even under the light microscope. The picture that we see in figure 1[6] appears to be rather unspectacular. It displays an in vitro culture of bacteria—*Escherichia coli* bacteria—in a petri dish that has been infected by bacteriophages.

Let me briefly lay out the biological principle behind this experiment: bacteriophages, or phages, in short—that is, virus particles specialized to infest bacteria—are essentially tiny molecular genetic packages containing a nucleic acid molecule wrapped in a protein shell that, in and out of themselves, are completely inert. In order to multiply, their nucleic acid part has to enter the bacteria. Once inside, it can induce the genetic apparatus of the bacterial cell to occupy itself with the multiplication of the molecular parasite instead of growing and multiplying itself. As a consequence, the bacterium gets filled with virus particles until it bursts. The phages are released and can then enter the neighboring bacterial cells, and the cycle is repeated over and over again.

Now this behavior can be exploited in order to make these minute molecular packages visible. First, a petri dish is prepared on which an evenly distributed lawn of bacteria on a nutrient agar plate has been grown. Because of their regular distribution, the bacteria remain invisible. Next, appropriately diluted virus particles are spotted on the plate. The dilution has to be high enough so that single viruses, well separated from each other, can fall onto the bacterial lawn. When the dishes are put in an incubator, the phages begin to do their work. They adhere to the cell wall of the neighboring bacterium, inject their nucleic acid into it, start to multiply, and finally burst the bacterium. The released viruses repeat the cycle and thus slowly form holes in the smooth bacterial lawn.

Experts call these holes "plaques." They grow around the very spot where the multiplication of a single virus started. The holes, when big enough, can be seen by the naked eye. Depending on the type of phage and its genetic variants, these holes can assume widely different shades, granulation,

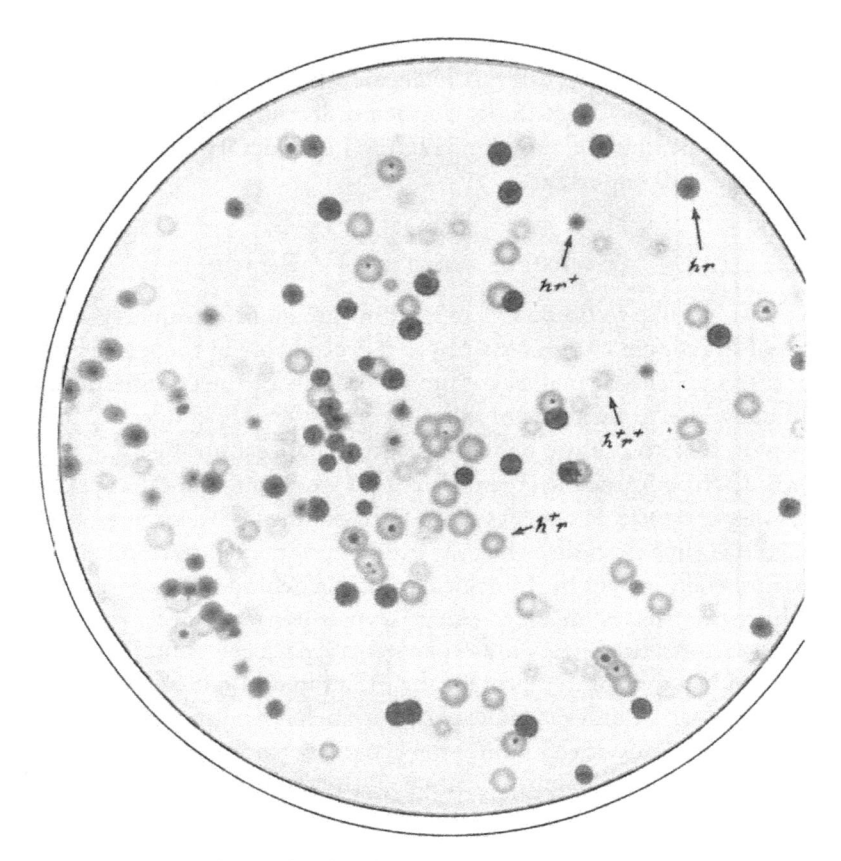

FIG. 1. In vitro culture of *Escherichia coli* bacteria in a petri dish that has been infected by bacteriophages (Stent, *Molecular Biology of Bacterial Viruses*, 177).

fringes and even colors. Accordingly, the plaques vary in their morphology, and these differences in turn can be correlated with particular types of phages. Identifying them, however, requires experience. One must be able to recognize minute differences, and one can acquire such experience only by repeating these experiments many times and getting acquainted with the material over a long period of time.

In the present case, *Escherichia coli* bacteria were infected by bacteriophages of the variant T2 and one of its genetic mutants. This is the phage type on which Max Delbrück and Salvador Luria, the US-based researchers who established the technique, specialized in the 1940s and 1950s.[7] They and their coworkers deliberately focused on this type of phage only, thereby acquiring an intimate knowledge of how to handle it and interpret the visible traces that it left. This deliberate restriction allowed them to develop what can be called a peculiar kind of microperformativity. It consists in allowing one to intuitively feel one's way through the traces that the experiment leaves behind and to become able to distinguish between meaningful tiny differences that can tell something about the experiment and its stakes from mere noise. And it helped establish an ambience in which, as Luria put it in his autobiography, "a broken test tube" can make all the difference.[8]

The image reproduced here is taken from an early 1960s textbook on the molecular biology of bacterial viruses by molecular genetician Gunther Stent. Stent had studied with Delbrück at the California Institute of Technology in Pasadena after World War II and had become immersed in phage work before he moved on to Berkeley. Two different T2 phages—type hr and h⁺r⁺—were used in the experiment carrying either two wild-type genes h⁺ and r⁺ or the two mutated genes h and r, respectively, "h" standing for "host range" and "r" for "rapid lysis." Interestingly, if we now have a look at the dish, four different types of plaques can be identified on the plate instead of two standing for the input phages. Thus, the preparation indicates that in the course of multiplication, to the surprise of the researchers who did the experiment, a reciprocal exchange of genes between the two types of phages could be assumed to have taken place. This was indicated by the fact that in addition to the two types hr and h⁺r⁺ that had been used in the experiment, two mixed types—h⁺r and hr⁺—appeared and could be identified because they caused slightly different plaques. It was one of these events where what at first looked like a negligible impurity of the experiment turned into a major signal that would have massive consequences for the manipulation of phage nucleic acid in emerging molecular genetics in the years to come.

Let us now have a look at this experiment from a micrological perspective that allows us to pay attention to the minutiae surrounding it. In addition to the particular kind of microperformativity already mentioned, four further aspects can be differentiated: The first concerns the kind of action involved. The second is connected to visualization. The third hangs together with the structure of experimentation. Finally, the fourth is to be sought in the specific dynamics of the molecular genetic research process.

First is *action*. What we have here is a very particular kind of performance of the agents involved, the bacteria and the phages. It is, we could say, a game of life and death. What we can see, the plaques, are the consequence of a devastation. The result of the myriads of procreated viruses is billions of dead bacteria in each of these spots. The microagents of the game are the bacteria and the phages themselves. They simultaneously multiply and destroy one another. We do not see the tiny individual acts; what we see are their massive consequences.

Second is *visualization*. The example also has to do with a very special form of visualization. Micropreparations of this kind lend macroscopic visibility to the presence of minuscule molecular entities that cannot be seen even under the best traditional microscope. But the form in which these entities (the viruses) become visible has nothing to do with their individual shape in space. Their visibility is the result of their action. They leave a heap of devastated bacteria of a defined and recurrent form. The image is not iconic but thoroughly indexical, to use Charles Sanders Peirce's semiotic vocabulary.[9] And there is more to the visualization at stake here. It is also, and essentially so, a visualization of *variants* of molecules. Not only can virus particles take on a visible form as a result of this game of life and death; tiny point mutations of DNA molecules can as well.

Third, we can also look at the dish from the point of view of *strategies of experimentation*. What happens here is characteristic for many forms of the technical enhancement on which scientific experimentation rests. In its simplest form, it can be put as follows: what is too small to be directly observed and manipulated must be enlarged. What we have is a principle of "dilatation," of blowing up.[10] Microscopic enlargement is probably the most obvious form of doing so. But experimenters have invented ingenious forms of enlargement or enhancement that are by no means restricted to the regime of the iconic. One of them we have here before our eyes, and more of that sort will be seen below. The reverse, of course, holds an important place in experimentation as well and can frequently be observed: what is too big to be handled must be downsized according to a principle of "compression"—in other words, what is too voluminous must be reduced. This principle can as well take on various and variegated shapes and is by no means restricted to the realm of the iconic. For instance, it can take the form of "what is too complex to be analyzed must be simplified" or "what is too thick to be penetrated must be thinned." And what holds for the dimension of space also holds for the dimension of time: what is too quick to be observed in detail must be slowed down, and what is too slow to be observed must be accelerated. These axes span a complete coordinate system of experimental action.

The fourth point concerns the *fine structure* of *molecular genetic research*. The virus preparation that we see here embodies and carries with it a rather complex package of knowledge on the genetic constitution and behavior of bacterial viruses. It is an arrangement that represents genetic knowledge in a compact, tangible, and manipulable form. It is characteristic of experimental arrangements that we have to look at their parts as an epistemic ensemble. They make sense in terms of the epistemic objects they serve to investigate. They are carriers of knowledge so far accumulated, but at the same time, they are also devices for the identification of new knowledge: in the present case, mutants of a virus arising from an alleged recombination of different virus DNAs. By the identification of events of this kind—unexpected but consequential—the experimental process of genetic knowledge acquisition is driven forward. The characteristic morphologies of the plaques are thus the result of not only an interaction between virus and bacterium but also, as it turns out, the interaction of virus strains among each other. What is made use of in this particular experiment are only different types of *viruses*. But the representational arsenal of this kind of preparation—we could call it a molecular analytics—can be further expanded: *bacteria* with different susceptibility toward the virus and even bacterial mixtures can also be taken advantage of. In this way, and in a permanent iteration of the process, new differences can be created that lead to new characterizations that in turn determine the ensuing course of the experiments. The rendering procedures accompanying them have, in principle, nothing in common with the traditional pictorial idea of representation. And yet molecular processes that otherwise would defy all imagery are made plainly visible in this way. The structural characteristics of visible contours and areas of spots come to stand for certain molecular structures and processes—genetic recombination events in the example.

In sum, we see here two forms of the minute interact. On the one hand, it takes the shape of an epistemic principle: that of iterativity in the sense of tiny but impactful turns in a given experimental regime. This is, as it were, a typical feature of productive experimental systems. On the other hand, the minute is inherent in the vision-enhancing technologies that are being used in order to make the infinitely small accessible.

Micropreparations 2: DNA Sequencing

Let us move on to a second example. In this way, we can get an impression of the variety of forms that molecular microperformance brings with it. It is the combined use of the polymerase chain reaction (PCR) and sequencing for the amplification and sequence analysis of deoxyribonucleic acid. Here, we have the case that molecular actions have become packaged into powerful technical sets with a wide range of uses in research experiments as well as in biotechnological production processes.

The polymerase chain reaction was worked out in the early 1980s.[11] It makes use of the capacities of an enzyme, the thermostable DNA polymerase

(Taq polymerase) of a bacterium called *Thermus aquaticus*. The bacterium is found in hot springs and can live and multiply in temperatures up to 70° Celsius (158° Fahrenheit). Kary Mullis, then a researcher at Cetus, one of the early biotech start-ups, realized that he could use the enzyme to amplify even the tiniest amounts of DNA. In principle, a single DNA molecule would suffice to start with. All that is needed is a sample of DNA; the Taq polymerase; a short, sequence-specific deoxyribonucleic acid primer molecule that tells the polymerase where to start the multiplication process; and a little thermocycler for the cyclic heating and cooling of the samples. Appropriately handled and fine-tuned, the process yields highly pure quantities of DNA that are suitable for sequencing afterward. We have here a nice example of the aforementioned strategy of dilatation, again in a particular form. Here, a sample with a concentration too low to be experimentally investigated is enriched so that it can be used in further experiments or else for sequencing purposes.

DNA sequencing was developed in the early 1970s already.[12] It depends on the availability of purified DNA in suitable amounts. When it was elaborated, what was most conveniently available in purified form were sufficient amounts of isolated virus or phage DNA. A few years later, DNA amplification became possible by DNA plasmid cloning in bacteria. Frederick Sanger and his group at Cambridge demonstrated the principle of the sequencing procedure with DNA from phage phi X 174.[13] The example exposes yet another form of visualization of the indexical kind already encountered in the first figure. We can address them as molecular preparations.

Let us look at figure 2.[14] What we see looks like a somewhat distorted barcode. The four lines correspond to the four bases to be found in DNA: adenine (A), thymine (T), cytosine (C), and guanine (G). Here again, a DNA polymerase plays a key role. On a single strand of DNA to be sequenced, it synthesizes a complementary strand that reflects its order of nucleotides. Statistically, however, a fraction of the reaction mixture is terminated after every addition of a new building block. This effect is generated by the appropriate addition of modified bases that can be incorporated but do not allow further elongation. The result is a mixture of DNA strands that differ in length from each other by one building block. For each of the four bases, a different line is generated. The resulting fragments can be separated from each other in each of the four lines as they are driven through a porous sequence gel by an electrical current. The bars mark the base that terminates the fragment. To make the bars visible, the modified bases added to the mixture are radioactively labeled. Once the gel is ready, it can be looked at under UV light that is absorbed by the DNA fragments. Alternatively, a photo plate

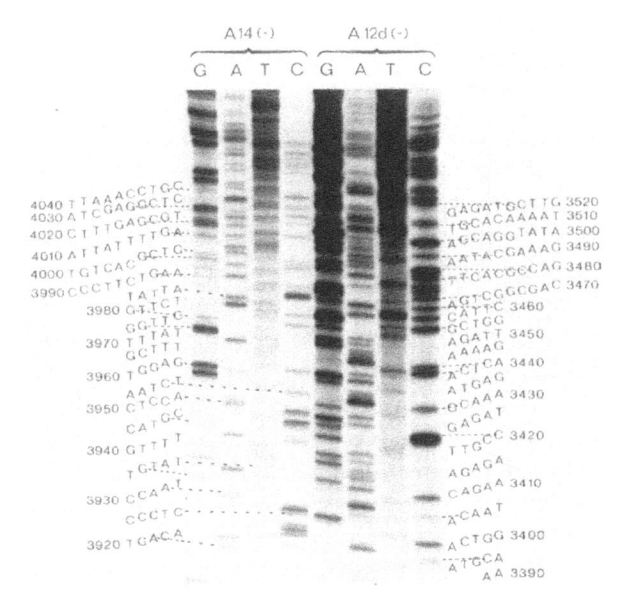

FIG. 2. The four bases to be found in DNA: adenine (A), thymine (T), cytosine (C), and guanine (G) (Sanger, Nicklen, and Coulson, "DNA Sequencing," fig. 1, 5464).

can be packed on it that comes to reflect, when developed, the bars as black stripes. Ideally, the sequence can be directly "read" from the autoradiogram. In contrast to the polymerase chain reaction that simply multiplies, here, we have to do with *analysis* through molecular synthesis.

Again, the representation is definitely indexical, not iconic. It represents but does not depict. This is decisive here. A mere magnification of the minute, the visualization of the *shape* of the DNA molecule, would not reveal anything about the *order* of its molecular building blocks and, with that, the information it contains for the construction of a protein. The macroscopic visualization of the microscopic goes along with a qualitative transformation. What we *see* in the end is a molecular order. And again, like in the first example, it goes hand in hand with a mass accumulation.

Experimental Environments

FINALLY, WE must have a look at the experimental environments—that is, the milieus in which all the aforementioned things find their place. This implies a move from what could be called the infraexperimental intricacies of the laboratory, with its production of traces, to experimentation in its integrity and the historical trajectories it creates. I would like to do this through the example of protein biosynthesis research that I have followed closely in my earlier book *Toward a History of Epistemic Things*.[15] It is, at the same time, an example of an in vitro system as described at the beginning of this essay. The minutiae of experimentation that concatenate in such a trajectory, with its options and decisions, its openings and its dead ends, can of course not be followed here in their details. A summary must do.

Around 1945, immediately after World War II, the experimental endeavor of unraveling the mechanism of protein biosynthesis started with a problem rooted in cancer research in an oncological laboratory at the Massachusetts General Hospital in Boston, in a group led by the oncologist Paul Zamecnik. The initial question was whether cancer cells would differ in their protein synthesis behavior from healthy cells. That question was, however, never answered unambiguously. Instead, the eventual addition of a little chemical—dinitrophenol—to their rat liver slices in the test tube gave a differential signal: the separation of respiration and amino acid incorporation, leading the dynamics of experimentation in a different direction. Under the hands of the researchers, the system completely changed to in vitro and, with that, into a device in which it became possible to analyze the energetic activation of amino acids, the building blocks of proteins. At the same time, this led to the abandonment of the theoretical framework that until then had dominated speculations about the mechanism of protein synthesis, derived from classical enzymology: protein synthesis as a reversal of proteolysis.

Around the middle of the 1950s and to the surprise of all those involved, the in vitro system, now separated into neat fractions by high-speed centrifugation and distinguished from each other by their activity through the use of radioactive labels, led to the identification of a hybrid molecule that so far had had no place in the universe of biochemistry. It was a combination of a small ribonucleic acid molecule with an activated amino acid. It was the completely unexpected result of an experimental cross-control, a tiny move in a protocol that otherwise left all things equal. It catapulted the system into the center of emerging molecular genetics because it opened the perspective to shed some light on the relation that was then beginning to be assumed to exist between nucleic acids as the carriers of genetic information and proteins as the main carriers of biological function.

Until this point, the experimental system was based on biological materials derived from rat liver, a legacy of its emergence in an oncological laboratory. The attempt to build a homologue based on the cell sap of a bacterium, that of *Escherichia coli*—the preferential model organism of molecular genetics at that time—led to yet another surprise, triggered again by a tiny experimental move: a preliminary incubation step prior to the addition of the test materials. In the hands of Marshall Nirenberg and Heinrich Matthaei, the system led, at the beginning of the 1960s, to the identification of the first genetic code word—that is, a triplet sequence of ribonucleic acid that codes for a particular amino acid. With that, the golden age of early molecular biology reached its height.

In between, there lies the trajectory of one and the same experimental system and its material continuity. But along with all the experimental fallout that this continuity and its incremental moves created, the theoretical assumptions changed radically. If, in 1945, the assumptions of classical enzymology prevailed with its dogma of the reversibility of every enzymatic process, in 1960, the mechanism was described in terms of a translation of genetic information—that is, in an entirely different conceptual horizon and linguistic idiom.

From the perspective of a micrology of knowledge, we might try the following summary. The major turning points in the development of the system came unexpectedly, and they happened, as a rule, to be connected to minute adjustments of its parameters and the technologies involved. At one point, it was a slight change in the way of homogenizing the tissue that had a decisive influence on the activity of the components. At another time, it was the addition, or the change of concentration, of a component of the artificial buffer system needed to make possible and to keep going the synthetic activity to be investigated. At yet another point, it was the deliberate but by no means necessary introduction of a control that actually aimed at controlling the background and produced a signal that, when followed, led to the characterization of an entirely novel molecular compound. And at yet another point, a component that was thought to be an impurity caused by the inadvertent breakdown of a macromolecule but could not be purified away led the researchers on the track of an essential intermediate involved in the synthetic process.

The instances could be multiplied. All these shifts were unprecedented and rested on a particular assemblage of a model organism (first the rat, then a bacterium), a style of experimentation (the test-tube experiment), and a number of research technologies (among them ultracentrifugation and radioactive tracing) that made them possible. If we really want to understand the generation of scientific knowledge, attention not only to the big things involved—big theories, big apparatus—but to these minutiae turns out to be essential. On the technical level, as exemplified by the viral plaque method and nucleic acid sequencing, the minute appears as the microscopic generator of forms of macroscopic visualization. On the epistemic level, as demonstrated by the trajectory of an experimental system, the minute appears as the gatekeeper for events that can induce major shifts in an experimental agenda.

Notes

I thank Christiane Frey for letting me borrow the term *micrologies of knowledge*.

1. Claude Bernard, *An Introduction to the Study of Medicine* (1865). Translated by Henry Copley Greene. Dover, New York 1957, 14–15.
2. For more details, see Hans-Jörg Rheinberger, "Cultures of Experimentation." In Karine Chemla and Evelyn Fox Keller (eds.), *Cultures without Culturalism: The Making of Scientific Knowledge*. Duke University Press, Durham 2017, 278–95.
3. Herbert C. Friedmann, "From Friedrich Wöhler's Urine to Eduard Buchner's Alcohol." In Athel Cornish-Bowden (ed.), *New Beer in an Old Bottle: Eduard Buchner and the Growth of Biochemical Knowledge*. Universidad de València, Valencia 1997, 67–122, 108.
4. Claude Bernard, *Leçons sur les phénomènes de la vie, communs aux animaux et aux végétaux* (1878). Librairie J. Vrin, Paris 1966, in particular 113–24.
5. For a preliminary study, see Hans-Jörg Rheinberger, "Preparations, Models, and Simulations." *History and Philosophy of the Life Sciences* 36 (2015), 321–34.
6. Taken from Gunther S. Stent, *Molecular Biology of Bacterial Viruses*. Freeman, San Francisco 1963, 177.
7. Ernst Peter Fischer and Carol Lipson, *Thinking about Science: Max Delbrück and the Origins of Molecular Biology*. W. W. Norton, New York 1988; see also John Cairns, Gunther S. Stent, James D. Watson (eds.), *Phage and the Origins of Molecular Biology*. Cold Spring Harbor Laboratory Press, Cold Spring Harbor NY 1992.
8. Salvador Luria, *A Slot Machine, a Broken Test Tube: An Autobiography*. Harper & Row, New York 1984.
9. Charles Sanders Peirce, "Logic as Semiotic: The Theory of Signs." In Justus Buchler (ed.), *Philosophical Writings of Peirce*. Dover Publications, New York 1955, 98–119.
10. Hans-Jörg Rheinberger, "Sichtbar Machen: Visualisierung in den Naturwissenschaften." In Klaus Sachs-Hombach (ed.), *Bildtheorien: Anthropologische und kulturelle Grundlagen des Visualistic Turn*. Suhrkamp, Frankfurt am Main 2009, 127–45.
11. See Paul Rabinow, *Making PCR: The Story of Biotechnology*. University of Chicago Press, Chicago 1997.
12. Miguel García-Sancho, *Biology, Computing, and the History of Molecular Sequencing: From Proteins to DNA, 1945–2000*. Palgrave Macmillan, London 2012.

13. Frederick Sanger, Steve Nicklen, and Alan R. Coulson, "DNA Sequencing with Chain-Terminating Inhibitors." *Proceedings of the National Academy of Sciences of the United States of America* 74 (1977): 5463–67.
14. Sanger, Nicklen, and Coulson, fig. 1, 5464.
15. Hans-Jörg Rheinberger, *Toward a History of Epistemic Things: Synthesizing Proteins in the Test Tube*. Stanford University Press, Stanford 1997.

PART THREE

Philological Minima

✦ ✦ ✦

Donne's Things
Epistemologies of the Small in John Donne's Love Poetry

ANDREAS MAHLER

✦ ✦ ✦

Things and/in Contexts

A WELL-KNOWN PORTRAIT of the Elizabethan poet John Donne, made in the 1590s by an unknown contemporary artist, depicts him as a young man in the pose of melancholic lover.[1] Apart from showing the upper half of the man himself, with reddish lips, a tiny mustache, an absent-minded—if not distracted—gaze, and his arms crossed, the painting highlights above all four more or less "small" things: a soft, dark broad-brimmed hat; an embroidered and partially diaphanous collar made of very fine lace; an ostentatious fur cuff belonging to a tight black glove on the man's right hand and wrist; and a bare left hand displaying four very long, elegant, slim, nonworking fingers, leaving the thumb invisible. In the painting, the representation of the four objects is quite obviously used to connote, if not seriously denote, their bearer's distinction and extravagance: in addition to staging him as rejected lover, they self-fashioningly display him as a kind of noble rake;[2] as the notorious "great visiter of Ladies, ... great frequenter of Playes, ... great writer of conceited Verses,"[3] for which he was known among his contemporaries at the time; as an eccentrically wild Inns of Court man who, not least in his official role as master of the revels at Lincoln's Inn beginning in February 1593, was lavishly spending all the money he had inherited this same year from his father—and one year later from his dead brother—on festivities and entertainments within the intellectual bohemian ambience of England's main law school and so-called third university.[4]

In a first instance, the four objects in the picture—hat, lace collar, cuff, and hand—are above all things in their own right; with the exception of the last, they are artisanal products displaying their makers' art and expertise, testifying to the times' advanced standards in matters of material treatment and cultural refinement.[5] In a second, as noted in the reading of the picture, these things find themselves used as signs, semiotically standing for something else—for example, signaling their bearer's intellectual attitude and social stance. And in a third, they also act as triggers for stories, tellingly evoking how the young man in the picture went about wasting a fortune to win popularity and acclaim among his peers and make himself a name—and career—in late Elizabethan London. In other words, things are things before they become signifiers and before they turn into narratives:[6] they are first of all nothing but syntactic elements contiguously and contingently brought together in one and the same material context; they then turn into semantic entities, endowed with meaning through the cultural context in which they are used; and they end up by developing a pragmatic force that relates these general meanings to the very personal contexts of their individual users.[7] As one can see, it is thus mainly the contexts or, in terms of this volume, the *milieus* that turn things, even and above all small things (details, trivia, *minutiae*),[8] from mere objects into meaningful entities by first placing them—hat, cuff, collar, hand—next to each other and then by making them culturally "distinguished, extravagant, noble, arrogant" or personally signify "interesting, handsome, and somewhat snobbish-looking bright young man wasting his money as well as time pretending to be in a state of unrequited love."

Donne's Things: A Flea

SMALL OBJECTS not only characterize Donne's portrait; they also figure prominently in his love poetry. His *Elegies* and *Songs and Sonnets* feature, among other things, a "bracelet," a "perfume," a "picture," a "blossom," a "primrose," a "broken heart" or a "jet ring sent," and, more generally or unspecifically, a "token," a "message," a "curse," a "book," a "will," and a "bait." From all we know, Donne's poems were themselves heavily contextualized; circulating in manuscript form from the early 1590s, they were used not so much, as would have been expected in the prevalent Petrarchan tradition, to address some distant and unreachable female object of love but mainly to impress precisely the coterie-like in-group of lascivious young intellectual males lingering in the halls and gardens of the inns, incessantly waiting for yet another "witty" thing to amuse them and bind them together as a group.[9] In other words, Donne's love poetry was not so much geared to female persuasion than to male bonding; where his speakers-in-the-poem pretend to convince fictitious women, the texts themselves try to ingratiate themselves with real men.[10]

That Donne's poems were material part of the carnivalesque Inns of Court celebrations is, for example, documented by the schedule of the 1597–98 Middle Temple's Prince d'Amour's Revels, which on Tuesday, January 3, 1598, significantly announced a Donneian "commendation of woman's inconstancy" that in the end, however, for some reason or other, was "not delivered."[11]

Among Donne's "thing poems," there is also his notorious song or sonnet on "The Flea."[12] The flea is the only animate thing in the list; from the point of view of scientific or literary observation, it is quite obviously one of those "trifling rarities" or epistemic inanities, such as insects or *animalcula*, that, as Lorraine Daston has pointed out, someone like Joseph Addison still in 1710 could ironize as being the objects of interest to "a little genius" only, traditionally tying the small to the small and stylistically linking it, in accordance with the precepts of the *rota Virgilii*, to the comparatively "little" worth of the ridiculous.[13] The poem apparently takes its cue from a whole row of flea poems going back to a pseudo-Ovidian late medieval piece entitled "Carmen del pulice" and ultimately triggered by a widely imitated 1582 poem composed by the French poet Étienne Pasquier under the title "La Puce de Madame des Roches" based on an alleged actual encounter three years before between the poet and said lady—a poem that Donne, who, from his mother's side, belonged to the Catholic Heywood family, may have come across in his puberty during a possible stay in exile with the Jesuits at Douai at some time in the 1580s.[14]

John Donne's "The Flea" very much looks like an exercise in situation building and contextualization.[15] Like other poems in the collection, it heavily exploits the faculty of deictics, not so much to anchor a written text in a situation extratextually already given, but to create a newly invented situation through their mere intratextual use.[16] As in "Break of Day," where he creates a dialogue between a cross-gendered female speaker urging her partner to stay in bed and a muted and presumably exhausted male lover eager to leave, Donne makes use of this fictionalizing device in "The Flea" three consecutive times, creating for the three nine-line stanzas of the poem three slightly differing consecutive situations between a male speaker and his lady, triggered each by some implied "event" before each stanza, involving the also present flea.[17] So what the poem presents is a kind of miniature narrative. It syntactically aligns man, woman, and flea and uses them for the suggestion of a rather melodramatic semantic sequence of "discovery," "threat," and "murder." As a small thing or mere *animalculum*, the flea thus initiates three changes of situation, which the speaker-in-the-poem in turn immediately contextualizes and semanticizes into pragmatic arguments meant to help him persuade the lady to sleep with him.

This rhetorical negotiation between things and contexts—*minutiae* and *milieus*—begins with the first stanza. That the undertaken enterprise will be a semiotic one is signaled right from the start in the repetition of the attention-seeking imperative "mark" as well as of the situation-building deictic "this" in the very first line:

> Mark but this flea, and mark in this,
> How little that which thou deny'st me is;
> Me it sucked first, and now sucks thee,
> And in this flea, our two bloods mingled be;
> Confess it, this cannot be said
> A sin, or shame, or loss of maidenhead,
> Yet this enjoys before it woo,
> And pampered swells with one blood made of two,
> And this, alas, is more than we would do. (ll. 1–9)[18]

The initial imperative "mark," additionally highlighted by a strong accent in a metrically weak syllable position, immediately turns the flea into a sign standing for something that, in the eyes of the speaker, does not rectify his rejection, since, according to his argument, it already demonstrates the fait accompli of the seemingly sinful "mingling" of his and the lady's "two bloods," signaling, if not proving by its mere existence, the "little" consequence this obvious union of the juices has had on them. In consequence of this, the speaker then goes on to make the addressee concede ("Confess it") by jubilantly rejecting ("cannot be said") all the moral meanings conventionally attributed to this fact ("sin," "shame," "loss of maidenhead") as well as by euphemistically playing down their own role in comparison to what the flea seemingly has already accomplished ("more than we would do").[19] Yet instead of being touched by this sophisticated—if not, to repeat Richard Baker's words, "conceited"—act of persuasion, the lady, as is suggested by the interruption between the two stanzas, rather bluntly reaches out for the flea in order to get rid of it as mere small, noisome thing as soon as possible.

This is where the second stanza sets in. But where stanza 1 is bent on belittling the flea's semantic value, stanza 2 adopts a very different strategy in maximally inflating it:

> Oh stay, three lives in one flea spare,
> Where we almost, nay more than married are.
> This flea is you and I, and this
> Our marriage bed, and marriage temple is;

> Though parents grudge, and you, we'are [sic] met,
> And cloistered in these living walls of jet.
>   Though use make you apt to kill me,
>   Let not to this, self murder added be,
>   And sacrilege, three sins in killing three. (ll. 10–18)

Against the moral playing down of what the flea allegedly stands for in the first argument, the second one now resorts to an extremely dense isotopy of the sacred and religious ("temple," "cloistered," etc.) in order to exaggerate the fact of the blood-mingling from a metonymy for the already consummated sexual union (the "marriage bed" where the two "are met") into a metaphor for the sacrament of marriage (the "marriage temple"),[20] with the lady's threat of killing the flea shifting into a "sinful" accumulation of not only "killing" the speaker (as is the Petrarchan lady's "use") but also inadvertently adding "self murder" by killing her own self in the "bed" and "sacrilege" in ruthlessly destroying the "temple." The moral invalidation of the first step thus contrasts with a kind of theological "overkill" in the second. Pragmatically, however, the speaker-in-the-poem rather seems to be on the defensive now; where in the first stanza his speech acts were initiating commands asking the addressee to do what he himself has in mind (to "mark" and to "confess," if not to cede), in the second, they are reactive pleas asking her not to do what she seems to have in mind in turn ("Oh stay, three lives in one flea spare"), which clearly shifts the agency from him to her. And even yet, the speaker's attempt at persuasion fails again.

The third stanza begins after the flea's death perpetrated by the lady's rash and determined hand:

> Cruel and sudden, hast thou since
> Purpled thy nail, in blood of innocence?
> In what could this flea guilty be,
> Except in that drop which it sucked from thee?
> Yet thou triumph'st, and say'st that thou
> Find'st not thyself, nor me the weaker now;
>   'Tis true, then learn how false, fears be;
>   Just so much honour, when thou yield'st to me,
>   Will waste, as the flea's death took life from thee. (ll. 19–27)

Pragmatically, stanza 3 starts with an exhortation and two rhetorical questions reproachfully accusing the lady of senselessly and pointlessly killing the animal, thus showing herself, as the initial adjectival apostrophe has it, as both "cruel and sudden"—that is, without reflection or feelings or a bad conscience. This can be read as an attempt on the part of the speaker to get himself out of the defense again; more importantly, however, the argument at the same time—with the lady's consent, too—surreptitiously begins to withdraw all semantic investment from the small object around which the entire poem is constructed: all of a sudden, the flea no longer means harm ("innocence") nor sinfulness ("guilt"), nor is it possible to attribute to its death any "weakening" effect, nor will it—as the act it seems to symbolize—in the end stand for any loss of "honor." To some degree, it almost looks as if the speaker, with his bombastic exaggeration in stanza 2, has placed a snare that in a first move slyly enables the lady—in inadvertently accepting the speaker's initial argument of "littleness"—to "triumph" by rhetorically (and literally) deflating the exaggeration (and the flea), only to win himself the upper hand again, as is testified by the last positively didactic (and paternalistic) imperative leading back to the beginning of the first stanza ("then learn").

The argument thus shifts from "little" meaning to, as it were, a world-encompassing "cosmological" meaning to no meaning at all; it moves the flea from trifle to sacrament back to mere thing or, even more, from "something," small or big, to "nothing."[21] Yet as has already been hinted at, while the speaker-in-the-poem is making use of the flea to create three contexts that are supposed to help him overcome the fictive lady's resistance, the author of the text makes use of it to win acclaim in the masculine context of the rhetorically and, above all, disputationally alert young members of the law schools by arguably nipping any female's counterargument outwittingly in the bud, proving his—and their—seemingly natural "superiority" and irresistibility.[22] At least latently, the fictitious conventional Petrarchan melancholy, as depicted in Donne's portrait, begins to tilt into, to some extent, "real" licentious, hedonistic, and exuberantly self-confident joy.[23]

Epistemologies of/in "The Flea"

FOLLOWING THE analysis, John Donne's poem presents a kind of perpetual metamorphosis. His flea is a thing that turns into a sign that turns into a story that becomes a thing again. It is part of the things of the world that we perceive/experience/see and then put into words/inscribe/talk or write about in order to find out what they might mean to the world—and to us. This epistemological process of bringing together the visible and the sayable has been called "diagrammatical"; it is one of our main pathways for acquiring, storing, and negotiating knowledge.[24] Donne's poetry, as has been suggested, seems to be mainly concerned with tracing these pathways by exploring how meaning and/or knowledge are culturally/socially/individually produced and questioned, established and relativized, constructed and deconstructed, or, as for that, contextually "territorialized" and "deterritorialized"; it is in this way that his poems can be understood as diagrammatical experiments or epistemological—if not, as will soon be explicated, epistemic—"undertakings."[25]

Accordingly, what Donne "undertakes" in "The Flea" is, as has been shown, first, the report of a "discovery"; second, the elaboration of an "interpretation"; and third, a spectacular annulment of both. The poem starts with the detection of something visible, it proceeds with its transposition into something sayable, and it ends by withdrawing both—without excluding, however, that all this, despite the flea's death, could just as well go on forever.[26] As a consequence, what we witness is a process of thing-based knowledge production and its annihilation. At the same time, the poem's shift from metonymy (stanza 1) to metaphor (stanza 2) to no trope at all (stanza 3) also seems to situate it on the threshold between the two epistemic formations characterizing Elizabethan England in the 1590s. Michel Foucault has called this "the age which, above all in England, saw the emergence of an observational, affirmative science" with "a new form of the will to knowledge"[27] and has described this process as a long and gradual shift from a thinking in analogies, with language being part of the things of the world (*verba* as *res*), to a method of finding truth through empirical observation only, with language as its mere external transmitter.[28] Diagrammatically, this can be seen as a shift from a comparative reading of the visible with an analogical interpretive expression of the sayable to a classificatory registering of the visible with its neutral and transparent rendering in words.[29]

As can be seen from our reading, stanza 1 seems to engage in the second procedure. In placing first the word "mark," it asks its addressee to discern and discriminate an object that "empirically" seems to have come into view and immediately makes use of it to classify and "translate" the speaker's wish as being of negligibly "little" value, since what the lady is still refusing—the "mingling" of the juices—has already become crude and factual reality in the visible *animalculum*'s swollen body, which, according to the speaker, can be "said to be" irrefutable empirical evidence—a kind of *demonstratio ad oculos*—for what the speaker's and the addressee's sexual union would bring about. This can be taken as an act of epistemological simulation. What the speaker is trying to do is "sell" the flea as a kind of preliminary test case or laboratorial experiment for what he has in mind to do with the lady, reaching the conclusion that, considering the visible evidence, one can say that it is "safe" to go ahead: if nothing much has happened to the flea, the argument goes, nothing much will happen to them either.

But as we know, the lady—somewhat literally—smells a (laboratorial) "rat" and moves her hand. Since his "scientific" way of reasoning does not convince her, the speaker in stanza 2 now resorts to the older analogical way of finding truth and winning knowledge as another possible strategy of simulation. Michel Foucault has characterized the episteme of analogism as a thinking in resemblances and similitude, specifying four different types:

> *Convenientia, aemulatio, analogy,* and *sympathy* tell us how the world must fold in upon itself, duplicate itself, reflect itself, or form a chain with itself so that things can resemble one another. They tell us what the paths of similitude are and the directions they take; but not where it is, how one sees it, or by what mark it may be recognized.... In order that we may know that aconite will cure our eye disease, or that ground walnut mixed with spirits of wine will ease a headache, there must of course be some mark that will make us aware of these things.... There are no resemblances without signatures.[30]

This epistemological interplay between marks or signatures and resemblances is what Foucault describes as a superimposition of "hermeneutics and semiology in the form of similitude," explaining that, in this thought system, "to search for a meaning is to bring to light a resemblance."[31] Accordingly, instead of further empirically observing the flea and translating its truth, which has already ended in a failure, the speaker-in-the-poem now seizes on its signature and looks for an analogy, which, from his point of view, he finds in another element standing for the union of the lovers, which is—sympathetically if not bombastically again—the "marriage temple": since both, the argument now goes, bring the lovers together, this may, or must, be interpreted in the way that they actually belong together if not that they designate, and factually are, the same. This is the speaker's switch from the metonymical to the metaphorical. If, as Foucault says, "to know" in this episteme is "to interpret: to find a way from the visible mark to that which is being said by it and which, without that mark, would lie like unspoken speech, dormant within things,"[32] the speaker in stanza 2 no longer tries to make believe that the blood metonymically stands for the prospective couple as in stanza 1 but alternatively insinuates that, in accordance with the sympathies in the great chain of being,[33] the entire flea now metaphorically stands, on a lower scale, for the sacredness of their prospective union, on a higher scale, with the consequence that the killing of the vehicle, the flea, would entail the destruction of the tenor—the sanctity of the "marriage temple" or the sacrament of the "marriage," if not the whole world order, hence the "sacrilege."

This untimely and counterfactual epistemological simulation of an analogical path of truth-finding has, in stylistic terms, been called "baroque,"[34] and this is precisely what it seems to look like to the lady—like a scintillating fake argument intellectually attempting to coax her into a consent she, at the moment, at least, does not feel like giving. Accordingly, what she does is act pragmatically and, as we know, kill the flea. Diagrammatically, from the point of view of knowledge production, there is in stanza 3 now nothing to be seen—neither any seemingly analogical evidence or "mark" nor any seemingly empirical evidence to "mark"—and, consequently, nothing to convert into meaning (to interpret or translate) any longer. This lack of a thing or even its loss, however, finds itself immediately instrumentalized again by the indefatigable speaker who sees in it an opportunity to shift his argument from epistemological simulation to dissimulation by all of a sudden projecting the unconditional wish for meaning-making onto the lady, insinuating that it is not he (as in the first two stanzas) but above all she who insists on seeing something in the flea and declaring—in a conceding verse that in a metrical anomaly seems to have strong accents on almost every syllable—that she shouldn't: "'Tis true, then learn how false, fears be." The final couplet summarizes this in a negative equation: no flea, no loss of life in killing the flea, no loss of honor in sleeping with the speaker. And yet the lady's reaction might, again, be nothing but an amused "Nice try."

Truth, Simulation, and the Act of Fictionalizing

JOHN DONNE's poem takes us, as has been shown, on an epistemological trajectory leading from an empirical episteme via an analogical episteme to a (more or less tongue-in-cheek) warning against falsity. It focuses on a small thing and gauges its potential meanings: it first uses the flea to dismiss the lady's fear that sleeping with the man means losing her reputation, it then uses the flea to instill in her the fear that killing the flea means committing a sin and acting against God, and it concludes by pointing out that fears like these, man-/woman-made as they are, are "false" and, as a consequence, should not be heeded. The speaker-in-the-poem makes use of truth-finding devices to simulate "truths" that rhetorically suit his purposes and to drop them as soon as he sees that they don't; the author of the poem makes use of elements of the "real"—man, woman, flea, nay Petrarchan love—to "cast" his imaginary "as a form" that "allows us" to become aware of knowledge- and truth-bound meaning-making processes and of the ways in which they are made and manipulated through linguistic anomalies and indirections such as paradoxes, allusions, implications, presuppositions, and whatnot, as would have fascinated any intellectually minded young lawyer-to-be.[35] It is in this sense that John Donne's "The Flea" can be understood, as has been suggested, as yet another knowledge-oriented diagrammaticopoetical "undertaking"; in ceaselessly creating imaginary *milieus* in which even trifles or almost invisible *minutiae*—in this case, a flea—can infinitely be invested with new meanings, the poem explores and negotiates the epistemological ways and means, as well as the epistemic conditions, by and under which we all continuously make, and abandon, cultural truths.

Notes

1. This portrait can, e.g., be found on the cover of the Penguin edition of Donne's poetry: John Donne, *The Complete English Poems*, ed. A. J. Smith, Harmondsworth: Penguin, 1975 (Penguin English Poets).
2. For the early modern idea of instrumentalizing things and the self for making an impression as well as for signaling social meanings, see Stephen Greenblatt's seminal study *Renaissance Self-Fashioning: From More to Shakespeare*, Chicago: University of Chicago Press, 1984, esp. 8–9; for its Italian prehistory in creating what some have called a *società spettacolo*, or a "facade culture," based, among other things, on Erving Goffman's concept of "impression management" as well as Thorstein Veblen's notion of "conspicuous consumption," see Peter Burke, *The Historical Anthropology of Early Modern Italy*, Cambridge: Cambridge University Press, 1987.
3. The words are Richard Baker's as quoted in R. C. Bald, *John Donne: A Life*, Oxford: Clarendon Press, 1970, 72; see there for further biographical detail; for an informative overview, cf. also the rather detailed "Table of Dates," in Donne, *Complete English Poems*, 17–25.
4. For the Inns of Court as leading intellectual centers of early modern—not least literary—life in the growing metropolis of London, see Philip J. Finkelpearl, *John Marston of the Middle Temple: An Elizabethan Dramatist in His Social Setting*, Cambridge, MA: Harvard University Press, 1969, esp. 1–80; for what he calls "Donne's circle" within the intellectual milieu of the universities and law schools, see A. Alvarez, *The School of Donne*, London: Chatto & Windus, 1962, esp. 187–95 for members' names; see also Arthur F. Marotti, *John Donne: Coterie Poet*, Madison: University of Wisconsin Press, 1986. For a Latour-inspired attempt at describing Donne's circle as an intellectually minded elitist "constellation" within the larger homosocial "network" of the Inns of Court, see my "Netzwerke, Konstellationen, intellektuelle Denkräume: John Donne und die Inns of Court," in *Religiöser Nonkonformismus und frühneuzeitliche Gelehrtenkultur*, ed. Friedrich Vollhardt, Berlin: Akademie-Verlag, 2014, 51–70.
5. Arguably, the extensive care visibly devoted to the portrayed person's hand (as well as his lips, mustache, and eyebrows) could also be seen as the result of a veritable artist's cultural activity.

6. For views asserting the "thingness" of things, see, as one of the triggers of the debate, Arjun Appadurai's edited volume *The Social Life of Things: Commodities in Cultural Perspective*, Cambridge: Cambridge University Press, 1986, esp. the editor's introduction; as well as Bill Brown, "Thing Theory," *Critical Inquiry* 28.1 (2001): 1–22; for the latent (or even manifest) "narrativity" of and in things, see Mieke Bal, "Telling Objects: A Narrative Perspective on Collecting," in *The Cultures of Collecting*, ed. John Elsner and John Cardinal, London: Reaktion Books, 1994, 97–115; for the early modern period, see Peter Burke, "*Res et Verba*: Conspicuous Consumption in the Early Modern World," in *Consumption and the World of Goods*, ed. John Brewer and Roy Porter, London: Routledge, 1993, 148–61 (I wish to thank Sabine Schülting for help and decisive hints).
7. For the syntactic, the semantic, and the pragmatic as the three aspects or dimensions of the semiotic, relating (1) signs to other signs, (2) signs to meanings, and (3) signs to their users, see, classically, Charles William Morris, "Esthetics and the Theory of Signs" (1939), in *Writings on the General Theory of Signs*, The Hague: Mouton, 1971, 415–33.
8. For further development of the relation between minutiae and milieus, with regard to perception and relevance, see the introduction to this volume, where it is also quite rightly pointed out that it is both (as it is "coemergently") the things that constitute the context and the contexts that constitute them as things (with meanings).
9. This function of creating some kind of "phatic community" is precisely what Marotti has in mind with his sobriquet of "coterie poet." For Donne's careful avoidance of publication, see Andrew MacColl, "The Circulation of Donne's Poems in Manuscript," in *John Donne: Essays in Celebration*, ed. A. J. Smith, London: Methuen, 1972, 28–46; for a more general discussion of this under the label of a "stigma of print," see, classically, James W. Saunders, "The Stigma of Print: A Note on the Social Bases of Tudor Poetry," *Essays in Criticism* 1 (1954): 139–64.
10. For an attempt at differentiation between a real (written) enunciation by an author such as John Donne (outer pragmatic level 1: L_1) producing a text (syntactic materiality: L_2) evoking a fictitious (oral) enunciation by a (homodiegetic) first-person speaker-in-the-poem (inner pragmatic level: L_3) producing, as enounced, a world-within-the poem (semantic result: L_4), see my "Towards a Pragmasemiotics of Poetry," *Poetica* 38.3–4 (2006): 217–57, esp. 221–25.
11. For a reconstruction of the schedule and further detail, see Finkelpearl, *John Marston of the Middle Temple*, 50–51; for the quotes, p. 50; for Donne, esp. note 15; for a useful list of "revels criteria," see W. R. Elton, *Shakespeare's Troilus and Cressida and the Inns of Court Revels*, Aldershot: Ashgate, 2000, 7.
12. My use of the term *thing poem* here is slightly anachronistic and oblique; the term usually refers to seemingly speakerless descriptive poems mainly coming to the fore in the nineteenth century with the Parnassian movement. "The Flea" was apparently the opening poem in the first printed editions of Donne's *Songs*

and Sonnets from 1635, four years after his death; for contextual and textual details, see the editor's notes in Donne, *Complete English Poems*, 376.

13. For this hierarchy of attitudes, following the style/object-relating laws of decorum and linking the high to the high (and to the important and serious) and the low to the low (as well as to the unimportant and ridiculous), which corresponds to—and somehow also seems to be a consequence of—the widespread medieval and early modern idea of the "chain of being," see, with reference to the Addison quote, Lorraine Daston, "Attention and the Values of Nature in the Enlightenment," in *The Moral Authority of Nature*, ed. L. D. and Fernando Vidal, Chicago: University of Chicago Press, 2004, 100–126; for a cogent differentiation between small things tout court and the hardly visible—and hence seemingly negligible—"minute body," see Christiane Frey, "The Art of Observing the Small: On the Borders of the 'Subvisibilia' (from Hooke to Brockes)," *Monatshefte* 105.3 (2013): 376–88; as well as Christiane Frey, "Bacon's Bee: The Physiognomy of the Singular," in *Exemplarity and Singularity: Thinking through Particulars in Philosophy, Literature, and Law*, ed. Michèle Lowrie and Susanne Lüdemann, London: Routledge, 151–65.

14. A lot of the biographical details are still uncertain, but for all we know, there is a strong French as well as Spanish connection to Donne's life. Donne's mother was the daughter of the Elizabethan author John Heywood and great-niece of Sir Thomas More.

15. For a first pioneering analysis of what he calls Donne's "inferred situations," see Geoffrey N. Leech, *A Linguistic Guide to English Poetry* (1969), London: Longman, 1977, 191–93; a systematization of this can be found in Jonathan Culler, "Poetics of the Lyric," in *Structuralist Poetics: Structuralism, Linguistics and the Study of Literature* (1975), London: Routledge, 1989, 161–88, esp. 164–70.

16. In terms of reference, this is a shift from an external exophoric (and somehow backward-looking anaphoric) use to an internal endophoric (and, to some extent, forward-looking cataphoric) use; for the terms, see Gillian Brown and George Yule, *Discourse Analysis*, Cambridge: Cambridge University Press, 1983 (Cambridge Textbooks in Linguistics), 190–222. For situation-building as one of the main devices of fictionalizing, see Wolfgang Iser, *The Act of Reading: A Theory of Aesthetic Response* (1976), trans. David Henry Wilson, Baltimore, MD: Johns Hopkins University Press, 1980, 62–68.

17. For a context-oriented, pragmatic description of the poem's inner contextuality, see Richard Bradford, *A Linguistic History of English Poetry*, London: Routledge, 1993 (Interface), 40–48. Formally, each stanza consists of alternating iambic tetrameters (ll. 1, 3, 5, and 7) and pentameters (ll. 2, 4, 6, 8, and 9) arranged in three couplets followed by a *volta* leading to a concluding, and surprising, triplet.

18. For the poem, see Donne, *Complete English Poems*, 58–59.

19. The flea both takes its pleasure "before" wasting its time with a lengthy process of "wooing" and "swells" without risking pregnancy. Stanza 1 uses the

situation-building word *this* more than any other single word six times, referring to the flea throughout (ll. 1, 4, 5, 7, and 9) as well as, in extension, to its metonymical acts of "mingling" (l. 5) and "swelling" (l. 9); for the intricacies of the deictic *this* in aesthetic contexts, see my "The Case Is '*This*': Metareference in Magritte and Ashbery," in *Metareference across Media: Theory and Case Studies*, ed. Werner Wolf, Amsterdam/New York: Rodopi, 2009 (Studies in Intermediality 4), 121–34.

20. For this shift from metonymy (stanza 1) to metaphor (stanza 2) back to metonymy (or arguably no trope at all; stanza 3), see Bradford, *Linguistic History of English Poetry*, 42.
21. In the tradition of early modern love poems, the flea was envied (1) because of its nearness to the lady and (2) for the fact that it was given the license to die a "love death," whereas the male lover was not.
22. For the institutionalized practice of "mock dispositions" at the inns, such as the so-called *moots* or the more dialogic variant of *bolting*, see Finkelpearl, *John Marston of the Middle Temple*, 9 (also the quote); for a detailed analysis of disputational activities in medieval and early modern universities, cf. Anita Traninger, *Disputation, Deklamation, Dialog: Medien und Gattungen europäischer Wissensverhandlungen zwischen Scholastik und Humanismus*, Stuttgart: Steiner, 2012 (Text und Kontext 33).
23. There is no fourth stanza, which means that the "fiction" does not continue: there is no further rejection, but there is no acceptance either. For the characteristic intertextual recombination of mutually exclusive love discourses—the Petrarchan, the Platonistic, and the hedonist variety—in one and the same poem (or, as for that, drama), especially in the 1590s and 1600s, see, with regard to the difference between Chaucer's and Shakespeare's *Troilus and Cressida*, my "'Potent Raisings': Performing Passion in Chaucer and Shakespeare," in *Love, History and Emotion in Chaucer and Shakespeare: Troilus and Criseyde and Troilus and Cressida*, ed. Andrew James Johnston, Russell West-Pavlov, and Elisabeth Kempf, Manchester: Manchester University Press, 2016, 32–45.
24. For the "diagrammatical" as the attempt at bringing together the regime of "looking" (the "dia" element) with a regime of "spelling it out" (the "grammatical"), see Gilles Deleuze's eulogistic further development and differentiation of Foucault's archaeology of knowledge in his *Foucault*, trans. Seán Hand, Minneapolis: Minnesota University Press, 1988; for the discovery of the minute, or the tiny detail, as an explicitly "epistemic thing" in the seventeenth and eighteenth centuries, see Hans-Jörg Rheinberger, *Toward a History of Epistemic Things*, Stanford, CA: Stanford University Press, 1997; as well as Frey, "Art of Observing the Small."
25. For a highly suggestive Deleuze-inspired reading of Donne's poetry under precisely this label, which, by the way, is also the title of one of Donne's *Songs and Sonnets*, see André Otto, *Undertakings: Fluchtlinien der Exklusivierung in John Donnes Liebeslyrik*, Munich: W. Fink, 2014; for the poems as "undertakings,"

see esp. 46–50; for a theoretical description of their territorializing and, above all, deterritorializing negotiation of knowledge under the notion of the "diagrammatical," see the remarks on 304–15. The term *undertaking* is also one of the words used to refer to Deleuze's essayistic philosophical negotiations and reflections.

26. In a way, this very much looks like an early modern equivalent to the late modern prose experiments of suggestion and negation, evocation and erasure, by Samuel Beckett; see my "From Nothing to Nothing: Emergence(s) and Residue(s) in Beckett's Prose," *Comparatio* 10.1 (2018): 19–38.

27. See Michel Foucault, "Orders of Discourse: Inaugural Lecture Delivered at the Collège de France," trans. Rupert Swyer, *Social Science Information* 10.2 (1971): 7–30, 25.

28. For this shift from analogism to a thinking in taxonomies, see Michel Foucault, *The Order of Things: An Archaeology of the Human Sciences* (1966), trans. Alan Sheridan Smith, London: Routledge, 2002 (Routledge Classics), 19–85; for an insightful explication and interpretation of Foucault's approach with regard to this shift, see Gary Gutting, *Michel Foucault's Archaeology of Scientific Reason*, Cambridge: Cambridge University Press, 1989 (Modern European Philosophy), 139–79.

29. For the role of language in this epistemic shift, see esp. Foucault, *Order of Things*, 62: "The written word ceases to be included among the signs and forms of truth; language is no longer one of the figurations of the world, or a signature stamped upon things since the beginning of time. The manifestation and sign of truth are to be found in evident and distinct perception. It is the task of words to translate that truth if they can; but they no longer have the right to be considered a mark of it. Language has withdrawn from the midst of beings themselves and has entered a period of transparency and neutrality."

30. See Foucault, 29 (emphasis in original); the aconite, whose seed has the signature of looking like an eye covered by an eyelid, thus shows its resemblance to the eye and discloses as its meaning that it will be useful for a cure against eye diseases, just as much as the walnut with its signature of a brain protected by a skull signals its affinity to anything to do with the head. For the topos of the readability of the world, see Hans Blumenberg, *Die Lesbarkeit der Welt* (1981), Frankfurt a.M.: Suhrkamp, 1986; the internet curiously refers to an English translation of this important book as *Legibility of the World*, Chicago: University of Chicago Press, 1990, which in all likelihood has never seen the day.

31. Foucault, 33; cf. also 64 (emphasis in original), where he summarizes the difference between the two episteme as one between a ("classical" and "modern") differentiating taxonomic "order" and a ("premodern") comparison-based analogical "interpretation": "This relation to *Order* is as essential to the Classical age as the relation to *Interpretation* was to the Renaissance. And just as interpretation in the sixteenth century, with its superimposition of a semiology

upon a hermeneutics, was essentially a knowledge based upon similitude, so the ordering of things by means of signs constitutes all empirical forms of knowledge as knowledge based upon identity and difference." For a cogent critique of Foucault, insisting on the knowledge finding (i.e., epistemological aspect) and not so much the knowledge founding (i.e., epistemic aspect of analogism) with language, according to the formula of *res et verba*, playing an important part in the finding process, see Stephan Otto, *Das Wissen des Ähnlichen: Michel Foucault und die Renaissance*, Frankfurt a.M.: Lang, 1992, 77–127; for a discussion of this in further detail, see my "Don Quixote, Hamlet, Foucault: Language, 'Literature,' and the Losses of Analogism," in *Nominalism and Literary Discourse: New Perspectives*, ed. Hugo Keiper, Christoph Bode, and Richard J. Utz, Amsterdam / Atlanta, GA: Rodopi, 1997 (Critical Studies 10), 251–68.

32. See Foucault, *Order of Things*, 36.
33. For the cosmological phantasm of a chain of being as an intricately linked hierarchical system of order determining the social imaginary from the Middle Ages right into the eighteenth century, see Arthur O. Lovejoy, *The Great Chain of Being: A Study of the History of an Idea* (1936), Cambridge, MA: Harvard University Press, 1964.
34. See Foucault, *Order of Things*, 56; for a comprehensive epistemological account of the truth-finding mechanisms from the medieval to the early modern epoch, with what he calls "mannerism" as no longer an epistemological but merely a nostalgic and wistful "aestheticist" use of analogism, see Joachim Küpper, *Diskurs-Renovatio bei Lope de Vega und Calderón: Untersuchungen zum spanischen Barockdrama*, Tübingen: Narr, 1990 (Romanica Monacensia 32), esp. 290–304. For an argument against using the term *baroque* in English cultural history, see my "Jahrhundertwende, Epochenschwelle, epistemischer Bruch? England um 1600 und das Problem überkommener Epochenbegriffe," in *Europäische Barock-Rezeption*, ed. Klaus Garber et al., 2 vols., Wiesbaden: Harrassowitz, 1991 (Wolfenbütteler Arbeiten zur Barockforschung 20), vol. 2, 995–1026. Donne wisely resists the temptation of making use of the analogically motivatable but nevertheless facetious pun of linking the signifier *flea* to the signifier referring to what the lady quite obviously does: *flee*.
35. The indirect quotes refer to Wolfgang Iser's suggestion to understand the "fictive," as he puts it, not—as in a binary—as simply opposed to the "real," but as a "triad" operating with a reality-inspired "imaginary" in the sense that "the fictionalizing act converts the reality reproduced into a sign, simultaneously casting the imaginary as a form that allows us to conceive what it is towards which the sign points"; see Wolfgang Iser, *The Fictive and the Imaginary: Charting Literary Anthropology* (1991), trans. David Henry Wilson, Baltimore, MD: Johns Hopkins University Press, 1993, 2. For the emphatic rediscovery of the art of fictionalizing in late sixteenth-century England, see my "New Ways of Worldmaking: English Renaissance Literature as 'Early Modern,'" in *Handbook of English Renaissance Literature*, ed. Ingo Berensmeyer, Berlin/Boston:

de Gruyter, 2019 (Handbooks of English and American Studies 10), 66–88; for the rhetorical practice of casuistic disputation as part of a prehistory of fictionalizing and the "aesthetic," see also my "Allegorie und Aisthesis: Zur Genealogie von Alteritätsagenturen," in *Allegorie*, ed. Ulla Haselstein, Berlin/Boston, MA: de Gruyter, 2016 (DFG-Symposion 2014), 356–81.

"Of a Parenthesis"

CYNTHIA WALL

✦ ✦ ✦

THE OXFORD ENGLISH DICTIONARY (OED) offers a peculiarly contradictory definition of *minutia*: "A precise detail; a small or trivial matter or point."[1] We might reconsider the triviality of a small point when we remember astronaut and US senator John Glenn's feelings about climbing into a rocket on a NASA launchpad in 1962: "I felt exactly how you would feel if you were getting ready to launch and knew you were sitting on top of 2 million parts—all built by the lowest bidder on a government contract."[2] The parenthesis—both verbal and typographical—is, in fact, a tiny detail carving out precision, modestly cloaking itself as a small or trivial matter, not meaning to interrupt the *important sentence*—but of course its very existence interrupts the *important sentence* and, like a wee David, can topple its Goliath. The "milieux" for my little minutia—its "environment; surroundings, esp. social surroundings" (OED, s.v. *milieux*)—are actually multiple: rhetorical, grammatical, social, textual, and even architectural.

In 1782, the landscape gardener Lancelot "Capability" Brown told Hannah More that "he compared his art to literary composition. Now *there*, said he, pointing his finger, I make a comma, and there pointing to another spot where a more decided turn is proper, I make a colon: at another part (where an interruption is desirable to break the view) a parenthesis—now a full stop, and then I begin another subject."[3] The page of any text is three-dimensional in the sense that it has a readerly topography, with a modeled surface and a historical underground; it shares its textures with the textures of its world.[4] (Just think of how the 1960s, for example, spawned not only the perception-changing psychedelic, the buildings of Louis Kahn and I. M. Pei, and far-out prose but also the fonts Filmsense and Benguiat Zenedipity; see fig. 1.) Virginia Woolf said that every great work is "based ... on the sentence that was current at the time,"[5] and every sentence has an identity made up of

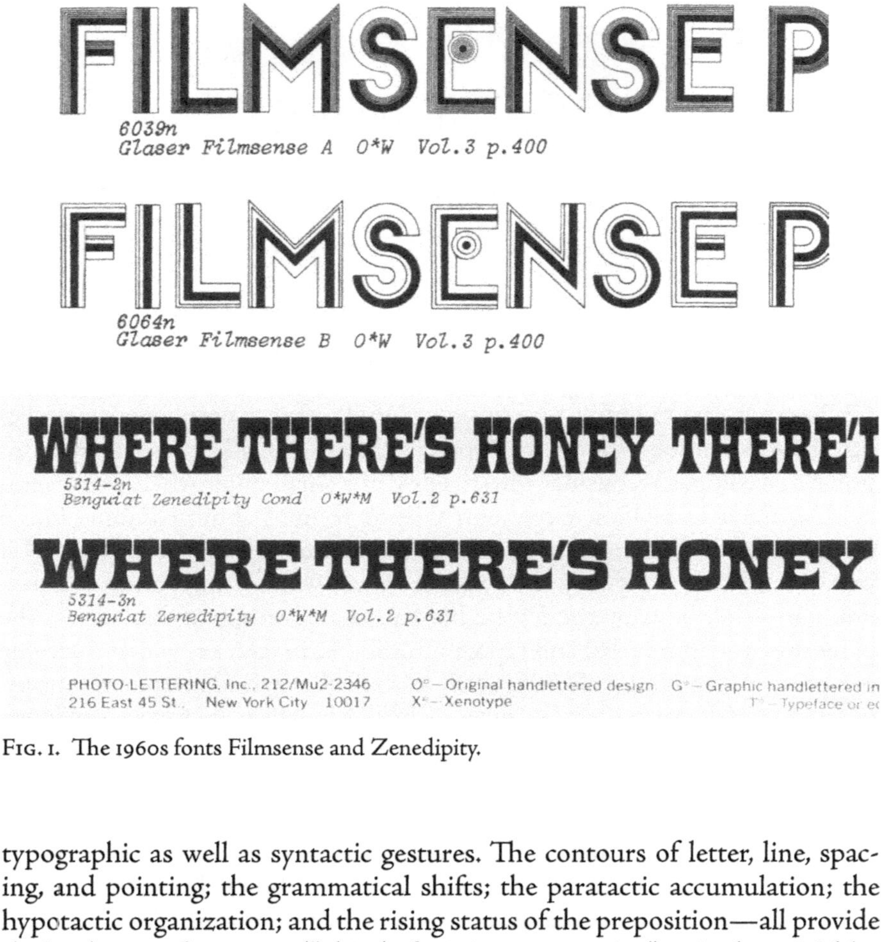

FIG. 1. The 1960s fonts Filmsense and Zenedipity.

typographic as well as syntactic gestures. The contours of letter, line, spacing, and pointing; the grammatical shifts; the paratactic accumulation; the hypotactic organization; and the rising status of the preposition—all provide the landscape of *narrative* ("a kind of miniature narrative," as Andreas Mahler puts it in the previous essay) of meaning, their forms differing in and traveling over time. Punctuation, as well as typography and grammar (not to mention actual landscapes), underwent a major landscaping in the eighteenth century, minutely captured in the difference between Edward Cocker's definition of the comma in 1704—"the least note of distinction in writing or printing, marked thus (,)"[6]—to Joseph Robertson's *fifty-eight pages* on the subject in his 1785 *An Essay on Punctuation*.[7] This essay explores the expressive potentialities of parentheses—both the rhetorical version and the typographical marks themselves—through a pair of two lonely mariners: the fictional Robinson Crusoe of Defoe's 1719 novel and the historical Lieutenant William Bligh of the 1789 mutiny on the HMS *Bounty*. Capability Brown was topographically clever in interrupting his views with landscaped parentheses; Crusoe at one end of the eighteenth century and Bligh at the other provide a pair of views on the interruptive powers of rhetorical parentheses. John Lennard has revived Erasmus's term for those minute typographical markers: "lunulae," or "little moons."[8] In prose as well as verse, those little moons can illuminate strange spaces and curious fissures in their psychological and narrative land- and seascapes. I will treat Crusoe and Bligh as open and close parens,[9] the very isolation of each metaphorically enacting the spatial milieus between the little moons.

Parentheses ("Lunulae")

EDITORS AND textual scholars will be familiar with the biography of punctuation, but for many, it remains, as John Lennard laments, "too readily invisible."[10] Punctuation in general, said Ephraim Chambers in 1728, "is a modern Art—the Antients were entirely unacquainted with the Use of our Commas, Colons, &c. and wrote not only without any Distinction of Members, and Periods, but also without Distinction of Words. . . . There is much more difficulty in *Pointing*, than People are generally aware of—In effect, there is scarce any thing in the Province of the Grammarians so little fix'd and ascertained as this. The Rules usually laid down, are impertinent, dark, and deficient; and the Practice, at present, perfectly capricious; Authors varying not only from one another, but from themselves too."[11] The eighteenth-century printers and grammarians went straight to work on those impertinent dark deficiencies. As John Smith explains in his 1755 *Printer's Grammar*, "It will not be labour in vain for a Compositor, to examine his Copy, and to observe in what manner it is pointed, whether properly, or at random: for some Gentlemen who have regard to make the reading of their Works consonant with their own delivery, point their Copy accordingly, and abide thereby, with strictness: which, were it done by every Writer, Compositors would sing, *Jubile!*"[12] (It should be underscored here that Smith gives credit to both authors and compositors for attention to the details of pointing.) Joseph Robertson confidently declared in 1785, "Some imagine, that punctuation is an arbitrary invention, depending on fancy and caprice. But this is a mistake. It is founded on rational and determinate principles."[13] And it is an *art*, an art of "infinite consequence in WRITING; as it contributes to the perspicuity, and consequently to the beauty, of every composition."[14]

The particular mark of the parenthesis was the invention of humanist scribes to isolate interpolations; Aldus Manutius introduced it to print in

1494, along with the comma, semicolon, and question mark.[15] Angel Day's sixteenth-century *English Secretorie* defines the parenthesis as "an intercluding of a sentence ... commonly set betweene two halfe circles."[16] It was not the most respected of punctuation marks (though it was received by printers more readily than the semicolon, and it's easier to argue authorial or compositorial intention, as it requires *two* strokes of the pen, *two* pieces of type).[17] It comes seventh in the typical hierarchy of definitions:

> At present, all European writers make use of the following points, as marks of division, which take their name from the sentence, or the clause, which they are respectively employed to distinguish.
>
> | 1. | A comma | , |
> |---|---|---|
> | 2. | A semicolon | ; |
> | 3. | A colon | : |
> | 4. | A period | . |
> | 5. | An interrogation | ? |
> | 6. | An exclamation | ! |
> | 7. | A parenthesis | ( ) |
>
> To these may be added the dash — (Robertson 14)

As Chambers put it in 1728, "The politest of our modern Writers avoid all *Parentheses*; as keeping the Mind in suspence, embarrassing it, and rendering the Discourse less clean, uniform and agreeable."[18] Robertson defines it as "a clause, containing some necessary information, or useful remark, introduced into the middle of a sentence obliquely, which may be omitted, without injuring the construction." Thus a parenthesis is by definition lonely, isolated, outcast—a little island in the text, best removed. (It is even spoken differently from the rest of its sentence: "In Speaking, *Parentheses* are to be pronounced with a different Tone ... to distinguish them from the rest of the Discourse").[19] To many grammarians and stylists, the parenthetical almost belongs in the French sense of "milieu," implying the "criminal underworld" (*OED* 2b): Hugh Blair's "third rule, for preserving the unity of sentences" is "to keep clear of all parenthesis in the middle of them.... For the most part, their effect is extremely bad; being a sort of wheels within wheels; sentences in the midst of sentences; the perplexed method of disposing of some thought, which a writer wants art to introduce in its proper place."[20] Indeed, in 1808 Caleb Stower noted in his own *Printer's Grammar* that "parentheses are not now so generally used as formerly: authors place their intercalations between commas, which make them equally as intelligible as though they were inclosed between parentheses, and look much neater in print."[21] Thus, in the enchanting book by Mr. Stops, *Punctuation Personified; or, Pointing Made Easy* (1824),[22] while Counsellor Comma (fig. 2), Ensign Semicolon, and the distinctly lugubrious Exclamation Point ! all have their own full page and image, the parentheses are decimated to thighbones and boneless legs (fig. 3)—the caboose to "A Dash – Circumflex ^ Breve ˘ . Diæresis ¨ Hyphen - Acute Accent ´ Grave Accent ` Parenthesis ( )."[23]

But parentheses have at the same time always been appreciated for their interpretive agility.[24] It is worth noting that the second definition of "milieu"

Fig. 2. "Counsellor Comma." *Punctuation Personified; or, Pointing Made Easy* (London: J. Harris and Son, 1824). Joseph Downs Collection of Manuscripts and Printed Ephemera, Courtesy of Winterthur Museum, Garden & Library.

Fig. 3. "A Dash – Circumflex ^ Breve ˘ . Diæresis ¨ Hyphen - Acute Accent ´ Grave Accent ` Parenthesis ( )." *Punctuation Personified; or, Pointing Made Easy* (London: J. Harris and Son, 1824). Joseph Downs Collection of Manuscripts and Printed Ephemera, Courtesy of Winterthur Museum, Garden & Library.

in the OED includes a parenthetical: "A group of people with a shared (cultural) outlook" (OED 2a); we could paraphrase and say that one happy milieu for parentheses includes a group of writers with a shared (punctuational) outlook. Lennard has memorably extrapolated, for example, the "intelligent nastiness" of a written parenthetical by Goneril in *King Lear*—"Your (Wife, so I would say) affectionate Seruant"—and Hero's "vnawares (Come hither)" in Marlowe's *Hero and Leander*;[25] Thomas Keymer has shown how, in Roger L'Estrange's translation of *Five Love-Letters from a Nun to a Cavalier* (1678), as "the parentheses [of Mariane's prose] become more intensive[,] they emphatically destabilise the surface content, like a visual heckling of what she wishes to say."[26] John Evelyn is quite liberal with parentheses in his *Kalendarium*—there is something significant embedded in lunulae in almost every entry, from the thickness of his father's hair to his account of his mother's death: "(I shall never forget it)."[27] *The Spectator* essays are riddled with parentheses (and accordingly criticized by Blair[28]); both Joseph Addison and Richard Steele were particularly fond of the opportunity for snark afforded by lunulae: "(as every rich Man has usually some sly way of Jesting, which would make no great Figure were he not a rich Man)."[29] Samuel Taylor Coleridge confessed, "Of Parentheses I may be too fond—and will be on my guard in this respect—. But I am certain that no work of empassioned & eloquent reasoning ever did or could subsist without them—They are the *drama* of Reason."[30] Even Blair acknowledged, "On some occasions, [parentheses] may have a spirited appearance; as prompted by a certain vivacity of thought, which can glance happily aside, as it is going along."[31] Robert Lovelace, the rake villain of Samuel Richardson's *Clarissa* (1747–48), has a way with parentheses as well as women, as in this letter to his friend John Belford:

> What! How! When!—And all the monosyllables of surprize.
>
> [Within parenthesis let me tell thee, that I have often thought, that the little words in the republic of letters, like the little folks in a nation, are the most significant. The *trisyllables*, and the *rumblers* of syllables more than *three*, are but the good for little *magnates*.][32]

(Lovelace [Richardson] here uses square brackets as parentheses rather than the more usual round brackets to cordon off his parenthesis; even today there is a tendency in the UK to call parentheses "brackets."[33] [Tsk.]) But Lovelace is not quite done with parentheses, rhetorical or typographical, as he continues to recount for Belford a scene he has stage-managed from behind the curtains:

> Too much in a hurry for good manners [Another parenthesis, Jack! Good manners are so little natural, that we ought to be *compos'd* to observe them: Politeness will not live in a storm], I cannot stay to answer questions, cries the wench—tho' desirous to answer [A third parenthesis—Like the people crying proclamations, running away from the customers they want to sell to]. This hurry puts the lady in a hurry to ask [A fourth, by way of embellishing the third!] as the other does the people in a hurry to buy. And I have in my eye now a whole street raised, and running after a proclamation or express crier, as if the first was a thief, the other his pursuers.[34]

Lovelace's parentheses themselves spring up among the hurry, hurrying until the whole paragraph is raised and running. He is Coleridgean in his love for the species. He considers himself an elegant writer, and his entire nature, if not his entire prose, is cast in obliquity, plots within plots, "wheels within wheels, sentence in the midst of sentences." He himself is like a line from Maurya Simon's poem "Parentheses: A Bestiary": "a scythe smiling sideways to itself in the mirror."[35]

Simon's poem proves in fact a lovely way to canvass the myriad roles of lunulae, which according to Parkes were employed in England "in the sixteenth and seventeenth centuries . . . much more freely than at any other time."[36] Besides their older appointments separating out *lemmata*, encircling invocations, isolating New Testament digressions, enclosing page numbers and vizzes, they also directed reading and meaning:

☞ "In reading or speaking, a parenthesis requires a moderate depression of the voice, and a quicker pronunciation, with a pause something longer than a comma."[37]

☞ "When the intervening clause is a deviation from the general tenor of the sentence, and particularly when it consists of many terms, it may be enclosed as a parenthesis."[38]

☞ "When the incidental sentence is not only introduced obliquely, but happens to be interrogatory or exclamatory, and, in that respect, different from the context, the parenthetical marks seem to be necessary, and to have a considerable use in directing the eye and the voice of the unexperienced reader."[39]

☞ "The parenthetical marks are sometimes necessary to prevent CONFUSION, or an AMBIGUITY in the construction of a sentence."[40]

Or to recast the functions within Simon's metaphors, modulating the voice ("a mini-corral for neighing vowels" [Simon l. 14]); extrapolating terms ("a way of drawing shadows into light" [Simon 1.6]); inserting obliquity ("a scythe smiling sideways" [Simon l. 2]); directing the eye ("the sentence keeping abreast of antimatter" [Simon 1.17]); and "necessary to prevent CONFUSION, or an AMBIGUITY"[41]—"twin hemispheres of gladness" (Simon l. 9)!

As Lennard elegantly puts it, "[Lunulae can function] not only epistemologically, as cues for the reader, but also ontologically, altering the status of the words which they enclose."[42] Simon's poem is only one (though perhaps the most beautiful) of the twentieth and twenty-first centuries' own metaphorical interpretations of the parenthesis. Think of Virginia Woolf's *Mrs. Dalloway*, prosodically (and characters, psychologically) destabilized by its *champagne* of parenthetical bubbles: "(She was like a poplar, she was like a river, she was like a hyacinth, Willie Titcomb was thinking. Oh how much nicer to be in the country and do what she liked! She could hear her poor dog howling, Elizabeth was certain.) She was not a bit like Clarissa, Peter Walsh said."[43]

A wonderful... picture book? illustrated poem? performance art?... called *This Is Me, Full Stop.*,[44] has a section on "brackets":

(This is us, brackets.) ( )[45]

Some say brackets, some say parentheses. [{(<>)}]

We don't mind what you call us. (!)

For some reason you only see us / from the side.

We do love a bit of gossip. {"**∴{{,~,}}∴**,,}
{"**∴{{'~'}};∴**"}

(Between you and me... well, everything's between you and me.)

(

We need each other (like the day / needs the night).

)

Once we open something, we always close it.
() [] {} {} [] ()
() [] {} {} [] ()
() [] {} {} [] ()
() [] {} {} [] ()
() [] {} {} [] ()

Where Simon's poem joyfully renders the metaphors of parentheses, "(This is us, brackets)" playfully *performs* their more ordinary functions.

And, of course, Stephen King has long plumbed the parenthetical depths of what and how we really think, as in this story of a little boy shamed by his teacher in front of his classmates because he's used the euphemism "go to the basement":

He walked down toward the boys' bathroom
(basement basement basement IF I WANT)
dragging his fingers along the cool tile of the wall, letting them bounce over the thumbtack-stippled bulletin board and slide lightly across the red
(BREAK GLASS IN CASE OF EMERGENCY)
fire-alarm box.[46]

For humiliated little Charles, parentheses enclose desire, speak truth, reject euphemism, record the world. (And Miss Bird gets eaten by a tiger in the "basement.")

Parentheses have led a dual life . . . all their lives. They are an eternal pair who contrive both to be invisible and to render their little packaged contents invisible, and yet it is *both* [( ] their agenda *and* [ )] their performance to open up the syntactical, psychological, epistemological, even, as Lennard says, *ontological* spaces on the page and in the mind.

Defoe and the Islands of Lunulae

THE YEAR 2019 was the 300th anniversary of the publication of Daniel Defoe's protonovel *Robinson Crusoe*; it was also the 230th anniversary of the mutiny on the *Bounty* and of Lieutenant William Bligh's remarkable four-thousand-mile journey from Tahiti to Java in a twenty-three-foot open launch, with eighteen men. Both Crusoe (through Defoe) and Bligh know how to tell a story of large things precipitated by small; of one parenthetical shoe waiting for the other to drop; of a figure lonely, isolated, outcast, marooned.[47] Part of the power of their stories is their textual craft, and part of their textual craft includes wielding the minutiae of parentheses. Defoe uses lunulae sparely but effectively; Bligh uses them even more rarely but every bit as effectively. The rhetorical parentheticals of both indicate stories submerged, stories poised to surface. These two maritime examples of men in isolation—one marooned on an island, the other on a boat, both parenthesized from the rest of their world—reveal some of the ways minute marks of punctuation can perform extraordinary tasks. As with punctuation more generally, each typographical mark, each piece of type, each rhetorical gesture simultaneously unites *and* divides, carving out and enacting psychological as well as geological islands of isolation *or* community. The Latin extension of the term "punctuation" indicates a point in space as well as time; parentheses can split the world apart and bring it back together.[48]

Defoe generally is quite nimble with parentheses. His Robinson Crusoe is quite fond of them: of his sixty-eight pairs, forty-four are *substantive* (versus "accidental"), to play on bibliographers' terms, overpowering the twenty-four vizzes, item numbers, and *I-says*.[49] They most often act as little rhetorical echo chambers to Crusoe's islandic isolations—literally and epistemologically. Examples begin in the "Editor's" Preface: "The Wonders of this Man's Life exceed all that (he thinks) is to be found extant; the Life of one Man being scarce capable of a greater Variety."[50] The sentence moves perfectly without the parenthetical, yet visually, "(he thinks)" behaves emblematically, serving as the kind of constant niggle inside many of our human assertions. (Was I right? Did I mean that? What did she think I meant? Oh god, is that how you pronounce it????) When Crusoe and his fellow-slave-cum-escapee Xury are pursued by the weird land-sea "Creatures," the parentheses widen the terror by underscoring the unknown: "(we knew not what to call them)"; "(whatever it was)" (*RC* 27). When Crusoe first builds his cave, at the moment, he says, "I began now to think my Cave or Vault finished," he discovers that "on a Sudden, (it seems I had made it too large) a great Quantity of Earth fell down from the Top and one Side, so much that in short it frighted me, and not without Reason too; for if I had been under it I had never wanted a Grave-Digger" (*RC* 86). The descriptive sentence is hollowed out by the hollowing out of the cave itself. (Note the typical Defovian wit at the end, which, like a number of his sharp little pointy bits, *could* have been lunulized but instead remains aboveground.)

One of the effects of looking closely at something tiny for a long time is that it suddenly seems *every* environment is a potential milieu allowing this minute particular to act as a little lever unearthing larger similitudes of itself. I will not actually go through all forty-four substantive parentheticals in *Robinson Crusoe*, but I will assert that for most if not all of them, an argument can be made that the texture of the parenthesis conforms to the form of the thought. (For example, Crusoe's weapons are either enclosed or reduplicated in parentheses; he also uses parentheses to distinguish between speaking aloud and speaking to himself.) For me, Maurya Simon's metaphorical parentheticals vividly recall and then recast scenes in *Robinson Crusoe*. In "Two hands upholding an invisible melon" (l. 4), I see Crusoe and his beloved limes: "Tandem heels of bread" (l. 5) . . . "(Cakes I should rather call them)" (*RC* 116). After his shipwreck, Crusoe holds up "two arms encircling longing" (l. 11): "I walk'd about on the Shore, lifting up my Hands" (*RC* 53), and again, after witnessing another shipwreck, he cries out, "That but one Man had been sav'd! *O that it had been but One!* I believe I repeated the Words, *O that it had been but One* [a close paren to the first?] a thousand Times; and the Desires were so mov'd by it, that when I spoke the Words, my Hands would clinch together" (*RC* 222). Crusoe essentially lives in "a still cove nestled among archipelagos" (l. 20); he ventures out from "the quays at either end of half-moon bay" (l. 38) and finds "a moat for transcendence" (l. 22); he sends "cryptograms to God" (l. 24); he discovers in himself, in his solitude, in his conversion, in his survival, "the other (another?) side of logic" (l. 32).

I have written elsewhere about how Defoe's narratives shape themselves around his plots, arguing against nineteenth- and early twentieth-century assumptions that the most masterful narratives are linear, that Defoe was constitutionally unable to tell a straight story, and that he as well as his characters digressed and repeated themselves in the *same* ways over and over. A closer and more historicized reading of Defoe's works suggests on the contrary a distinctive formal artistry in those digressions, asides, and parentheticals.[51] Just as the texture of Crusoe's parentheticals conforms to his thoughts, so in Defoe's plots, form follows function. The narrative digressions in *A Journal of the Plague Year* follow the ebb and flow of the plague as well as the venturing in and out of the narrator H. F. Moll Flanders's obsessive street indexing of her criminal escapes charts a route around Newgate Prison. Roxana's narrative pattern is archaeological: what's really going on is textually buried under what she chooses to tell us is going on.[52] I am thus prepared to argue to the death that the plot of *Robinson Crusoe* has the shape of lunulae. Let me put it this way: As Jonathan Lamb has observed, early mariners had very little to rely on: "They had no base, no longitude, no system of cognitive mapping, and no loyal home audience. The islands they made for were themselves mobile. The Portuguese in the Pacific called the coordinates of their dead reckoning the *punto de fantasia*."[53] Crusoe's markers of what Coleridge called the "drama of reason" are the coordinates of *un milieu du point de fantaisie*.

Crusoe's very dwelling is an architectural *milieu des parenthèses*:[54]

> Before I set up my Tent, I drew a half Circle before the hollow Place, which took in about Ten Yards in its semi-diameter from the Rock, and Twenty Yards in its Diameter, from its Beginning and Ending.
>
> In this half Circle I pitch'd two Rows of strong Stakes ... the two Rows did not stand above six Inches from one another. (RC 68)

The half circle drawn is repeated by two more half-circles of stakes, and the description itself is repeated in Crusoe's journal—where, as Crusoe confesses in parentheses, "(tho' in it will be told all these Particulars over again)" (RC 81):

> Jan. 3. I began my Fence or Wall; which, being still jealous of my being attack'd by some Body, I resolv'd to make very thick and strong.
>
> N. B. *This Wall being describ'd before, I purposely omit what was said in the Journal;* ... *perfecting this Wall, tho' it was no more than about* 24 *Yards in*

Length, being a half Circle from one Place in the Rock to another Place about eight Yards from it, the Door of the Cave being in the Center behind it. (RC 88)

During the rainy season, Crusoe works away on enlarging his cave, expanding the close paren: "And by Degrees work'd it on towards one Side, till I came to the Outside of the Hill, and made a Door or Way out, which came beyond my Fence or wall, and so I came in and out this Way" (RC 121). Later on he repeats the spiky semicircles of protection (*which could be another line from Simon's poem*) in his country bower, which he then repeats yet again back home. The hedge around his bower is actually a full circle, "firm and entire," and he's delighted to find that the stakes have shot out willowlike branches and enclosed the whole: "This made me resolve to cut some more Stakes, and make me a Hedge like this in a Semicircle round my Wall; I mean that of my first Dwelling, which I did; and placing the Trees or Stakes in a double Row, at about eight Yards distance from my first Fence, they grew presently, and were at first a fine Cover to my Habitation, and afterward serv'd for a Defence also, as I shall observe in its Order" (RC 123). The wall's semicircles have almost as many possible functions as their textual lunulae.

There is of course the canoe (something between a bracket and a lunula?—or a "smoky [hoop] halved by fire" [Simon l. 26]), but what is really the embedded lunulae of plot is Crusoe's journal, the main and peculiar source and cause of his digressions. Crusoe's journal is a textual parenthesis, the insertion of which interrupts (and sometimes confuses the boundaries of) the main narrative, the omission of which would not be noticed by any first reader. We can tell the journal because of the italicized dates. Except when we can't—when we suddenly realize that the journal narrative has been spilled on by the later Crusoe's recollections, and we often don't realize that until we suddenly hit the close paren: "But to return to my journal" (RC 92); "This was the 21st" (RC 96); "But I return to my Journal" (RC 106); "But leaving this Part, I return to my Journal" (RC 113). (*Unlike* the typeset lunulae, Crusoe doesn't always supply both parts, although he does occasionally give us the opening paren: "No one, that shall ever read this Account" [RC 102]; "This was the first Time I could say" [RC 113].)

Typographically, Crusoe carves out paragraphical islands—a visual interruption of the topographical page. He decides to "[draw] up the State of [his] Affairs in Writing, not so much to leave them to any that were to come after me, for I was like to have but few Heirs" (RC 76)—note the *non*parenthetical witty aside—and in order to comfort himself, he sets out "very impartially, like Debtor and Creditor, the Comforts [he] enjoy'd, against the Miseries [he] suffer'd" (fig. 4). On the left-hand column are the isolations rendered

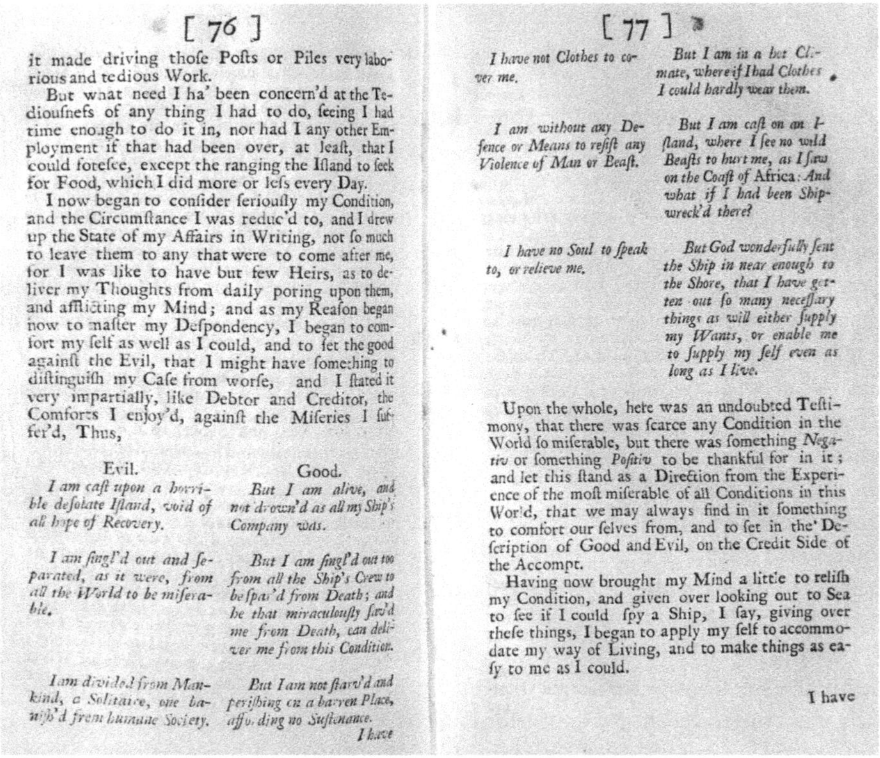
[76]

it made driving thoſe Poſts or Piles very laborious and tedious Work.

But what need I ha' been concern'd at the Tedioufneſs of any thing I had to do, ſeeing I had time enough to do it in, nor had I any other Employment if that had been over, at leaſt, that I could foreſee, except the ranging the Iſland to ſeek for Food, which I did more or leſs every Day.

I now began to conſider ſeriouſly my Condition, and the Circumſtance I was reduc'd to, and I drew up the State of my Affairs in Writing, not ſo much to leave them to any that were to come after me, for I was like to have but few Heirs, as to deliver my Thoughts from daily poring upon them, and afflicting my Mind; and as my Reaſon began now to maſter my Deſpondency, I began to comfort my ſelf as well as I could, and to ſet the good againſt the Evil, that I might have ſomething to diſtinguiſh my Caſe from worſe, and I ſtated it very impartially, like Debtor and Creditor, the Comforts I enjoy'd, againſt the Miſeries I ſuffer'd, Thus,

Evil.	Good.
I am caſt upon a horrible deſolate Iſland, void of all hope of Recovery.	*But I am alive, and not drown'd as all my Ship's Company was.*
I am ſingl'd out and ſeparated, as it were, from all the World to be miſerable.	*But I am ſingl'd out too from all the Ship's Crew to be ſpar'd from Death; and he that miraculouſly ſav'd me from Death, can deliver me from this Condition.*
I am divided from Mankind, a Solitaire, one baniſh'd from humane Society.	*But I am not ſtarv'd and periſhing on a barren Place, affording no Suſtenance.*

I have

[77]

I have not Clothes to cover me.	*But I am in a hot Climate, where if I had Clothes I could hardly wear them.*
I am without any Defence or Means to reſiſt any Violence of Man or Beaſt.	*But I am caſt on an Iſland, where I ſee no wild Beaſts to hurt me, as I ſaw on the Coaſt of* Africa: *And what if I had been Shipwreck'd there?*
I have no Soul to ſpeak to, or relieve me.	*But God wonderfully ſent the Ship in near enough to the Shore, that I have gotten out ſo many neceſſary things as will either ſupply my Wants, or enable me to ſupply my ſelf even as long as I live.*

Upon the whole, here was an undoubted Teſtimony, that there was ſcarce any Condition in the World ſo miſerable, but there was ſomething *Negativ* or ſomething *Poſitiv* to be thankful for in it; and let this ſtand as a Direction from the Experience of the moſt miſerable of all Conditions in this World, that we may always find in it ſomething to comfort our ſelves from, and to ſet in the Deſcription of Good and Evil, on the Credit Side of the Accompt.

Having now brought my Mind a little to reliſh my Condition, and given over looking out to Sea to ſee if I could ſpy a Ship, I ſay, giving over theſe things, I began to apply my ſelf to accommodate my way of Living, and to make things as eaſy to me as I could.

I have

FIG. 4. Crusoe's paragraphical islands. Daniel Defoe, *The Life and Strange Surprizing Adventures of Robinson Crusoe, of York, Mariner*, 2nd ed. (London: Printed for W. Taylor, 1719). General Collection, Beinecke Rare Book and Manuscript Library, Yale University.

by the "horrible desolate Island": "singled out and separated"; "divided from Mankind, a Solitaire"; with no clothes, no weapons, and "no Soul to speak to" (*RC* 76–77). But from each small textual island of isolation, he can instantly cross to an equal or larger milieu of comfort: not drowned like everybody else, isolated to be *spared* death, not starved, in no need of clothes or weapons, and with *so much stuff*—eventually a table, chairs, shelves, baskets, kiln, pots, raisins, limes, melons, bread, milk, butter, *chevre* and *chevon*, an impressive costume, and a functioning umbrella. Crusoe's answers to his own complaints textually refashion the islands of isolation into islands of refuge: the open paren of "EVIL" is emphatically closed by "GOOD."

William Bligh and Navigating Isolation

SEVENTY YEARS after the publication of *Robinson Crusoe*, on April 29, 1789, the HMS *Bounty's* master's mate Fletcher Christian and "in all 25 hands, . . . the most able men of the ship's company" mutinied and forced Lieutenant William Bligh and eighteen of his men into an open launch 23 feet long, 6'9" wide, and 2'9" deep.[55] In this boat they traveled over twelve hundred leagues (some four thousand miles) in forty-one days from Tofoa, one of the Friendly Islands (now Tofua, in Tonga), to the Dutch settlement in Coupang, Java. Bligh's narrative, taken almost verbatim from the log he rigorously (and neatly!) kept on the voyage and published on his return to London in 1790, is as compelling as a novel. Of course, it is utterly unlike a novel, though it has points of narrative intersection with *Robinson Crusoe*—not surprisingly, given Defoe's familiarity with seamen's journals. Defoe complained in the *New Voyage Round the World* (1724) that such journals were nothing more than "long . . . tedious Accounts of their Long [sic]-work, how many Leagues they sail'd every Day; where they had Winds, when it blew hard, and when softly; what Latitude in every Observation, what Meridian Distance, and what Variation of the Compass."[56] As Maximilian Novak puts it, "In a ship's journal events must be recorded even if nothing happens" (he then quotes a parody attributed to Defoe, noting, "It is doubtful that anyone ever made an entry like 'the Captain came upon Decke, looked up at the Vane, put his Right Hand in his Pocket, and said Nothing.'"[57]) But there *is* a way of saying nothing that can be very powerful. Bligh's log and narrative do track the leagues sailed, the winds' directions, the boat's coordinates, the food rations, but the austerity is punctuated with horror; the spareness has a grim rhythm; what is left out presses ominously against what is recorded.

In some ways, there is almost no "characterization" in the novelistic or even the memoirist sense. Bligh's favorite pronoun is the first-person singular but functions to superimpose him on his ship or the launch: "I SAILED from Otaheite on the 4th of April 1789"; "I had passed ten islands"; "I resumed my course"; "I had only four feet water" (*Narrative* 1, 26, 27; April 4, May 6, May 7). The eighteen men with him in the boat, whom he lists with their position at the beginning (Thomas Ledward, Acting Surgeon; David Nelson, Botanist; Robert Tinkler, A boy [*Narrative* 6]), remain narratively silent. Bligh rarely records any conversation, apart from Christian's famous muttered cry during the mutiny: "I am in hell—I am in hell" (*Narrative* 8). ("The (visible) witness who speaks sotto voce" [Simon, l. 7].) None of the men emerges as an individual—which makes it a bit more striking and almost novelistic when, in the log, Bligh gives physical "Description[s] of the Pirates"[58] that have a novelistic savagery:

> Age   F In
> Fletcher Christian 24   Masters Mate 5..9. High. Blackish or very dark brown Complexion Dark brown Hair. Strong made. a Star tatowed on his left breast and tatowed on the backside. His knees stand a little out and he may be called a little bow legged. He is subject to violent perspiration, particularly in his hand so that he soils anything he handles. (*Log* 359)

Fletcher Christian, of course, becomes the fallen star of the narrative (although the narrative of the launch quickly leaves that disaster in its wake); Christian's betrayal of Bligh's early affection and mentoring infects the post hoc description—soiling everything in hindsight.

Bligh's narrative measures space: "At noon I considered my course and distance from Tofoa to be WNW¾. 86 miles, my latitude 19° 28′ S. I directed my course to the WNW, that I might get a sight of the islands called Feejee" (*Narrative* 24; May 3). It measures time: "I was fully determined to make what provisions I had last eight weeks, let the daily proportion be ever so small" (*Narrative* 24; May 3). It measures food: "I now served a tea-spoonful of rum to each person, (for we were very wet and cold) with a quarter of a bread-fruit, which was scarce eatable, for dinner" (*Narrative* 24; May 3). Calculating his limited provisions, Bligh allows each crew member "one 25th of a pound of bread, and a quarter pint of water, at sun-set, eight in the morning, and at noon" (*Narrative* 31; May 10). Often, the measurements of space, time, and food occupy the same paragraph, the same narrative milieu, with nothing interrupting their interconnectedness: "[May 4] Just before noon we discovered a small flat island of a moderate height, bearing W S W, 4 or 5 leagues. I observed in latitude 18° 58′ S; our longitude, by account, 3° 4′ W from the island Tofoa, having made a N 72° W course, distance 95 miles, since yesterday noon. I divided five small cocoa-nuts for our dinner, and every one was satisfied" (*Narrative* 25). Geography, clocks, and coconuts combine. Everyone was satisfied. For now. The combination of endless measurements, endless repetitions of weather and water, and a narrative trapped in Bligh's metonymic isolation, would seem to earn Defoe's satiric criticism. And yet there is nothing tedious about Bligh's narrative or his log. The horror grows with the details and without any details. It is compounded by measurement as well as explanation. Syntax changes with circumstances.

Part of the drama lies, in fact, in the bare, ceaseless repetition. On rations: "I issued the 25th of a pound of bread, and a quarter of a pint of water, as yesterday" (*Narrative* 32; May 11); "The twenty-fifth of a pound of bread, and water I served as usual" (*Narrative* 33; May 13); "Served the usual allowance of bread and water" (*Narrative* 33; May 14). On actions: "The sea ran higher than yesterday, and the fatigue of baling, to keep the boat from filling, was exceedingly great" (*Narrative* 24; May 3); "We continued constantly shipping seas, and baling, and were very wet and cold in the night" (*Narrative* 33; May 13); "Constantly shipping water, and very wet, suffering much cold and shiverings in the night" (*Narrative* 33; May 14); "Two people constantly employed baling" (*Narrative* 33; May 15); "Always baling" (*Narrative* 37; May 20). The repetition is all day, every day. And that repetition is rhythm, a ceaseless beat hammering home the weariness. It has a cumulative effect:

> Tuesday, May the 12th. Strong gales at S E, with much rain and dark dismal weather, moderating towards noon, and wind varying to the N E.
> Having again experienced a dreadful night, the day showed to me a poor miserable set of beings full of wants, without any thing to relieve them. Some complained of a great pain in their bowels, and all of having but very little use of their limbs. What sleep we got was scarce refreshing, we being covered with sea and rain. Two persons were obliged to be always baling the water out of the boat. I served a spoonful of rum of day-dawn, and the usual allowance of bread and water, for supper, breakfast, and dinner. (*Narrative* 32)

Date; weather; cold, starvation, scurvy, pain; a spoonful of rum and the usual allowance, "for supper, breakfast, and dinner."

The deadly rhythm of repetition thus dramatizes any details of *change*:

> [Tuesday, May 26] In the evening we saw several boobies flying so near to us, that we caught one of them by hand. This bird is as large as a good duck; like the noddy, it has received its name from seamen, for suffering itself to be caught on the masts and yards of ships. I directed the bird to be killed for supper, and the blood to be given to three of the people, who were the most distressed for want of food. The body, with the entrails, beak, and feet, I divided into 18 shares, and with an allowance of bread, which I made a merit of granting, we made a good supper, compared with our usual fare" (*Narrative* 41).[59]

It turns out that eating raw bird—even if you're allotted the beak or the feet (the lucky ones get the blood)—has become a celebratory event. The horror of the celebration (to the comfortable reader) deepens the horror of what it interrupts. The moonlit ray of cheer illuminates the pools of darkness outside its brief enclosure.

Bligh is rhetorically powerful on the horrors of darkness, literal and metaphorical. (His *Narrative* was, of course, a bestseller.) The mariners are always conscious of their isolation from land and people and safety and comfort: "We had now all but two or three things in the boat, when I took Nageete by the hand, and we walked down the beach, every one in a silent kind of horror" (*Narrative* 20; May 3); "The sight of these islands served but to increase the misery of our situation" (*Narrative* 34; May 15). At its worst, isolation penetrates into the boat itself: "In the night we had very severe lightning, but otherwise it was so dark that we could not see each other" (*Narrative* 37; May 19); "The misery we suffered this day exceeded the preceding. The night was dreadful. The sea flew over us with great force, and kept us baling with horror and anxiety" (*Narrative* 28; May 23). Daylight would discover its own darkness: "At dawn of day, some of my people seemed half dead: our appearances were horrible; and I could look no way, but I caught the eye of some one in distress" (*Narrative* 37; May 20); "At dawn of day I found every one in a most distressed condition, and I now began to fear that another such a night would put an end to the lives of several. . . . Every one complained of severe pains in their bones" (*Narrative* 39; May 23).

The log entries compress the horror: "Cold and Wet"; "Very Wet and Cold"; "Very Wet and Cold" (*Log* n.p.; June 7–10). Pronouns disappear; sentences hunch down: "Thursday, May the 14th. Fresh breezes and cloudy weather, wind southerly. Constantly shipping water, and very wet, suffering much cold and shiverings in the night" (*Narrative* 33; May 14); "Wednesday, May the 20th. fresh breezes E N E with constant rain; at times a deluge. Always baling" (*Narrative* 37). I agree with Jonathan Lamb, who takes issue with the kind of reading of ships' logs that characterizes them as inscribing "the imperatives of eighteenth-century discovery": "'shoot! classify! name! describe!'" (Lamb 9).[60] "So much is taken for granted in these generalizations," he points out. "It is as if no alarms or mistakes disturbed the collection of specimens, and no terrors smudged the cartographic grid. How easy it is to read the log of George Anson's *Centurion*, perhaps the best known of the British navigations before Cook, and to assume that the laconic references to tempests, freezing rain, and broken equipment, and the daily list of names terminating in D.D. (discharged dead), are heroic reductions of dangers to the level of inconveniences, an impassive assertion of the majesty of a higher purpose" (Lamb 10). The compressed brevity of Bligh's account earns its power through the horror of repetition, the apparent sidelining into an account column, the understatement of what is. It has its textual and experiential corollary in Crusoe's journal, where labor and pain are also compressed: the compression, caught as it were (parenthetically) between narratives, gives us a tight sense of its boundaries, of what is pulsing against abbreviation: "June 24. Much better" (*RC* 101). (????)

And that brings us back to the beginning: Bligh's own lunulae. There are only five, but they appear in the published narrative identical to those in his manuscript log. The first appears early, when he is bound by the mutineers: "Isaac Martin, one of the guard over me, I saw, had an inclination to assist me, and, as he fed me with shaddock, (my lips being quite parched with my endeavours to bring about a change) we explained our wishes to each other by our looks" (*Narrative* 4; April 1789). Bligh almost never mentions his own weaknesses or deprivations during the rest of the narrative, often remarking how he doesn't feel the pain or hunger (although recounting—for the future court-martial?—that one of his crew hinted he looked the worst of them all); the parenthetical underscores his ongoing efforts and commands our sympathy. The second is during the same period, and it makes sure the reader knows the *true* hero of the argument: "Some swore 'I'll be damned if he does not find his way home, if he gets any thing with him,' (meaning me); others, when the carpenter's chest was carrying away, 'Damn my eyes, he will have a vessel built in a month'" (*Narrative* 5; April 1789). The third is a translation, as Bligh is strategizing to escape an attack from the inhabitants of Tofoa: "Maccaackavow [a chief] then got up, and said, 'You will not sleep on shore? then Mattie,' (which directly signifies we will kill you) and he left me" (*Narrative* 19–20; May 3). The chief speaks directly; Bligh understands at least some of the language; in this case, the meaning of the sentence *would* in fact be materially altered for most readers without the translation in parentheses. The parenthesis *is* the shock. The fourth occurs toward the beginning of their real sufferings, on May 3: "I now served a tea-spoonful of rum to each person (for we were very wet and cold) with a quarter of a breadfruit, which was scarce eatable, for dinner" (*Narrative* 24). Not much later, there would be no need for the explanation of the extra; later, "Very Wet and Cold" stays out in the syntactical open. In all the rest of the narrative—another sixty large quarto pages—*everything* is on the surface, nothing is lunulaed—*except* for one brief moment, when they land on an island full of blackberries and plums and grapes and beans and ferns and hearts of palm and oysters *and no more raw bird bits*: "This being the day of the restoration of king Charles the Second, and the name not being inapplicable to our present situation (for we were restored to fresh life and strength), I named this Restoration Island" (*Narrative* 51; May 29). It is almost as if Bligh wants to swallow the parallel to digest it in satisfaction.

As with *Robinson Crusoe*, one can start seeing little moons everywhere. Let's call this a touch of magical realism in literary criticism: on the way toward Restoration Island, Bligh charts a course between reefs described and drawn almost like semicircles. "On the outside, the reef inclined to the N E for a few miles, and from thence to the N W; on the south side of the entrance, it inclined to the S S W as far as I could see it" (*Narrative* 44; May 28). He then describes "being now happily within the reefs, and in smooth water" (*Narrative* 45). The chart he draws of the northeast coast of New Holland (i.e., Australia) rather nicely features a series of little moons that are replicated where the ink has offset onto the facing page, creating a literal topography of the page, a milieu of shadow text (see fig. 5).

The arrival of the *Bounty* launch to the Dutch settlement in Java is the restoration to the island of community, and once there, Bligh preserves the small community forged in his small boat; the men, islanded together for forty-one days, threatened or isolated or restored by islands along the way, are restored by the island in Indonesia. There is a very moving communal moment when Bligh is shown to the house reserved for him by the governor of Coupang; Bligh decides to override social hierarchy and instead keep his little boat community together (albeit reinscribing a little architectural hierarchy):

> The house consisted of a hall, with a room at each end [*architectural parentheses!*], and a loft over head; and was surrounded by a piazza, with an outer apartment in one corner, and a communication from the back part of the house to the street. I therefore determined, instead of separating from my people, to lodge them all with me; and I divided the house as follows: One Room I took to myself, the other I allotted to the master, surgeon, Mr. Nelson, and the gunner; the loft to the other officers; and the outer apartment to the men. The hall was common to the officers; and the men had the back piazza. Of this I informed the governor, and he sent down chairs, tables, and benches, with bedding and other necessaries for the use of every one. (*Narrative* 81; June 14)

Although Bligh preserves naval and social hierarchy in the architectural arrangements at Coupang, it is worth remarking on his decision "to lodge them all with me." I began this section by considering Bligh's narrative in terms of his penchant for the first-person singular, which is, after all, the prerogative of a ship's master. Bligh appeared completely flabbergasted by the mutiny—though later documents (the proceedings of the court-martial, Edward Christian's investigations, the Nordhoff and Hall trilogy, and of course Marlon Brando and Trevor Howard, Mel Gibson and Anthony

Hopkins) give a rather fuller context. But I expect it was probably the same qualities in Bligh that sparked the mutiny that then got the boat across those four thousand miles: the single-mindedness, the *bloody*-mindedness, the iron concentration, the iron will. Yet the pressure of those events—the shock of the mutiny; the grim limits of supplies; the dark, cold, and wet; the isolation on the seas; and the intensified necessity of community ("all baling")— emerge, almost parenthetically, in his prose. On the same day that "I" directed "my" course, "I" discovered two islands, and "I" passed ten more, we see "*Our* supper, breakfast, and dinner consisted of a quarter of a pint of cocoa-nut milk" and "To *our* great joy we hooked a fish" (*Narrative* 26–27; May 6 [my emphases]); "our health," "our cloathes" (*Narrative* 84; June 14). Much like Crusoe, Bligh the person seems fundamentally isolated, inherently alone, but also like Crusoe, he ends up with a community (albeit one each shapes and rules). And also like Crusoe, he knows how to shape the textual realities of islands and isolations. Each has a compelling narrative, the parentheticals of which disrupt, upheave, undermine, redirect, clarify—and restore.

In all the multitudes of milieus of minutiae, the lowly parenthesis has acquired powers beyond its early wildest dreams. Where critics and compositors were often annoyed, writers quickly and creatively exploited its possibilities. If the "politest of our modern Writers avoid all *Parentheses*," as Chambers claimed, the greatest of modern writers embrace them (judiciously). Wheels within wheels, perhaps, but "wheels within wheels" has come to connote "hidden or unknown things that influence a particular situation, making it more complicated than it at first seems";[61] any machinery worth its salt (say, John Glenn's Atlas rocket) multiplies its wheels within wheels. Shakespeare, Marlowe, L'Estrange have parentheses visually, interpretively destabilizing their syntactical hosts. *The Spectator* found rich possibilities for inserted asides. Capability Brown saw the parenthesis as a metaphor for "(where an interruption is desirable to break the view)." Richardson's Lovelace employs parentheses to champion little words. Stephen King has parentheses vivisect the subconscious. Maurya Simon's poem finds thirty-nine metaphors of function. While Mr. Stops's parentheses do look boneless, they are, in fact, the legs balletically supporting the body ("A perfect plié" [Simon l. 1]). The minute parenthesis is an Archimedean lever: with the right syntactical fulcrum, it can move worlds.

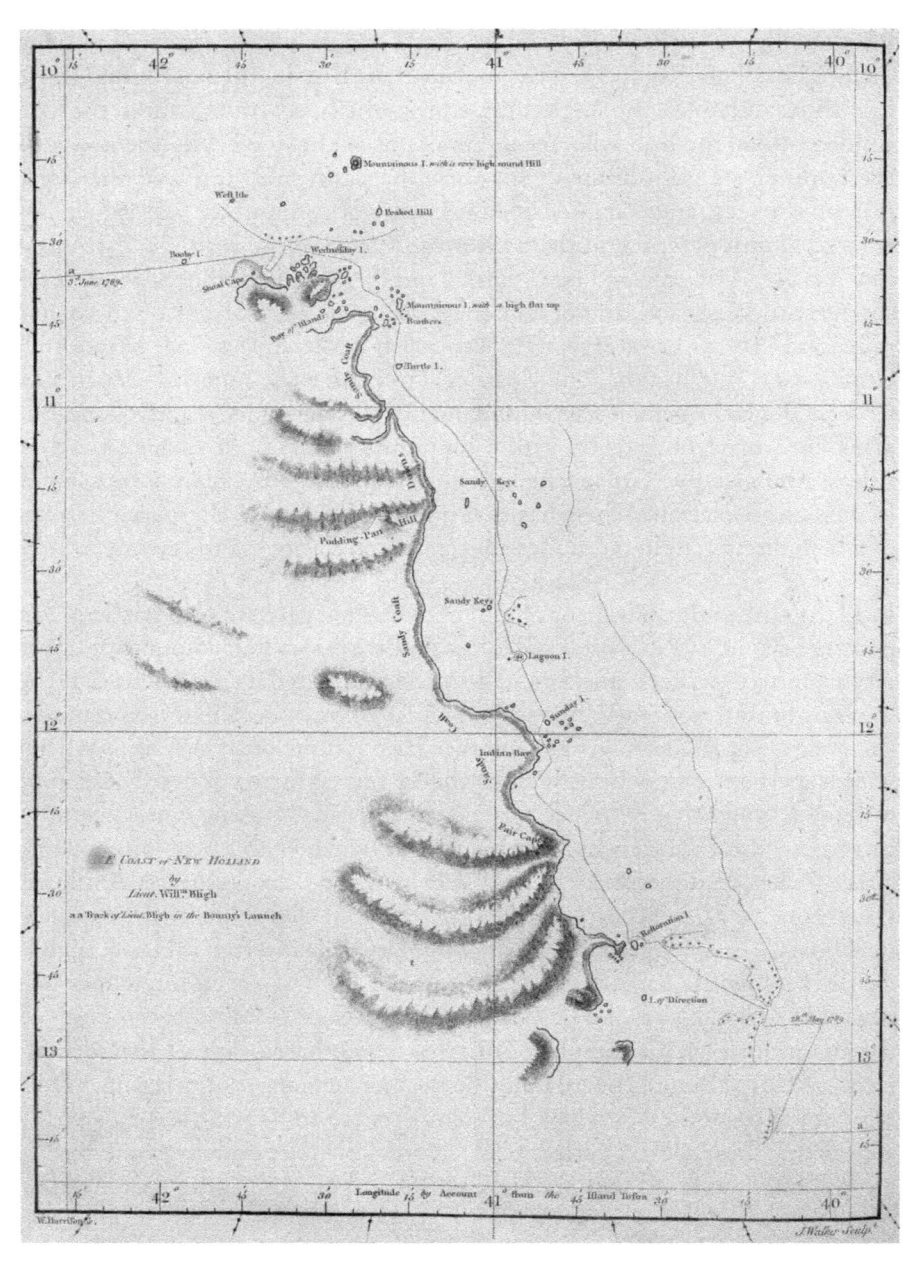

FIG. 5. Coast of New Holland and its inked closed paren. William Bligh, *A Narrative of the Mutiny, on Board His Majesty's Ship Bounty* (London: Printed for George Nichol, 1790). Reproduced by permission of the British Library (shelfmark 212.d.17).

[45]

lies in, but I confidered it to be within a few miles of this, which is fituate in 12° 51′ S latitude.

Being now happily within the reefs, and in fmooth water, I endeavoured to keep near them to try for fifh; but the tide fet us to the N W; I therefore bore away in that direction, and, having promifed to land on the firft convenient fpot we could find, all our paft hardfhips feemed already to be forgotten.

At noon I had a good obfervation, by which our latitude was 12° 46 S, whence the foregoing fituations may be confidered as determined with fome exactnefs. The ifland firft feen bore W S W five leagues. This, which I have called the ifland Direction, will in fair weather always fhew the channel, from which it bears due W, and may be feen as foon as the reefs, from a fhip's maft-head: it lies in the latitude of 12° 51′ S. Thefe, however, are marks too fmall for a fhip to hit, unlefs it can hereafter be afcertained that paffages through the reef are numerous along the coaft, which I am inclined to think they are, and then there would be little rifk if the wind was not directly on the fhore.

My longitude, made by dead reckoning, from the ifland Tofoa to our paffage through the reef, is 40° 10′ W. Providential channel, I imagine, muft lie very nearly under the fame meridian with our paffage; by which it appears we had out-run our reckoning 1° 9′.

We now returned God thanks for his gracious protection, and with much content took our miferable allowance of a 25th of a pound of bread, and a quarter of a pint of water, for dinner.

Friday, May the 29th. Moderate breezes and fine weather, wind E S E.

As we advanced within the reefs, the coaft began to
fhew

1789.
M A Y 28.

Friday
29.

Notes

1. *Oxford English Dictionary*, 2nd ed. (Oxford: Oxford University Press, 2004), s.v. *minutia*. Hereafter cited parenthetically in the text.
2. "That Famed John Glenn Quote about a Rocket's Millions of Lowest-Bidder Parts," MarketWatch, December 12, 2016, https://www.marketwatch.com/story/that-famed-john-glenn-quote-about-a-rockets-2-million-lowest-bidder-parts-2016-12-11.
3. Hannah More, *Memoirs of the Life and Correspondence of Mrs. Hannah More: By William Roberts, Esq.*, 3rd ed. (London: R. B. Seeley and W. Burnside, 1825), 1:267.
4. For an extended analysis of the concept of the topography of the page, see Cynthia Wall, *Grammars of Approach: Landscape, Narrative, and the Linguistic Picturesque* (Chicago: University of Chicago Press, 2019), particularly chap. 3. Some of the introduction to this essay paraphrases that recent book.
5. Virginia Woolf, *A Room of One's Own* (London: Hogarth, 1929), 115.
6. Edward Cocker, *Cocker's English Dictionary: Interpreting the Most Refined and Difficult Words* (London: Printed for A. Back and A. Bettesworth, 1704).
7. Joseph Robertson, *An Essay on Punctuation* (London: Printed for J. Walter, 1785), 18–76. The title of this essay, "Of a Parenthesis," is taken from Robertson (115). Impressive as these fifty-eight pages might be, they don't compete with the ninety-some pages devoted to "glossing the word 'milieu'" mentioned in the subsequent essay by Christopher Johnson.
8. John Lennard, *But I Digress: The Exploitation of Parentheses in English Printed Verse* (Oxford: Clarendon Press, 1991), 1.
9. "1948 Words into Type 282 (note) The terms curves, brackets, and round brackets are never used in printing offices for parentheses. The term there used is parens, separately designated open paren and close paren" (*OED*).
10. John Lennard, "In/visible Punctuation," *visible language* 45, no. 1/2 (2011): 123.
11. Ephraim Chambers, *Cyclopædia; or, An Universal Dictionary of Arts and Sciences*, 2 vols. (London: Printed for James and John Knapton et al., 1728), 2:911.
12. John Smith, *The Printer's Grammar . . . With Directions to Authors, Compilers, &c. how to Prepare Copy, and to Correct their own Proofs: The Whole calculated for the Service of All who have any Concern in the Letter Press* (London: Printed for the Editor, 1755), 88. But, he sighs, "where Authors think otherwise" (than the printer, compositor, or corrector), "they ought not to be thwarted in their judgment, especially if they express it in their Copy."
13. Joseph Robertson, *An Essay on Punctuation*, 2nd ed. (London: Printed for J. Walter, 1786), preface (n.p.). He adds, "In this circumstance, books are no certain guides; for most of them are carelessly and irregularly pointed; and many pauses are necessary in reading, where no point is inserted by the printer."
14. Robertson, n.p.
15. M. B. Parkes, *Pause and Effect: An Introduction to the History of Punctuation in the West* (Aldershot, Hants: Scholar Press, 1992), 51.

16. From *OED*: "1592 A. Day 2nd Pt. Eng. Secretorie sig. N2v, in Eng. Secretorie (rev. ed.)."
17. Aldus himself said, in his *Interpungendi Ratio*, "Unum hoc tacere non possum, ineptè facere, qui hæc parenthesi includunt: ut puto, ut res indicat, ut à majoribus accepimus, quod equidem facilè intellexerim, et similia; quæ si semicirculo distinguantur, aut saltem puncto et semicirculo, satis erit" (Manutius, *Interpungendi Ratio*, 5; qtd. in Robertson, *Essay on Punctuation*, 119–20), or "One thing I can't keep quiet about is that it is inept to put in parentheses: 'as I think,' 'as the evidence indicates,' 'as most people agree,' 'with which indeed I would readily concur,' & the like; which if they were set off with a comma, or indeed with a period & a comma [i.e., Bembo's semicolon], that would be enough" (trans. Professor Gordon Braden, University of Virginia).
18. E. Chambers, *Cyclopædia; or, An Universal Dictionary of Arts and Sciences*, 2 vols. (London: Printed for James and John Knapton et al., 1728), 2:752. In 1767, James Buchanan noted, "A Parenthesis marked thus ( ); which is a Sentence inserted within another Sentence as an Illustration of the Sense; tho' it may be left out, and the Structure of the Sentence remain entire." James Buchanan, *A Regular English Syntax* (London: Printed for J. Wren, 1767), 189–90.
19. Chambers 2:752.
20. Hugh Blair, lecture 11, "The Structure of Sentences," in *Lectures on Rhetoric and Belles Lettres*, 2 vols. (London: Printed for W. Strahan, T. Cadell, W. Creech, 1783), 1:222.
21. Caleb Stower, *The Printer's Grammar* (London: B. Crosby, 1808), 87. At another point, he reinforces the matter: "Crotchets are so seldom made use of now, that they require little notice: both parentheses and crotchets were formerly used to inclose folios, &c.; but the modern method of putting folios in full-faced figures, unattended, leaves the crotchet scarce a duty to perform" (87).
22. Mr. Stops [John Harris], *Punctuation Personified; or, Pointing Made Easy* (London: J. Harris and Son, 1824).
23. Stops 10.
24. John Lennard points to the earliest reference to parentheses in an English book, Joannes Sulpitius Verulanus's *Opus Grammaticum* (1494), where Sulpitius says (in Latin, naturally), "Parenthesis is where a different utterance (so says Perottus) is introduced into an as yet incomplete utterance" (qtd. in Lennard 20). "Sulpitius interrupts himself with a parenthesis while defining the parenthesis as an interruption. Such self-instantiation is rarely evident in sixteenth-century uses, but became increasingly common thereafter" (Lennard 20). "Sulpitius's witty definition is important: his book went through many editions and influenced the Elizabethan schoolmen. Certainly both Mulcaster's and Puttenham's definitions derive from Sulpitius's (though both add to it); and the *Opus Grammaticum* stands at the head of a clear tradition of definition, but of the parenthesis, not lunulae" (Lennard 21).
25. Come thither; As she spake this, her toong tript,

> For vnawares (*Come hither*) from her slipt,
> And sodainly her former colour chang'd,
> And here and there her eies through anger rang'd. (*Hero and Leander* 1598, ll. 357–60)

> "To call Hero's 'mistake' a Freudian slip is psychologically fair to her, but does no justice to the skill with which Marlowe constructed the slip" (Lennard 50).

26. Thomas Keymer, "Novel Designs: Manipulating the Page in English Fiction, 1660–1780," in *New Directions in the History of the Novel*, ed. P. Parrinder et al. (London: Palgrave Macmillan, 2014), 29.
27. John Evelyn, *The Diary of John Evelyn*, ed. E. S. De Beer, 6 vols. (Oxford: Clarendon Press, 1955), 2:1, 13.
28. See particularly lecture 20, "Critical Examination of the Style of Mr. Addison, in N° 411. of the Spectator" (*Lectures* 2:67–68); and lecture 22, "Critical Examination of the Style of Mr. Addison, in N° 413. of the Spectator" (*Lectures* 2:119).
29. Joseph Addison and Richard Steele, *The Spectator*, ed. Donald Bond, 5 vols. (Oxford: Clarendon Press, 1965), 1:10 (no. 2, Friday, March 2, 1711).
30. Samuel Taylor Coleridge, *The Letters of Samuel Taylor Coleridge*, ed. Earl Leslie Griggs, 6 vols. (Oxford: Clarendon Press, 1956–71), 3:282 (no. 801). To Thomas Poole, January 1810.
31. Blair 1:281; Robertson 115.
32. Samuel Richardson, *Clarissa; or, The History of a Young Lady*, 7 vols. (London: Printed for S. Richardson, 1747–48), 4:206.
33. "The upright curves ( ) used to mark off a word or clause inserted parenthetically; round brackets. Originally as collective singular; now usually in plural (sometimes with singular form). in parentheses (also in (a) parenthesis): in brackets. Occasionally applied to square brackets [ ]" (*OED*).
34. Richardson 4:206–7.
35. Maurya Simon, "Parentheses: A Bestiary," in *A Brief History of Punctuation: Poems by Maurya Simon: Calligraphy by Cheryl Jacobsen* (Winona, MN: Sutton Hoo Press, 2002), l. 2. Hereafter cited parenthetically in the text.
36. Parkes 87.
37. Robertson 116.
38. Robertson 120.
39. Robertson 121.
40. Robertson 124.
41. Robertson 124.
42. Lennard 21.
43. Virginia Woolf, *Mrs. Dalloway*, foreword by Maureen Howard (1925; repr., San Diego: Harcourt, 1981), 188.
44. Philip Cowell and Caz Hildebrand, *This Is Me, Full Stop.* (n.p.: Particular Books / Penguin Random House, 2017).
45. These are my feeble approximations of the large, clever illustrations on the facing pages of the text.

46. Stephen King, "Here There Be Tygers," in *Skeleton Crew* (1968; repr., New York: G. P. Putman's Sons, 1985), 136.
47. An obsolete definition of *marooned*—though the example is from William Dampier (1699), one of the models for Crusoe—is "To be lost and separated from one's companions" (*OED* 1a).
48. Lennard points out that the Latin extension of the term *punctuation* indicates a point in time as well as a point in space ("In/visible" 123). All the early grammarians defined the different punctuation marks by the length of their pause in reading: the author of *Some Rules for Speaking and Action; To be observed At the Bar, in the Pulpit, and the Senate, and by every one that Speaks in Public* (1716) puts it succinctly: "A *Comma* stops the Voice while we may privately tell One, a *Semicolon* Two, a *Colon* Three, and a *Period* Four." These are what they called in "musical proportion" to each other; James Buchanan argued that the different marks of punctuation direct the "different Emphasis and tones of Voice" that in turn express "the different Passions and Emotions of the Soul" (Buchanan 189). There is, thus, poetry and music and passion in punctuation ("Parentheses are the *drama* of reason").
49. See W. W. Greg, "The Rationale of Copy-Text," *Studies in Bibliography* 3 (1950): 21: "substantives" include those matters of text "that affect the author's meaning or the essence of his expression"; "accidentals" are things like "spelling, punctuation, word-division, and the like, affecting mainly its formal presentation." I am arguing rather fiercely here that punctuation can affect meaning every bit as much as syntax or word choice. (And in *Grammars of Approach*, I take on spelling, capitalization, font, word division, "and the like," for good measure.)
50. Daniel Defoe, *The LIFE And STRANGE SURPRIZING ADVENTURES OF ROBINSON CRUSOE, Of YORK, MARINER: Who lived eight and twenty Years all alone in an un-inhabited Island on the Coast of AMERICA, near the Mouth of the Great River of Oroonoque; Having been cast on Shore by Shipwreck, wherein all the Men perished but himself. With an ACCOUNT how he was at last as strangely deliver'd by PYRATES. Written by Himself*, 2nd ed. (London: Printed for W. Taylor, 1719), "preface," recto. Hereafter cited parenthetically as *RC*.
51. For the interested few, there is a sustained analysis of Roxana's use of the summative prepositional phrase "In a Word" in *Grammars of Approach*, chap. 5, 183–85.
52. See, for example, my introduction to the Penguin Classics edition of *A Journal of the Plague Year* (New York: Penguin, 2003) and *The Literary and Cultural Spaces of Restoration London* (Cambridge: Cambridge University Press, 1998), chap. 4, 141–47.
53. Jonathan Lamb, *Preserving the Self in the South Seas 1680–1840* (Chicago: University of Chicago Press, 2001), 165; quoting Martin de Munilla, *La Austrialia del Espíritu Santo: The Journal of Fray Martin de Munilla, O.F.M., and Other Documents Relating to the Voyage of Pedro Fernández de Quirós to the South Sea (1605–1606) and the Franciscan Missionary Plan (1617–1627)*, trans. and ed. Celsus Kelly, intro. G. S. Parsonson, 2 vols. (Cambridge: Hakluyt Society, 1966), 1:51. Hereafter cited parenthetically in the text.

54. The *OED* offers an architectural example: "1592 T. Nashe Pierce Penilesse (Brit. Libr. copy) sig. C2 The Kitchin ... was no bigger than the Cooks roome in a ship, with a little Court chimney, about the compasse of a Parenthesis in Proclamation print" (*OED* 2a).
55. William Bligh, *A Narrative of the Mutiny, on Board His Majesty's Ship Bounty; and the Subsequent Voyage of Part of the Crew, in the Ship's Boat, From Tofoa, one of the Friendly Islands, to Timor, a Dutch Settlement in the East Indies* (London: Printed for George Nichol, 1790), 4. Hereafter cited parenthetically in the text under *Narrative*.
56. Daniel Defoe, *A New Voyage Round the World* (1724; repr., London: Printed for G. Read, 1725), 4.
57. Maximilian E. Novak, "'Simon Forecastle's Weekly Journal': Some Notes on Defoe's Conscious Artistry," *Texas Studies in Literature and Language* 6, no. 4 (Winter 1965): 438.
58. William Bligh, *The Log of H.M.S. Bounty 1787–1789, by Lieutenant W. Bligh*, foreword by Admiral of the Fleet the Earl Mountbatten of Burma (Guildford, Surrey: Genesis Publications, 1975), 359–62. Hereafter cited parenthetically in the text under *Log*.
59. Bligh describes his method of fair distribution in a footnote: "One person turns his back on the object that is to be divided: another then points separately to the portions, at each of them asking aloud, 'Who shall have this?' to which the first answers by naming somebody. This impartial method of division gives every man an equal chance at the best share" (*Narrative* 41; May 25). This note is, of course, a form of parenthesis.
60. Lamb 9. Lamb is quoting from Jonathan Raban's *Passage to Juneau: A Sea and Its Meanings* (New York: Pantheon Books, 1999), 25.
61. https://dictionary.cambridge.org/us/dictionary/english/wheels-within-wheels, accessed December 5, 2020.

"Love" of Detail

Auerbach, Spitzer, and Curtius

CHRISTOPHER D. JOHNSON

✦ ✦ ✦

> I am like a point in Euclid, explained the Good Fairy, position but no magnitude, you know.
> —Flann O' Brien, *At Swim-Two-Birds*

Introducing the volume *World Philology* (2015), Sheldon Pollock repeats Erich Auerbach's warning about the "imminent disappearance of philology."[1] By way of a response to this and after surveying the current dismal state of the field, Pollock urges a "maximalist" approach that "aims to rethink the very nature of the discipline, transhistorically and transculturally."[2] Drawing on both European and Southeast Asian philological traditions and keen to reverse philology's loss of cultural capital, Pollock broadly defines philology "as the discipline of making sense of texts" and, more ambitiously still, "as a unitary global field of knowledge."[3]

Worth pondering, then, is his comment that Edward Said's much-discussed invocations of Auerbach, Leo Spitzer, and Ernst Robert Curtius, "a very ill-sorted and discordant triumvirate, by the way," raises more questions about the state and direction of current philological practices than it answers.[4] The following will attempt to sort this "triumvirate" by tracing how they distinguish, on the one hand, between the potentially limitless parts of the textual corpus and, on the other, philological details containing significant minutiae, details that Auerbach treats as microcosmic "departure-points" for interpreting macrocosmic literary worlds. It is by parsing and ordering these *Ansatzpunkte* that these philologists construct milieus at once changing but also constant—indeed, as we shall see, Spitzer showcases his method by glossing the word *milieu* for some ninety pages. Perceived as an emblematic detail rather than as a mere part among countless other possible parts, the philological point can lead both synchronic and diachronic lives. Sometimes the detail seems almost to impersonate a Leibnizian monad or a Borgesian aleph, but then invariably, it is made to respond to historical contexts and contingencies. And if a philological detail has greater aesthetic and epistemological ambitions than a part does,[5] then, as Curtius's method confirms, this may come at the cost of anachronism. In any case, a detail by definition is already an object of scrutiny, already perceived and prized. It is "true" in the Viconian sense that to know details is to know when and how and perhaps even why they were fabricated. In the same vein, Ezra Pound urges the poet to pursue "the method of luminous detail" over "the method of multitudinous detail": "Any fact is, in a sense, 'significant'. Any fact may be 'symptomatic', but certain facts give one sudden insight into circumjacent conditions, into their causes, the effects, into sequence, and laws."[6] The "luminous detail," per Pound, condenses meaning in timeless and time-bound ways, or as his favorite formula has it, "DICHTEN=CONDENSARE."[7]

Our trio of philologists are empiricists insofar as they think inductively, gather detailed points intuitively; they are metaphysicians insofar as they thirst for synthesis. Like their predecessor, Goethe, they practice morphology: first they gather, intuitively or systematically, an inventory of particular forms or (data) points; then they start analyzing changes and continuities in these points; and eventually, they grasp after syntheses. Such gathering and grasping generates particular affects. Spurred especially by Auerbach, I want to argue that the *love* etymologically embedded in *philology* inspires both starting and end points for these philologists. They all variously prize Aby Warburg's methodological maxim, "Der liebe Gott steckt im Detail" (which is inverted and distorted in the English equivalent: "The devil is in the details"). Yet as they find and arrange philological details in the swiftly changing milieu of postwar literary studies, their "love of the logos" proves a bittersweet if not melancholy one.

Curtius's collection of topics and metaphors; Spitzer's explications of etymological, semantic, and stylistic facts; and Auerbach's "arbitrary" *Ansatzpunkte* all aim to endow philological minutiae, intuitively mined from the potentially limitless supply of such minutiae, the kinds of meanings that would enable an understanding of the whole (however this is variously defined). More particularly, Curtius cultivates a largely synchronic "phenomenology" that takes a "spiral" path through centuries and cultures to paint a synthetic picture of a vanishing European "tradition." For his part, Spitzer views semantic changes as an organic historical process inherent in the nature of language, literature, and culture. Linguistic "details" matter because they reveal interior and exterior worlds. They are constitutive of "milieus" where an abiding thirst for earthly and higher things thrives. Alternatively, Auerbach chooses exemplary "starting points," ambiguously labeled "arbitrary," to make sense of the literary representation of "reality" over the course of almost three millennia. This "reality," though, also includes the pathos of Auerbach's self-representation as an exiled philologist inhabiting a changed "earth." In brief, the networks of significant philological details fabricated by these philologists become so persuasive, so thick, so interlinked, that they yield synthetic, universalizing claims, claims shaped both by objective forces and by various subjective, even affective ones.

The first definition in Euclid's *Elements* reads, "The *point* is that which has no part."[8] From here the geometer proceeds to define "a *line*" as "breadthless length" and "the ends of a line" as "points" (definitions 2, 3). This soon leads to another cardinal figure: "A *circle* is a plane figure contained by one line such that all the straight lines falling upon it from one point among those lying within the figure equal one another; And the point is called the *centre* of the circle" (definitions 15, 16).[9] Like Euclid, who constructs inductive and deductive relations between the point and geometric figures (lines, circles, triangles, etc.) generated from points, our philologists endeavor to explicate the relations of the part or the parts to the whole. They tell narratives about these relations; make epistemic, phenomenological, and historical truth claims about them; and variously reflect on their own methods—though the hermeneutic spirals, circles, and paths they trace, given the materials they work with, never of course have the deductive certainty of Euclid's *Quod erat demonstrandum*.

Leibniz's monads, like Euclid's points, have no parts: "The monad . . . is nothing but a simple substance, which enters into composites; *simple*, meaning without parts."[10] Our philological points, however, are never allowed to lose sight of their past and future lives as significant details belonging to synthetic wholes. These "minimal things," to borrow Rudolf Gasché's term, resemble Leibnizian monads, but as they incorporate external relations and contingencies of various kinds, they afford less-than-ideal—more material but thoroughly dynamic—starting points for the inspection of complex, changing worlds and selves. Finding the point, the meaningful detail—be it a word, a verse, an excerpt, a short passage, an anecdote—is the first intuitive step of inductively gathering enough *points* to see a/the whole. Assembling these points together affords an experience (*Erlebnis*) that, to literalize the cliché, is more than the sum of its parts. And in this, they function like the fragment as Friedrich Schlegel and Novalis conceived it in their accounts of Romantic poetry and philosophy.

From a traditional philological perspective—that is, one centered in linguistics—the ontology of these points/details resides in material forms that constitute texts and textual traditions: manuscripts, books, and now digital media.[11] While from a new philological perspective, the point transcends the page: it represents the initial place from which a "tradition" (Curtius), "*milieux*" (Spitzer), and the "world" itself (Auerbach) can emanate. For once found (invented or induced), this detail allows the intuitive accumulation, description, and arrangement of other details that point beyond themselves. Conversely, inducing these other points threatens a dangerous *copia*, one already palpable in the word's lexicography. Among possible glosses, the *Deutsches Wörterbuch von Jacob Grimm und Wilhelm Grimm* (1854) cites Kant's "punct des anfangs, punct der berührung" and Wieland's "das ist der punkt, wo alles zusammentrifft." Then it adds, "Darnach ist der punkt (sinnlich oder geistig) die stelle als stand-, ruhe-, gesichtspunkt, ausgangs-, anfangs- oder endpunkt, höhe-, mittel-, brenn-, kern-, hauptpunkt, halt-, stützpunkt, bewegungs-, hebepunkt u. s. w."[12] The *OED* offers thirty-four definitions, each of which contains additional meanings. These compass temporal (1b) as well as spatial connotations (3), geometric and scientific (4) as well as conceptual or rhetorical ones ("the *point* is . . .") (10), but most relevantly, a *point* signifies (5a) "a separate or single item, article, or element in an extended whole (usually an abstract whole, as a course of action, a subject of thought, a treatise, a discourse, a set of ideas, etc.); an individual part, element, or matter; a detail, a particular; (sometimes) a detail of nature or character, a particular quality or respect; Something that is the focus of attention, consideration, or purpose."[13]

However, the philological point/detail is not just a semantic object. It can be an "epistemic thing," an object subject to technological scrutiny, as Hans-Jörg Rheinberger might say. Consisting in the first instance of an agglomeration of ink on the page, it can be known not just via reading. The first microscopic image in Robert Hooke's *Micrographia* (1665) shows the "point" of a "small and very sharp Needle." But then Hooke contemplates God's ability "to include as great a variety of parts and contrivances in the yet smallest Discernable Point, as in those vaster bodies (which comparatively are called also Points) such as the *Earth, Sun,* or *Planets.*" After explaining, though, that these points can became vast when seen through "a Mechanicall contrivance," he turns to another "*point* commonly so call'd, that is, the mark of a full stop, or period." Accompanied by a drawing of a single printed period made from his microscopical observations and by comments on the practices of printers and typesetters, the description wittily concludes, "We have now shown a *point* to be a point."[14]

In the late 1920s, Aby Warburg became friendly with the renowned *Neuphilologe*, Karl Vossler. Greatly influenced by Benedetto Croce, especially Croce's *The Aesthetic as the Science of Expression and General Linguistics* (1902), Vossler developed an "idealist" version of linguistics that focused more on the vitality of stylistic expression and its roots in cultural contexts than the analysis of rule-based grammatical phenomena. In his numerous books and articles, Vossler thus pushed philology to include "the cultural-historical and psychological development of language."[15] And though his approach to reforming linguistics did not gain large traction among traditional philologists, his focus on literary style profoundly influenced a handful of younger contemporaries, like Spitzer, as well as scholars in the next generation, who, like Auerbach, were also keen to redraw and expand philology's disciplinary borders and epistemological reach. More to the point, as Adi Efal and Anna Guillemin note, Warburg's work has great affinities with the "morphological direction" of the *Neuphilologie*.[16] An art historian who aspired to a *Kulturwissenschaft* (cultural science), Warburg invokes in a 1900 text the "philological gaze"—that is, a sober, historicist interpretation of the image that attends to details and leans heavily on textual sources—all the while never forgetting the more ecstatic, Platonic gaze.[17]

In "Dürer and Italian Antiquity" (1905), based on a lecture for the Hamburg Philological Society, Warburg borrows catachrestically from linguistics and philology to convey his novel ideas and methods concerning the "liminal values of mimetic and physiognomic expression."[18] Dürer, Warburg argues, gives masterly, self-conscious expression to the *pathos formula*—Warburg's cardinal term for a recursive symbolic form that, when knowingly manipulated by the artist, could coolly express extreme emotional energy—in the 1494 drawing *Death of Orpheus*.[19] This and other pathos formulas paradoxically capture how an artwork's style can entail both formal and affective qualities, how it can be at once a synchronic and diachronic phenomenon. Thus, Warburg's "philological gaze" would interpret how visual details are transformed over time by artists who give expression to their perceptions of transformation and repetition.[20] Such scrutiny of details was also meant to open larger perspectives—a balancing act epitomized by Warburg's "Der liebe Gott steckt im Detail." In other words, Warburg wagered that certain details—say, the "animated accessory" (*bewegtes Beiwerk*) of a nymph's fluttering clothes or the classical "language of corporeal gestures" (*Gebärdensprache*) that adorns Roman sarcophagi or animates Ovid's *Metamorphoses* and that is later remade by early modern artists like Domenico Ghirlandaio and Rembrandt—when juxtaposed and then placed in combinatory sequences, could foster immediate, synoptic insights into the afterlife of pathos-charged images.[21] This morphology of images structures Warburg's last project, the *Mnemosyne-Atlas* (1926–29), which "is intended," Warburg writes, "to be first of all an inventory of pre-coined classical forms that impacted the stylistic development of the representation of animated life in the age of the Renaissance."[22] By selecting and arranging such images and eschewing discursive arguments, by highlighting details rather than trying to exhaust the whole, the *Atlas* aims to make palpable the "ethical confrontation" that the polarities contained in pathos formulas occasioned in history and, self-consciously, in Warburg's own thought.[23] Thus, Giorgio Agamben regards the *Mnemosyne-Atlas* as a "paradigm" for what thinking in "the human sciences"—not just art history and *Bildwissenschaft* (visual studies)—might mean and do. Further, Agamben grapples with how Warburg's paradigmatic images are "undecidable in regards to diachrony and synchrony, unicity and multiplicity" and how this "immanent" undecidability both encourages and frustrates any synthetic attempt to reconcile such paradigmatic "phenomena" with "the laws of historical philology"—that is, with the "chronological archive."[24]

In 1928 Vossler, at Warburg's invitation, gave a lecture in Hamburg on the legacy of Seneca in Renaissance tragedy. When Warburg visited Croce in Rome the same year, he reported that while the philosopher and literary historian was intrigued by the maxim "Der liebe Gott steckt im Detail," which Vossler had discussed with him, he possessed "for the visual element no entryway-receiver-hook [*Vorplatzauffangehaken*]."[25] Warburg's strange neologism—his late writings abound with them—is symptomatic of how verbal and visual expression converge in his thinking. In this, he resembles Walter Benjamin, who, as Samuel Weber shows, regards the verbal and visual detail as a microstructure of the whole. More particularly, Weber invokes Benjamin's *The Origin of the German Trauerspiel* (1928), where, with the aid of Leibniz's *Monadology* and baroque allegory, the detail acquires astonishing metaphysical force. Focusing on propositions 12 and 13 in the *Monadology*, Weber concludes the Leibnizian detail becomes "plurivalent" in Benjamin's hands and asserts, "The detail is that principle through which variety is reconciled with immanence."[26] Yet while Leibniz ascribes divinity to this "immanence," Weber also entertains the notion, in the spirit of Benjamin, that the detail is in the devil, that it can express, in other words, the ruination caused by time and that is barely mediated by memory.

Alternatively, in his 1932 review essay "Strenge Kunstwissenschaft" ("Rigorous Study of Art"), Benjamin urges attention to what he calls "das Unbedeutende" (the insignificant). After objecting to Heinrich Wölfflin's formalist, universalizing version of art history, Benjamin quotes approvingly the art historian Hans Sedlmayr, who urges microscopic attention to individual artworks. He then praises three articles published in a new journal, *Kunstwissenschaftliche Forschungen*: "What these studies share is a convincing love for—and a no less convincing mastery of—their subject [eine überzeugende Liebe zur Sache und eine nicht minder überzeugende Sachkenntnis]."[27] Such "love" helps these articles avoid the trap of treating artworks as the expression of "'problems of form' as such."[28] Stressing the need to attend to historical contexts and merging the interpretations of visual and verbal artifacts, Benjamin elaborates:

> Also characteristic of this manner of approaching art is the esteem for the insignificant [Andacht zum Unbedeutenden] (which the brothers Grimm practiced in their incomparable expression of the spirit of true philology). But what animates this esteem if not the willingness to push research forward to the point where even the "insignificant"—no, precisely the insignificant—becomes significant? The bedrock that these researchers come up against is the concrete bedrock of past historical existence. The "insignificance" with which they are concerned is neither the nuance of new stimuli nor the characteristic trait, which was formerly employed to identify column forms much the way the Linné taxonomized plants. Instead it is the inconspicuous or also the offensive aspect (the two together are not a contradiction) which survives in true works and which constitutes the point where the content reaches the breaking point for an authentic researcher [der Punkt ist, an welchem der Gehalt für einen echten Forscher zum Durchbruch kommt].[29]

To exemplify then how this "point" can be made significant, Benjamin praises Carl Linfert's study of baroque architectural drawings, which reveals at once the historical "process of decay" and, methodologically, "the capacity to be at home in marginal domains [das Zuhausesein in Grenzgebieten]."[30] In other words, Linfert, like Benjamin himself, prizes how the detail can entail concrete, nonsynthetic worlds of meaning.

Curtius first met Warburg in Rome in late 1928, when the latter was preparing the lecture "Roman Antiquity in Domenico Ghirlandaio's Workshop," which, besides casting the late Quattrocento painter as an exemplary mediator of antiquity's pathos formulas, made the case for placing art history in dialogue with the disciplines of archaeology, anthropology, psychology, and literary criticism.[31] In offering a "comparative view" of various exemplary objects following the model of his *Mnemosyne-Atlas*, Warburg wanted to make visible "the process of energetic inversion, that is, the production of contrary interpretations."[32] One of the "inversions" or pathos formulas he discussed is how Botticelli's illustration of an episode in Dante's *Purgatorio* (10) is remade in "a relief of the emperor leaping over dead enemies under his horse's hooves." That this very same "inversion" is cited by Curtius in a 1950 article that argues that Warburg's pathos formulas are also viable in medieval literature confirms again how the philological detail tends to wander away from its origins.[33]

Curtius's *European Literature and the Latin Middle Ages* (1948; henceforth *ELLMA*) presents itself as an exemplary "synthesis" enabled by the "scientific technique" of "philology" but precipitated by "a concrete historical situation."[34] Dedicated to Warburg and the philologist Gustav Gröber (Curtius's teacher), *ELLMA* was written during Curtius's so-called inner emigration during the Nazi reign.[35] With its eighteen chapters and twenty-five *exkurses*, *ELLMA* aspires to be "a phenomenology of literature," whose bountiful material and grand historical scope would elicit the "unity" of European culture by showing how the topoi and metaphors fueling the invention and style of Latin medieval literature are rooted in antiquity but survive well into the 1800s.[36] Indebted theoretically to Henri Bergson's account of the creative spirit and Max Scheler's philosophical phenomenology, Curtius describes an inductive search for those philologically "'significant' facts" ("bedeutsamen" Tatsachen) that would gradually allow him to obtain a panoptic, cartographical, and deductive view of how "European literature is an 'intelligible unit' ['Sinneinheit']."[37] Yet as Curtius readily grants, these philological minutiae are not easily reconciled with the whole: "Specialization without universalism is blind. Universalism without specialization is inane."[38] Indeed, in grasping after axiomatic, universalist principles, *ELLMA* is fueled by a species of anachronism or synchronism that has drawn sharp critiques from Hans Blumenberg, Hans Robert Jauss, and Hans Ulrich Gumbrecht.[39] Spurning what he regards as facile literary history and overly theoretical *Geistesgeschichte*, Curtius seizes on the topical continuities he finds in Western literature to sketch a "comparative morphology of cultures" that would allow him "to see European literature as a whole."[40]

More to the point, the second chapter on medieval "Romania" ends by invoking Warburg: "Through Romania and its influences [*Ausstrahlungen*] the West received its Latin schooling. The forms and the fruits of that schooling are now to be considered. That is, we must now proceed from generalities to the concrete wealth [*Fülle*] of the substance of history. We must now go into details. But, as Aby Warburg used to say to his students, 'God is in detail.'" Fair enough, but as he proceeds to map, with his "technique of philological microscopy," exemplary topoi, and metaphors in subsequent chapters,

Curtius repeatedly admits that the abundance of compelling minutiae, "a surfeit of bits and pieces" (ein Übermaß von Kleinkram), threatens to obscure "contemplation of the whole."[41]

Representative of this precarious dialectic of parts and whole is the section on *The Book of Nature* in chapter 16. Here, beginning with verses by Alan of Lille ("Omnis mundi creatura / Quasi liber et pictura / Nobis est et speculum"), Curtius shows how the metaphor—Blumenberg will call it an *absolute metaphor*—is remade by a host of thinkers, from the "preacher and mystic Luis de Granada (1504–1588)" back to Hugh of St. Victor in the twelfth century, then to Cusanus in the early fifteenth, up to Paracelsus, Montaigne, Thomas Browne ("We carry with us the wonders we seek without us: there is all Africa and her prodigies in us; we are that bold and adventurous piece of Nature, which he that studies wisely learns in a compendium what other labour at in a divided and endless volume."), Galileo, Diderot, Voltaire, Jacob Grimm, and Goethe. In this fashion, Curtius prepares the way in the next section for a synthetic account of Dante's elaborate, "intellectual [*geistige*]" refashioning of the "imagery of the book."[42]

Then in the "Epilogue," Curtius steps back from inducing "examples" to contemplate his book's "arduous journey [*Wanderung*]."[43] He wanted to "give in the text only as many examples as were necessary to confirm the argument," and he wanted it to have both "aesthetic" proportion and conceptual rigor.[44] Thus, he refers back to the tenth and last of his "Guiding Principles" prefacing the book, Ortega y Gasset's maxim, "Un libro de ciencia tiene que ser de ciencia; pero también tiene que ser libro."[45] In other words, with his book's exemplary "facts" or details having been artfully arranged, Curtius can now induce a conceptual whole: "The course [*Gang*] of the presentation and the succession of the chapters are such as to result in a step-by-step progress and a spiral ascent. The first chapters present facts whose significance is illuminated later.... The structure is determined not by a logical order but by thematic continuity. The interweaving of threads, the reappearance of persons and motifs in different designs, reflects the concatenated historical relations."[46] Curtius's carefully constructed "Gang"—the last paragraph of the book proper begins "Wir schließen unsern Gang"—aims to be both objective and subjective: to find "'significant' facts," the philologist must "read a great deal" and have extraordinary intuition as to which matter when arranging them to make larger, synthetic claims.[47] Then to clarify all this and to secure his claim of systematicity, Curtius invokes Warburg:

> When we have isolated and named a literary phenomenon, we have established one fact [*ein Befund*]. At that point we have penetrated the concrete

structure of the matter of literature. We have performed an analysis. If we get at a few dozen or a few hundred such facts, a system of points [*Punkten*] is established. They can be connected by lines; and this produces figures. If we study and associate these [*Betrachtet und verknüpft man sie*], we arrive at a comprehensive picture. That is what Aby Warburg meant by the sentence quoted earlier: "God is in detail." We can put it: analysis leads to synthesis. Or: the synthesis issues from the analysis; and only a synthesis thus brought into existence is legitimate.[48]

In appropriating Warburg's motto to justify how philological "analysis" can yield a universal "synthesis," Curtius both affirms elective affinities and masks profound differences with his older colleague. Warburg's "Verknüpfungszwang" (compulsion to link), which informs his arrangement of art-historical details, is far less systematic, far more haptic, and a good deal more subjective than Curtius implies. Still, both believed that philology could (and should) traverse and thereby, in a certain sense, collapse historical time and geographic space in order to win new perspectives and meanings from the past. Both shared a contempt for disciplinary boundaries: earlier in the book, Curtius evokes Warburg to explain why he, too, does not fear the specialists, "the guardians of Zion," who would defend their fiefdoms instead of sharing his ambitious comparatist program.[49] Both prized their intuition over deduction. Spurning deductive "Problemgeschichte" and "Begriffsgeschichte," Curtius insists that topoi like "sapientia et fortitudo" and "the theater of the world" are enough to guide the literary historian in constructing an account of the continuous European tradition.[50] Invoking Bergson again, Curtius terms his ability to "divine" (*erraten*) the scattered "facts" that are "significant" (*bedeutsam*) a "psychological function," but when applied to literature, it goes by the name of "philology."[51]

Spitzer's richly ambivalent 1949 review of *ELLMA* is colored by historical and political concerns as well as intellectual ones. Spitzer, who famously makes the dialectic of "Erlebnis" (*experience*) and "Methode" the basis for his philology, reads *ELLMA* at least partially through his experience as an exiled Austrian German Jew who had escaped the Holocaust and who could, at the time he wrote the review, survey the critical landscape from his perch at Johns Hopkins.[52] In a nutshell, while praising *ELLMA* for its astonishing scope, Spitzer blames Curtius for failing to penetrate the "inward form" of individual works via close, sustained readings. Conversely, a good part of Spitzer's critique verges on the ad hominem. After tracing Curtius's "scholarly development" and noting that by 1930 he was being hailed as "a Goethe *redivivus*," Spitzer offers a finely ironic account of the genesis of *ELLMA*:

As early as 1932, he had become aware of the "perils" for the German mind which lay in its too easy, too lovingly-fostered irrationalism and which was able to engender such a barbarous movement such as Hitlerism. With his flair for the duty of the hour, Curtius turned toward "solid philology" and toward medieval philology where sobriety and discipline of mind had reached the greatest triumphs. It was logical that an aristocratic mind such as Curtius' should, before the onslaught of the plebeian hordes, retreat into the Latin past of Germany. ... The "European" Curtius could thus still preserve his scholarly integrity and also survive—in medieval garb.[53]

Furthermore, the review ends, in smaller type, with Spitzer regretting that Curtius's "resentments seem to include the German emigré scholars in Romance who have worked before him in the same direction."[54] In other words, Curtius fails to cite Auerbach, Helmet Hatzfeld, Leonard Olschki, and Spitzer himself—all Jews.

With this biographical-political frame in place, Spitzer evaluates the scholarly achievement of *ELLMA*. First, he lauds its mastery and reader-friendly arrangement of the "source material," reserving special praise for Curtius's discovery and arrangement of minutiae: "This encyclopedic book has become an inexhaustible mine of uncontroversial facts about the ultimate sources of European poetry. How happy we must be to possess such a vast grammar (or dictionary) of *topoi* in which a glance at the index may enable us to find all that is known to a unique connoisseur of classical, Neolatin and Romance philology." In sum, "No comprehensive book on medieval literature in the last fifty years is more epoch-making than this."[55] Yet while he admires how Curtius achieves "the realization of the historical continuity of our European civilization," Spitzer also chides him: "One feels as though the world-clock stood still: man appears here as a being consisting in continuity. And is not the insight into such basic conservatism of man an antidote against the feeling of helplessness engendered by the vista of chronic dismemberment and of the crumbling of tradition that the world of today offers us? Before the forces of barbarism that encircle us, Curtius has found an escape by immersing himself in the necropolis of the past that was alive as late as the eighteenth century."[56] In blaming such nostalgic synchrony, Spitzer also regrets how Curtius's "topology" undervalues the parts that compose that whole: "It is also platitudinously true that the sum total of the sources does not explain the inward form of a particular work of art. Does Curtius forget that the great work of art is always unique and that art strives for uniqueness?"[57] Essentially critiquing Curtius's thirst for an objective synthesis, Spitzer then points to other approaches, particularly that "admirable work of Erich Auerbach, *Mimesis*, written by a German in exile without any resentment against current German movements and interpreting, historically *and* aesthetically, individual texts that cover the same span of twenty-six centuries as does Curtius' book." Books like Auerbach's, Spitzer concludes, "lead us farther into the inner sanctum of medieval poetry. Such books truly interpret the individual work of art while Curtius informs us, more completely than his predecessors, about its general background."[58] Point taken, but Curtius (like Warburg) had good reasons for eschewing aesthetic interpretation. As suggested by the analogies he makes between his method and geometry, it was precisely the "inner sanctum" he wished to keep shut. For such subjectivism, Curtius believed, tended to cloud rather than clarify critical thought.

Exemplary of how Spitzer's own scholarship moves from minutiae to larger, even cosmic structures is the lengthy 1942 essay "Milieu and Ambiance: An Essay in Historical Semantics." Taking as its starting point Hippolyte Taine's use of the word *milieu*, which, Spitzer notes, was immediately adopted as a loan word by other languages but also matched with equivalents "to represent the same concept (German *Umwelt*; Spanish *medio*; Italian *ambiente*; English *environment*),"[59] the essay then regresses chronologically to ponder associated Greek and Latin words and only later progressively builds a thick history of ideas out of these and their vernacular derivatives. Along the detailed way, we are instructed about words/concepts signifying "climate," "air," "cosmos," "space," "place," "medium," and so on. The essay, then, is also an inquiry into the history of physics and cosmology.

An intellectual history of the first order, which could and should be a part of any history of human perceptions of the environment, "Milieu and Ambiance" is accompanied by copious footnotes filled with philological details. (The bottom-heavy mise-en-page resembles Pierre Bayle's *Dictionnaire* or certain texts by Derrida.) As it winds toward its conclusion, via Pascal's and Leibniz's figuration of man as positioned between two infinities, the essay circuitously returns to Taine's milieu, which Spitzer judges to be weirdly "anthropomorphic," divorced from material, spatial, and cosmic connotations.[60] But Spitzer also finds space to condemn Hitler's words and deeds and then to celebrate Válery.[61] Válery, in turn, prompts Spitzer to reiterate his belief that diachronic details and synchronic truths can be harmonized: "What admirable constancy of word-material, surviving through the flux of changing conceptions! ... In language the chrysalis can live on, along with the butterfly. We are forever carrying with us the chrysalises, the primeval 'eggshells' of human thought."[62] In this metaphoric, synthetic manner, Spitzer moves from minutiae to milieus and back again to other minutiae, at once "old" and "new."

Alternatively, Spitzer's programmatic essay, "Linguistics and Literary History," introducing *Linguistics and Literary History* (1948), begins with an autobiographical account of his *Bildung* as a linguist and literary historian. It celebrates how attention to the etymological detail, the "spiritual etymon"—accompanied by a brief, exemplary analysis of how the English words *conundrum* and *quandary* were formed—can lead to an understanding of "psychology and history of civilization."[63] Recalling his earlier interpretation of Rabelais's word "pantagruélisme," Spitzer insists the scholar must "work from the surface to the 'inward life-center' of the work of art" and then repeat this movement until an "account of the whole" is possible.[64] At stake, in other words, is the hermeneutic circle and what it takes to prevent it from becoming vicious, as the scholar moves "to and fro" between the linguistic and the literary, by induction and then deduction, by attending now to stylistic details, now to the ideas and insights furnished by *Geistesgeschichte*: "It is the basic operation in the humanities, the *Zirkel im Verstehen* as Dilthey has termed the discovery, made by the Romantic scholar and theologian Schleiermacher, that cognizance in philology is reached not only by the gradual progression from one detail to another detail, but by the anticipation or divination of the whole because 'the detail can be understood only by the whole and any explanation of detail presupposes the understanding of the whole.'"[65] Spitzer's credo regarding the value of the "observed detail" and how it enables "the all-too-human" philological "induction" is at once the empirical and almost mystical:

> But, of course, the attempt to discover significance in the detail, the habit of taking a detail of language as seriously as the meaning of a work of art ... this is an outgrowth of a preestablished firm conviction, the "axiom," of the philologian, that details are not inchoate chance aggregation of dispersed material through which no light shines. The philologian must believe in the existence of some light from on high. ... If he did not know that at the end of his journey there would be awaiting him a life-giving draught from some *dive bouteille*, he would not have commenced it: "Tu ne me chercherais pas si tu ne m'avais pas déjà trouvé," says Pascal's God. Thus, humanistic thought, in spite of the methodological distinction just made, is not completely divorced from that of the theologian as is generally believed; it is not by chance that the "philological circle" was discovered by a theologian, who was wont to harmonize the discordant, to retrace the beauty of God in this world. This attitude is reflected in the word coined by Schleiermacher: *Weltanschauung*: "die Welt anschauen": "to see, to cognize the world in its sensuous detail." The philologian will then continue the pursuit of the microscopic because he sees therein the microcosmic; he will practice that "*Andacht zum Kleinen*" which Jacob Grimm has

prescribed; he will go on filling his little cards with dates and examples, in the hope that the supernal light will shine over them and bring out the clear lines of truth.[66]

Striking is how Spitzer here variously transforms "the pursuit of the microscopic" into a vision that is as affective as it is immanent—immanent almost in the sense that Leibniz ascribes to monads. The philological detail enlightens not only because the hermeneut has revelatory powers but also because a writer like Rabelais has the inspiration and skill to fabricate a "dive bouteille."

This credo is formulated still more affectively when Spitzer, defending the extraordinary rigor with which the philologist approaches literature and playing implicitly on the "spiritual etymon" of *philologia*, writes in the essay's first footnote, "Love, whether it be for love for God, love for one's fellow men, or the love of art, can only gain by the effort of the human intellect to search for the reasons of its most sublime emotions, and to formulate them. It is only a frivolous love that cannot survive intellectual definition; great love prospers with understanding."[67] But how does such "love" translate into actual practice? While I cannot here delve into the book's chapters on *Don Quijote*, the writings of Diderot, and an ode by Claudel,[68] I do wish very briefly to indicate how chapter 3, "The 'Récit de Théramène,'" reads a line from the famous soliloquy in Racine's *Phèdre* (1677) to argue that stylistic details need to be interpreted as part of larger worldview—namely, that aspects of Racine's style point to "a baroque tragedy of *desengaño*."[69] Although warning the reader that this reading initially departs from his "usual procedure" of starting "from a detail (of style)" and then trying "to establish the meaning of the whole (the tragedy)," Spitzer does in fact pivot midway in the essay to a careful *explication de texte* of Théramène's pathos-filled description of what he saw when the monster emerges from the sea to slaughter the unfortunate Hippolyte.[70] He does so by arguing that the paradoxical verse that most offends French neoclassical tastes ("Le flot qui l'apporta recule épouvanté") not only expresses with enormous affect nature's aversion to the "monstre furieux," which it itself has engendered, but also points to larger, decidedly "baroque" stylistic traits. An early reader of Benjamin's *Trauerspiel* book, Spitzer thus ultimately would show how "rhetorical patterns mould and purify crude reality."[71]

Auerbach wrote two reviews of *ELLMA* in 1950. His review in English opens by celebrating "a monument of powerful, passionate and obstinate energy."[72] However, this is quickly qualified with the formalist reproach that the book's compelling *copia*, its instructive minutiae, overwhelm: "The material collected is not only admirable by its choice and overwhelming abundance, not only extremely suggestive as a model for further research, but very often delightful in itself. But it is too much. The material sometimes interrupts or blurs the leading ideas, and some of these scattered over different chapters, are not presented with the strength and consistency they deserve."[73] Curtius's book, "inspired by synthetic ideas—ideas which are, indeed, mostly ... based rather on a rational and as it were strategical vision than on insight into the living individual phenomena," fails, Spitzer contends, to place its innumerable details in tangible historical-cultural contexts. After regretting that Curtius neglects how Christian theology informs his "material," Auerbach also notes "that sometimes [he] seems to underestimate its popular trends. Thus, this magnificent book is a model of modern combinative and perspective scholarship, yet presents a somewhat incomplete and one-sided picture of the European Middle Ages."[74] *ELLMA* is either "too much" or not enough.

Auerbach's German review focuses more on the details. It begins, "This is one of the most nourishing, humanistic books that I have come across ... at once repertoire, in which one can look things up, and presentation, whose sharply delineated ideas prompt collaboration and elaboration."[75] Then, though, he reproaches Curtius for expecting his model reader to know intimately, like its author, an astonishing range of languages and a whole library of canonical and obscure texts, literary, philosophical, and otherwise.[76] Sardonically paraphrasing Curtius, Auerbach invokes Warburg's ghost: "Only then can specialized research become meaningful, in which the most exacting treatment of particulars must not be abandoned, for 'der liebe Gott steckt im Detail.' With all this one must agree. That the situation looks hopeless is no objection. The gods love him [*Den lieben die Götter*] who desires the impossible. And it entirely suffices, if only a few meet the challenge: they become models and preserve the tradition."[77] Even more than its perceived elitism, Auerbach most regrets the book's subjective ("einseitig") quality—namely, that the synthetic judgments Curtius makes based on his mastery of the textual tradition leave little room for the kind of philology in which psychological factors, historical contexts, and material conditions can also be valued.[78] The "topology" mapped in *ELLMA* is too synchronic and "the demotic, the individual, and the national recede, [while] the continuity of the transmission is stressed."[79] Put another way, Auerbach implicitly blames Curtius for not being more Viconian, for not attending to the *verum-factum* principle, which asserts that to know a thing to be true, however large or small, is to know how and when it was made.[80]

Despite these misgivings, in Auerbach's massively influential 1952 essay, "Philologie der Weltliteratur," Curtius is enlisted to make an entirely different point. Here Auerbach considers the impossibility of sustaining Goethe's ideal in light of the dizzying, almost Borgesian confluence of cultures and literary traditions after the war: "Should mankind succeed in withstanding the shock of so mighty and rapid a process of concentration—for which the spiritual preparation has been poor—then man will have to accustom himself to existence in a standardized world, to a single literary culture, only a few literary languages, and perhaps even a single literary language. And herewith the notion of *Weltliteratur* would be at once realized and destroyed."[81] In this bleak light, Auerbach then contemplates ways philology, with its historical perspectives, might adapt to changed circumstances and the familiar problem of "the superabundance [*Überfülle*] of materials, of methods and of points of view [*Anschauungsweisen*]."[82] How, he asks, can a philologist still tell "the inner history" of "humanity achieving self-expression"? Or less nebulously, "How can anyone, given such circumstances, ponder a scholarly and synthesizing philology of *Weltliteratur*?"[83] How, in brief, can the *Weltphilolog* still solve the conundrum of the many and the one?

Auerbach's primary response is to point to the ability of "personal intuition" and then "combinatory intuition" to find *Ansatzpunkte* that would help "solve the problem of synthesis."[84] In other words, his dilemma is ours as well—information overload is nothing new.[85] Nor are concerns about cultural or linguistic homogenization. But if his search for cultural "unity" and "continuity," for epistemological, formal, or symbolic syntheses, may nowadays be regarded as naive or worse, then his worry about literary studies being overwhelmed by excessive minutiae seems more urgent than ever. It is here, then, after pointedly dismissing "encyclopedic collecting" as inimical to synthetic thought and after doubting whether the "unity" offered by single-author studies is viable,[86] that Auerbach turns to Curtius's book, which he treats as a methodological rather than a conceptual triumph: "The most impressive recent book in which a synthesizing historical view is accomplished is Ernst Robert Curtius's book on *European Literature and the Latin Middle Ages*. It seems to me that this book owes its success to the fact that despite its comprehensive, general title, it proceeds from a clearly prescribed, almost narrow, single phenomenon: the survival of the rhetorical-school tradition. Despite the monstrosity of the materials it mobilizes, in its best parts [*Teilen*] this book is not a mere agglomeration [*Anhäufung*] of many items, but a radiation outward [*Ausstrahlung*] from a few items."[87] This last metaphor reveals a great deal about Auerbach's own preferred method, even as he calls Curtius's "the only method that makes it possible for us now to write a history-from-within against a broader background, to write synthetically and suggestively."[88] More particularly, Auerbach admiringly describes Curtius's "discovery" of starting points:

Whether Curtius's choice for a point of departure [*die Wahl des Ansatzes*] was satisfactory, or whether it was the best of all possible choices for his intention, is not being debated; precisely because one might contend that Curtius's point of departure [*Ansatz*] was inadequate one ought to admire the resulting achievement all the more. For Curtius's achievement is obligated to the following methodological principle: in order to accomplish a major work of synthesis it is imperative to locate a point of departure [*Ansatz*], as it were, by which the subject can be seized. The point of departure [*Ansatz*] must be the election of a firmly circumscribed, comprehensible set of phenomena whose interpretation must possess radiating power [*muß Strahlkraft besitzen*] and which orders a greater region than they themselves occupy.[89]

Curtius grasps topoi, philological details, only then to trace circles encompassing broader meanings. And though these *Ansätze* are not the ones Auerbach would have chosen himself, in terms of method, they are paradigmatic. Indeed, "the method also makes it possible for a younger scholar, even a beginner, to conceive" such syntheses, for "a comparatively modest general knowledge buttressed by advice can suffice once intuition has found an auspicious point of departure [glücklichen Ansatz]."[90] Conflating words and things, literature and reality, Auerbach enthusiastically elaborates the spatial-geometric conceit to insist that the point can almost be read monadically—if the scholar is willing and able to let its meanings emanate. But such "a history-from-within" must be grounded in what Auerbach redundantly, almost catachrestically, terms "Wirklichkeit" (*reality* or *the real*).[91] (The repetition of the words *Wirklichkeit, Wirkung, wirken*, and their cognates would be a fine *Ansatzpunkt* to interpret this essay, to say nothing of *Mimesis: Dargestellte Wirklichkeit*.) Essential, then, is to find "a partially apprehendable phenomenon [*Teilphänomen*] that is as circumscribed and concrete as possible, and therefore describable in technical, philological terms."[92] The "Ansatz" must be "exact and objective"; it must have "potential for centrifugal radiation [*Strahlkraft*]"—whether this be a "semantic interpretation, a rhetorical trope, a syntactic sequence, the interpretation of one sentence," or otherwise—what matters is it possess per se "radiating power, so that with it we can deal with world history [*Weltgeschichte*]."[93]

This last extraordinary claim brings Auerbach back to his other theme, *Weltliteratur*. With a gesture that has provoked responses of all kinds—one thinks, above all, of Said's loving but contrapuntal reception—Auerbach regrets the absence of a "synthetic philology of world literature," regrets that humanity no longer seems at home in the "universe," even while insisting that "our philological home is the earth: it can no longer be the nation,"[94] for as it continues to induce detailed points and to conceive milieus, philology must "expand" its reach even as "our earth grows closer together."[95] And what is the effect of all this? After intimating the exile's despair, with the help of Hugh of St. Victor (1097–1141), Auerbach concludes, "Hugo intended these lines for one whose aim is to free himself from a love of the world. But it is a good way also for one who wishes to earn a proper love [*rechte Liebe*] for the world."[96]

In the last part of this essay, I want to ponder what this "love" might mean in Auerbach's epochal *Mimesis: The Representation of Reality in Western Literature* (1946, 1949). With its twenty chronologically arranged essays, each of which seizes upon (at least) a single passage (usually) in its original language and then in translation, to then enable the unfolding of a panoptic view of a work's and, by extension, a culture's "Wirklichkeit," *Mimesis* microscopically examines exemplary parts of the West's literary legacy even as it deftly apprizes millennia of historical change.

Mimesis features no introduction proper—the book begins in medias res like Homer's *Odyssey*, the first text discussed. Homer's digressive, realistic description of Odysseus's scar and how he acquired it (book 19) serves as the book's first exemplary detail, which in turn yields an *Ansatzpunkt* for the entire book. Thus, after enabling Homer's epic and Genesis 22 to radiate meanings, Auerbach reflects at the end of the first chapter, "We have compared these two texts, and, with them, the two kinds of style they embody, in order to reach a starting point [*Ausgangspunkt*] for an investigation into the literary representation of reality in European culture."[97] However, it is only in the twentieth chapter on literary modernism and in the epilogue does Auerbach offer sustained methodological reflections. My own focus, then, concerns the nominally "random" or "arbitrary"—the word Auerbach uses repeatedly is *beliebig*—character of his *Ansatzpunkte* in terms of both the content studied and how that content is read. Grimm's etymology of *beliebig* has it originating in the substantive *Liebe*, which yields the verb *belieben* (to wish, to please) and, beginning sometime in the seventeenth century, the adjective *beliebig* (arbitrarily chosen).[98] An electronic word search of *Mimesis* yields forty-seven instances of *beliebig*. The first instance, in chapter 2, reads, "Petronius' literary ambition, like that of the realists of modern times, is to imitate a random [*beliebiges*], everyday, contemporary milieu with its sociological background."[99] Thus, unlike Spitzer in his "Milieu" essay, Auerbach is less interested in the fortunes of individual words and the shifting values in intellectual history than he is in the style that Petronius, like Balzac or Dostoevsky (both invoked in the next paragraph), employs to represent a distinctive historical "milieu," one that embraces the everyday myriad contingencies. Taking "Petronius' romance" as its not-so-arbitrary starting point, the chapter then subtly shifts attention to other *Ansätze*: a passage by Tacitus and a magisterial reading of Mark's account of Peter's denial of Christ, all to contend that "the antique stylistic rule according to which realistic imitation, the description of random everyday life [*beliebigen Alltäglichkeit*], could only be comic" is "incompatible" with the representation of random phenomena in the New Testament, since this always also points beyond—or better yet above—what is being represented.[100] Then tellingly, Auerbach pauses to offer a gloss: "The term random [*beliebig*] being here employed to designate people from all classes, occupations, walks of life, that is, who owe their place in the account exclusively to the fact that the historical movement engulfs them as it were accidentally, so that they were obliged to react to it one way or another."[101] In brief, *beliebig* pinpoints both what Auerbach's authors represent and how his method functions. It signals how one can mediate the literary world's plenitude by randomly (but not so randomly) selecting examples or *Ansätze* out of all-too-ample possibilities.

When it comes to Rabelais, "examples, both from the work as a whole and from sections of it, can be multiplied at will [*beliebig vermehren*]."[102] Or, in the Montaigne chapter, "The obligatory basis of Montaigne's method is the random life one happens to have [*das beliebige eigene Leben*]. But then this random life of one's own must be taken as a whole."[103] Though to conclude the chapter, Auerbach reflects, "The text we have analyzed is a good point of departure [*Ausgangspunkt*] for a conscious comprehension of the largest possible number of the themes and attitudes in Montaigne's undertaking, the portrayal of his own random [*beliebigen*] personal life as a whole."[104] In other words, given the caprice and heterogeneity intrinsic to Montaigne's *écriture*, one well-chosen but still random "point" can and must lead to other such "points" until Montaigne and Auerbach (and his readers) become "conscious" of the "whole." And while there are also random "Anhaltspunkte" that fail to emanate[105] and while "the systematic privileging of the textual fragment," with its embryonic function, often comes at the expense of the "various contexts" in which it occurs and thus may constrict rather than free the reader's imagination,[106] there can be little doubt that the starting points selected in *Mimesis* precipitate a form of literary criticism and history that illuminates and affects in ways that remain urgently immanent—if, as Said observes, we value "discursive continuity" and the effort to "enact the combination of past and future woven into the historical fabric." Alternatively, these *Ansatzpunkte* are places or perhaps events affording critics the chance to "endlessly" imagine or represent themselves.[107]

In the book's final chapter, "The Brown Stocking," Auerbach ambiguously, self-consciously, imitates yet also blames the experimental novels of Woolf, Joyce, Proust, and Mann. In summarizing the "technique" of his modernist contemporaries, he segues, almost by metonymy, to the most sustained reflections on his book's method:

> There is a confidence that in any random [beliebig] fragment plucked from the course of a life at any time the totality of its fate is contained and can be portrayed. There is greater confidence in syntheses gained through full exploration of an everyday occurrence than in a chronologically well-ordered total treatment which accompanies the subject from beginning to end, attempts not to omit anything externally important, and emphasizes the great turning points of destiny [Schicksalswendungen]. It is possible to compare this technique with certain modern philologists who hold that the interpretation of a few passages from *Hamlet*, *Phèdre*, or *Faust* can be made to yield more, and more decisive information about Shakespeare, Racine, or Goethe and their times than would a systematic and chronological treatment of their lives and works. Indeed, the present book may be cited as an illustration. I could never have written anything in the nature of a history of European realism; the material would have swamped me [ich wäre im Stoff ertrunken].[108]

Thus, like the exemplary realists studied earlier in the book, "modern philologists" like himself—presumably, Vossler, Curtius, and Spitzer are meant here as well—have found a way to parse the small parts, the fragments of an ungraspable whole, in order to tell a larger historical narrative that would be somewhat immune from the violent vicissitudes of history.[109] By contrast, Auerbach admits how "the modern writers here under discussion," with their heightened consciousness of experiencing the chaos of things small and large, with their penchant for "overlapping, complementing, and contradiction," offer "a challenge to the reader's will to interpretive synthesis."[110] In other words, when the last chapter ends on the same pessimistic note about "the approaching unification and simplification" that informs the "Weltliteratur" essay, it becomes undeniably clear that *Mimesis* is as much about Auerbach's *Erlebnis*, as Spitzer would say, as it is a methodical attempt to make sense of Western literature's "representation of reality."[111]

It is altogether fitting, then, that when Auerbach reflects further on method in the book's epilogue, the description of his "Absicht" (intention, purpose) seems self-consciously riddled by contradiction. He describes how his method permits him "some room to play [einigen Spielraum]," insists on the "random [beliebig]" nature of his starting points, notes how the limited scholarly resources in Istanbul shaped his text, and yet reiterates his belief that he could tell a synthetic narrative of Western literary realism to a radically changing world. As such, Auerbach's ultimate intention is not methodological, conceptual, or world historical. Instead it is—as in the "Philologie der Weltliteratur" essay—an affective one. The last sentence of the epilogue reads, "I hope that my study will reach its readers—both my friends of former years, if they are still alive, as well as all others for whom it was intended. And may it contribute to bringing together again those who have retained unclouded love for our western history [die die Liebe zu unserer abendländischen Geschichte ohne Trübung bewahrt haben]."[112] In this manner, the semantic relation between "beliebige" and "Liebe" emerges with real, almost Warburgian pathos.

Finally, I would end by mentioning one afterlife of our "triumvirate." In his 1992 essay, "Modern European Literature: A Geographical Sketch," Franco Moretti invokes Auerbach's review of *ELLMA*, thereby commencing an intellectual trajectory that will subsequently include the controversial notion of *distant reading* and his celebrated *Atlas of the European Novel, 1800–1900*. Initially hailing *ELLMA* as "the only scholarly masterpiece devoted to our subject," Moretti then tempers his praise, for "our" Europe has long ceased to be Curtius's retrospective "Romania"; therefore a different milieu, a "polycentric Europe," must be mapped if the notion of *Weltliteratur* is to be reconfigured and redeemed.[113] Put otherwise, the question that much of Moretti's later work is dedicated to addressing, namely, how to chart the innumerable individual points in the global, nonsynchronic, literary space, a space no longer dominated by European voices, is one that, as Pollock and others remind us, philology must have a part in answering.

NOTES

1. Pollock, *World Philology*, 4.
2. Pollock, 6.
3. Pollock, 22.
4. Pollock, "Future Philology," 959–61.
5. Schäffner, Weigel, and Macho, "Das Detail," 7. All translations are mine unless otherwise noted.
6. Pound, *Selected Prose*, 22.
7. Pound, *ABC*, 92.
8. Euclid, *Elements*, 155.
9. Euclid, 158–83.
10. Leibniz, "Monadology," 268.
11. Compare with Wall's indicative readings of parentheses as "points."

12. "PUNKT, m.," *Deutsches Wörterbuch von Jacob Grimm und Wilhelm Grimm*, digitalisierte Fassung im Wörterbuchnetz des Trier Center for Digital Humanities, https://www.woerterbuchnetz.de/DWB?lemid=P08547.
13. "point, n.1," *OED Online*, March 2021, https://www-oed-com.ezproxy1.lib.asu.edu/view/Entry/146609?rskey=hWWuSc&result=1&isAdvanced=false.
14. Hooke, *Micrographia*, 3–4. Wall's essay on the parenthesis in this volume also discovers worlds in small punctuation marks.
15. Vossler, *Frankreichs Kultur*, vii.
16. Efal, "Le 'regard philologique,'" paragraph 10.
17. Warburg, "Ninfa fiorentina," 203.
18. Warburg, GS I.2, 449.
19. Warburg, 446.
20. Warburg's *pathos formula*, Guillemin argues, "functions like a language and can be read as a linguistic principle" (Guillemin, "Style of Linguistics," 613).
21. Weigel, "'Nichts weiter,'" 91–93.
22. Warburg, "Absorption of the Expressive Values," 277–78; Warburg, GS II.1, 3.
23. Warburg, "Absorption of the Expressive Values," 283; Warburg, GS II.1, 3.
24. Agamben, "Paradigm," 28–32. Early in his career Warburg briefly eyes a synthetic "art history." See Warburg, GS I.1, 93–94.
25. Warburg, GS VII, 446.
26. Weber, "Gott und Teufel," 47. Weber notes that Adorno favored the phrase "Der liebe Gott wohnt im Detail."
27. Benjamin, "Rigorous Study," 86; Benjamin, "Strenge Kunstwissenschaft," 365.
28. Benjamin, "Rigorous Study," 86.
29. Benjamin, 86–87 (translation modified); Benjamin, "Strenge Kunstwissenschaft," 366.
30. Benjamin, "Rigorous Study," 90; Benjamin, "Strenge Kunstwissenschaft," 369.
31. Warburg, GS II.2, 306. Curtius attended the lecture and was astonished at its breadth and depth.
32. Warburg, GS II.2, 307.
33. Curtius, "Antike Pathosformeln," 23–27. For Curtius's misunderstandings of art history as practiced by the Warburgians, see Curtius, *ELLMA*, 14–15; Curtius, *Europäische*, 24–25; see also Pfisterer, "Die Bilderwissenschaft ist mühelos."
34. Curtius, *ELLMA*, 382, x; Curtius, *Europäische*, 386.
35. The phrase is Werner Krauss's. See Gumbrecht, *Vom Leben und Sterben*, 68.
36. Curtius, *ELLMA*, vii–ix.
37. Curtius, 382–83; Curtius, *Europäische*, 386–87.
38. Curtius, *ELLMA*, ix.
39. See Pfisterer, "Die Bilderwissenschaft ist mühelos," 25–26; Gumbrecht, *Vom Leben und Sterben*, 52–55. But Warburg is celebrated for his self-consciously "anachronic" (as opposed to an unthinkingly anachronistic) approach in Christopher Wood and Alexander Nagel's *Anachronic Renaissance* (New York: Zone Books, 2010).

40. Curtius, *ELLMA*, 11, 4, 12; Curtius, *Europäische*, 21, 14, 22.
41. Curtius, *ELLMA*, 228 (translation modified), 291, ix; Curtius, *Europäische*, 295.
42. Curtius, *ELLMA*, 319–26.
43. Curtius, 380; Curtius, *Europäische*, 384.
44. Curtius, *ELLMA*, 380.
45. Curtius, 380, unnumbered page.
46. Curtius, 381; Curtius, *Europäische*, 385. But Curtius also grants that the order of the chapters does not match his order of discovery.
47. Curtius, *ELLMA*, 401, 382; Curtius, *Europäische*, 404, 386–87.
48. Curtius, *ELLMA*, 382–83; Curtius, *Europäische*, 386–87.
49. Curtius, *ELLMA*, 13; Curtius, *Europäische*, 23.
50. Curtius, *ELLMA*, 390; Curtius, *Europäische*, 394.
51. Curtius, *ELLMA*, 383; Curtius, *Europäische*, 386–87.
52. A brilliant appreciation of Spitzer's life and work is Starobinski's "La stylistique et ses méthodes."
53. Spitzer, review of *ELLMA*, 425–26. For another extremely ambivalent but much more comprehensive review, see Lida de Malkiel, "Perduración," 99–131.
54. Spitzer, review of *ELLMA*, 431.
55. Spitzer, 428.
56. Spitzer, 428.
57. Spitzer, 429.
58. Spitzer, 430.
59. Spitzer, "Milieu and Ambiance," 3.1: 2.
60. Spitzer's thoughts on early modern expressions of anthropomorphized infinity could be profitably compared with Mahler's reading above of Donne. Particularly remarkable is how the exorbitant cosmographical figuration Donne undertakes of Elizabeth Drury's body and spirit in the two *Anniversary* poems often hinges on the nearly catachrestic repetition of the (small) word *all*.
61. Spitzer, "Milieu and Ambiance," 3.2: 169, 175, 191, 196ff.
62. Spitzer, 3.2: 198. As Spitzer pithily puts it in German, "Wortwandel ist Kulturwandel und Seelenwandel."
63. Spitzer, "Linguistics and Literary History," 10, 8.
64. Spitzer, 15–19.
65. Spitzer, 19–20.
66. Spitzer, 23–24. Spitzer cites Curtius's teacher, Gröber, as having "formulated the idea of the philological circle" (35).
67. Spitzer, 30.
68. Spitzer, *Linguistics and Literary History*, 41–86; 135–92; 193–236. Invoking Auerbach reading Dante while pondering the one and many (215), Spitzer describes the hermeneut's experience as a kind of *radiation*, the same metaphor Auerbach privileges (see below).
69. Spitzer, 89.
70. Spitzer, 87, 105ff.

71. Spitzer, 133, 124.
72. Auerbach, review of *ELLMA*, 348.
73. Auerbach, 349.
74. Auerbach, 349–50.
75. Auerbach, "Curtius," 330.
76. Auerbach, 331.
77. Auerbach, 331.
78. Auerbach, 338.
79. Auerbach, 332.
80. For his appropriation of Vico's thought, see Auerbach, *Literary Language*, 7–17.
81. Auerbach, "Philology and *Weltliteratur*," 3.
82. Auerbach, 8; Auerbach, "Philologie der Weltliteratur," 304.
83. Auerbach, "Philology and *Weltliteratur*," 5, 9 (translation modified).
84. Auerbach, 11 (translation modified). Analysis of what Auerbach means exactly by "intuition" or parallels with his friend Panofsky's version of "synthetic intuition" are unfortunately beyond this essay's scope. For meanings of *Ansatz* and *ansetzen*, see Holdheim, "Hermeneutic Significance," 627–31.
85. See Blair, *Too Much to Know*.
86. Auerbach, "Philology and *Weltliteratur*," 12–13.
87. Auerbach, 13 (translation modified); Auerbach, "Philologie der Weltliteratur," 307–8.
88. Auerbach, "Philology and *Weltliteratur*," 14.
89. Auerbach, 13–14 (translation modified); Auerbach, "Philologie der Weltliteratur," 308.
90. Auerbach, "Philology and *Weltliteratur*," 14; Auerbach, "Philologie der Weltliteratur," 308.
91. See Gallagher and Greenblatt, "Touch of the Real," 20–48.
92. Auerbach, "Philology and *Weltliteratur*," 14; Auerbach, "Philologie der Weltliteratur," 308–9.
93. Auerbach, "Philology and *Weltliteratur*," 15; Auerbach, "Philologie der Weltliteratur," 309.
94. Auerbach, "Philology and *Weltliteratur*," 17; Auerbach, "Philologie der Weltliteratur," 310.
95. Auerbach, "Philology and *Weltliteratur*," 16; Auerbach, "Philologie der Weltliteratur," 310.
96. Auerbach, "Philology and *Weltliteratur*," 17; Auerbach, "Philologie der Weltliteratur," 310.
97. Auerbach, *Mimesis*, trans. Trask, 23; *Mimesis*, 26.
98. "BELIEBIG," Deutsches Wörterbuch von Jacob Grimm und Wilhelm Grimm, digitalisierte Fassung im Wörterbuchnetz des Trier Center for Digital Humanities, https://www.woerterbuchnetz.de/DWB?lemid=B03712.
99. Auerbach, *Mimesis*, trans. Trask, 30; *Mimesis*, 34.

100. Auerbach, *Mimesis*, trans. Trask, 44; *Mimesis*, 47.
101. Auerbach, *Mimesis*, trans. Trask, 44; *Mimesis*, 47.
102. Auerbach, *Mimesis*, trans. Trask, 272; *Mimesis*, 259.
103. Auerbach, *Mimesis*, trans. Trask, 298; *Mimesis*, 283.
104. Auerbach, *Mimesis*, trans. Trask, 309; *Mimesis*, 294.
105. For example, see Auerbach, *Mimesis*, trans. Trask, 474; *Mimesis*, 441.
106. Brownlee, "Ideology of Periodization," 183–87.
107. Said, *Beginnings*, 68–73. On Said's *contrapuntal* readings of Auerbach, see Newman's "Nicht am 'falschen Ort.'"
108. Auerbach, *Mimesis*, trans. Trask, 547–48; *Mimesis*, 509.
109. See Auerbach, introduction to *Literary Language*, 6–7, 17–24.
110. Auerbach, *Mimesis*, trans. Trask, 549.
111. Auerbach, 553. See Porter, "Earthly (Counter-)Philology," 258–59.
112. Auerbach, *Mimesis*, trans. Trask, 557; *Mimesis*, 518 (translation modified). Barck and Treml read *Mimesis* as a "homily" (Barck and Treml, *Erich Auerbach*, 29).
113. Moretti, "Modern European Literature," 3–27.

Bibliography

Agamben, Giorgio. "What Is a Paradigm?" In *The Signature of All Things: On Method*. Translated by Luca D'Isanto with Kevin Attell, 9–32. New York: Zone Books, 2009.
Auerbach, Erich. "Ernst Robert Curtius: Europäische Literatur und Lateinisches Mittelalter." In *Gesammelte Aufsätze zur romanischen Philologie*, 330–38. Bern and Munich: Francke, 1967.
Auerbach, Erich. "Introduction: Purpose and Method." In *Literary Language and Its Public in Late Latin Antiquity and in the Middle Ages*. Translated by Ralph Mannheim, 5–24. New York: Bollingen, 1965.
Auerbach, Erich. *Mimesis: Dargestellte Wirklichkeit in der abendländischen Literatur*. Tübingen: Francke, 2015.
Auerbach, Erich. *Mimesis: The Representation of Reality in Western Literature*. Translated by Willard R. Trask with a new introduction by Edward W. Said. Princeton: Princeton University Press, 2003.
Auerbach, Erich. "Philologie der Weltliteratur." In *Gesammelte Aufsätze zur romanischen Philologie*, 301–10. Bern and Munich: Francke, 1967.
Auerbach, Erich. "Philology and *Weltliteratur*." Translated by Marie and Edward Said. *Centennial Review* 13.1 (1969): 1–17.
Auerbach, Erich. Review of *ELLMA*. *Modern Language Notes* 65.5 (1950): 348–51.
Barck, Karlheinz, and Martin Treml. "Einleitung: Erich Auerbachs Philologie als Kulturwissenschaft." In *Erich Auerbach: Geschichte und Aktualität eines europäischen Philologen*. Edited by Karlheinz Barck and Martin Treml, 9–29. Berlin: Kadmos, 2007.
Benjamin, Walter. "Rigorous Study of Art." Translated by Thomas Y. Levin. *October* 47 (1988): 84–90.

Benjamin, Walter. "Strenge Kunstwissenschaft." In *Gesammelte Schriften*. Vol. 3. Edited by Hella Tiedemann-Bartels, 363–69. Frankfurt am Main: Suhrkamp, 1991.

Blair, Ann M. *Too Much to Know: Managing Scholarly Information before the Modern Age*. New Haven and London: Yale University Press, 2011.

Brownlee, Kevin. "The Ideology of Periodization: *Mimesis* 10 and the Late Medieval Aesthetic." In *Literary History and the Challenge of Philology: The Legacy of Erich Auerbach*. Edited by Seth Lerer, 156–75. Stanford: Stanford University Press, 1996.

Curtius, Ernst Robert. "Antike Pathosformeln in der Literatur des Mittelalters." In *Gesammelte Aufsätze zur romanischen Philologie*, 23–27. Bern and Munich: Francke, 1960.

Curtius, Ernst Robert. *Europäische Literatur und lateinisches Mittelalter*. Tübingen: Francke, 1993.

Curtius, Ernst Robert. *European Literature and the Latin Middle Ages*. Translated by Willard R. Trask. Princeton: Bollingen, 1990.

Efal, Adi. "Le 'regard philologique' de Warburg." Special issue, *Images Re-vues* 4 (2013). doi.org/10.4000/imagesrevues.2853.

Euclid. *The Thirteen Books of the Elements*. Vol. 1: Books 1–2, 2nd ed. Edited by Thomas L. Heath. New York: Dover Publications, 1956.

Gallagher, Catherine, and Stephen Greenblatt. "The Touch of the Real." In *Practicing New Historicism*, 20–48. Chicago and London: University of Chicago Press, 2000.

Gasché, Rudolf. *Of Minimal Things: Studies on the Notion of Relation*. Stanford: Stanford University Press, 1999.

Guillemin, Anna. "The Style of Linguistics: Aby Warburg, Karl Voßler, and Hermann Osthoff." *Journal of the History of Ideas* 69.4 (2008): 605–26.

Gumbrecht, Hans Ulrich. *Vom Leben und Sterben der großen Romanisten: Karl Vossler, Ernst Robert Curtius, Leo Spitzer, Erich Auerbach, Werner Krauss*. Munich: Carl Hanser, 2002.

Holdheim, W. Wolfgang. "The Hermeneutic Significance of Auerbach's *Ansatz*." *New Literary History* 16.3 (1985): 627–31.

Hooke, Robert. *Micrographia; Or, Some Physiological Descriptions of Minute Bodies Made by Magnifying Glasses With Observations and Inquiries Thereupon*. London, 1665.

Leibniz, G. W. "Monadology." In *Philosophical Texts*. Translated and edited by R. S. Woolhouse and Richard Francks, 267–81. Oxford: Oxford University Press, 1998.

Lida de Malkiel, María Rosa. "Perduración de la literatura antigua en Occidente." *Romance Philology* 5 (1951): 99–131.

Moretti, Franco. "Modern European Literature: A Geographical Sketch." In *Distant Reading*, 1–42. New York: Verso, 2013.

Newman, Jane O. "'Nicht am 'falschen Ort': Saïds Auerbach und die 'neue' Komparatistik." In *Erich Auerbach: Geschichte und Aktualität eines europäischen Philologen*. Edited by Karlheinz Barck and Martin Treml, 341–70. Berlin: Kadmos, 2007.

Pfisterer, Ulrich. "'Die Bilderwissenschaft ist mühelos': Topos, Typus und Pathosformel als methodische Herausforderung der Kunstgeschichte." In *Visuelle Topoi*.

Edited by Ulrich Pfisterer and Max Seidel, 21–47. Munich and Berlin: Deutscher Kunstverlag, 2003.

Pollock, Sheldon. "Future Philology? The Fate of a Soft Science in a Hard World." *Critical Inquiry* 35 (2009): 931–61.

Pollock, Sheldon. "Introduction." In *World Philology*. Edited by Sheldon Pollock, Benjamin A. Elman, and Ku-Ming Kevin Chang, 1–24. Cambridge, MA, and London: Harvard University Press, 2015.

Porter, James. "Erich Auerbach's Earthly (Counter-)Philology." *Digital Philology: A Journal of Medieval Cultures* 2.2 (2013): 243–65.

Pound, Ezra. *ABC of Reading*. London: Faber and Faber, 1991.

Pound, Ezra. *Selected Prose, 1909–1965*. New York: New Directions, 1973.

Rheinberger, Hans-Jörg. *Toward a History of Epistemic Things*. Stanford: Stanford University Press, 1997.

Said, Edward W. *Beginnings: Intention and Method*. New York: Columbia University Press, 1985.

Schäffner, Wolfgang, Sigrid Weigel, and Thomas Macho. "Das Detail, das Teil, das Kleine: Zur Geschichte und Theorie eines kleines Wissen." In *"Der liebe Gott steckt im Detail": Mikrostrukturen des Wissens*. Edited by Wolfgang Schäffner, Sigrid Weigel, and Thomas Macho, 7–17. Munich: Wilhelm Fink, 2003.

Spitzer, Leo. *Linguistics and Literary History: Essays in Stylistics*. Princeton: Princeton University Press, 1948.

Spitzer, Leo. "Milieu and Ambiance: An Essay in Historical Semantics." *Philosophy and Phenomenological Research* 3.1 (1942): 1–42; and 3.2 (1942): 169–218.

Spitzer, Leo. Review of *ELLMA*. *American Journal of Philology* 70 (1949): 425–31.

Starobinski, Jean. "La stylistique et ses méthodes: Leo Spitzer." *Critique* 206 (1964): 579–97.

Vossler, Karl. *Frankreichs Kultur im Spiegel seiner Sprachenentwicklung: Geschichte der französischen Schriftsprache von den Anfängen bis zur Klassichen Neuzeit*. Edited by W. Meyer-Lübke. Heidelberg: Carl Winter, 1921.

Warburg, Aby. "The Absorption of the Expressive Values of the Past." Translated by Matthew Rampley. *Art in Translation* 1.2 (2009): 272–83.

Warburg, Aby. *Bilderreihen und Ausstellungen*. Edited by Uwe Fleckner and Isabella Woldt. *Gesammelte Schriften* II.2. Berlin: Akademie, 2012.

Warburg, Aby. *Der Bilderatlas: Mnemosyne*. Edited by Martin Warnke with Claudia Brink. *Gesammelte Schriften* II.1. Berlin: Akademie, 2008.

Warburg, Aby. *Die Erneuerung der heidnischen Antike: Kulturwissenschaftliche Beiträge zur Geschichte der europäischen Renaissance*. Edited by Horst Bredekamp and Michael Diers. *Gesammelte Schriften* I.1–2. Berlin: Akademie, 1998.

Warburg, Aby. "Ninfa fiorentina." In *Werke in einem Band*. Edited by Martin Treml, Sigrid Weigel, and Perdita Ladwig, 198–210. Berlin: Suhrkamp, 2010.

Warburg, Aby. *Tagebuch der Kulturwissenschaftlichen Bibliothek Warburg*. Edited by Karen Michels and Charlotte Schoell-Glass. *Gesammelte Schriften* VII. Berlin: Akademie, 2001.

Weber, Samuel. "Gott und Teufel—im Detail." In *"Der liebe Gott steckt im Detail": Mikrostrukturen des Wissens*. Edited by Wolfgang Schäffner, Sigrid Weigel, and Thomas Macho, 43–51. Munich: Wilhelm Fink, 2003.

Weigel, Sigrid. "'Nichts weiter als . . .' Das Detail in den Kulttheorien der Moderne: Warburg, Freud, Benjamin." In *"Der liebe Gott steckt im Detail": Mikrostrukturen des Wissens*. Edited by Wolfgang Schäffner, Sigrid Weigel, and Thomas Macho, 91–111. Munich: Wilhelm Fink, 2003.

PART FOUR

Minimilieus of Modernity

✦✦✦

Architecture and Ambience
From Walter Benjamin to the French Avant-Garde

MALTE FABIAN RAUCH

✦ ✦ ✦

ON MARCH 21, 1959, Guy Debord sent a letter to the artist and architect Constant Nieuwenhuys, to which he appended a series of notes titled "Ecology, Psychogeography and Transformation of the Human Milieu." The tone of the text is programmatic. Questions are asked and answers given. In the opening lines, Debord declares, "Psychogeography is the playful part in contemporary urbanism. Through this ludic apprehension of the urban milieu, we will develop perspectives of an interruptive construction of the future."[1] This ludic apprehension is concerned not with classical parameters of urban space or architecture but with the elusive notion of "ambience," which Debord and Constant also describe as "microambiences" and "*microclimats*" (translated as "microatmospheres") in the structure of the urban tissue. Thus, in an essay from 1958, Constant and Debord claim that "the construction of a situation consists in the edification of transitory micro-ambiances."[2] Two years later, in 1960, Constant presented a lecture in Amsterdam that outlines the plan for a new city, New Babylon, where "atmosphere" itself would become an "artistic medium" through which social space in its entirety could be reorganized.[3] What are such microambiences? And how could ambience and atmosphere, the most minute and ephemeral things, be conceived of as the basis of a radical politics and architecture?

Since the early 1950s, Constant and Debord, spokesmen of the avant-garde groups Lettrist International (LI; 1952–57) and the Situationist International (SI; 1957–72), developed a critique of established theories of urbanism and architecture as well as an experimental "unitary urbanism." Constant came to be interested in architecture through a trajectory that led him from the avant-garde group Cobra to the Mouvement international pour un Bauhaus imaginiste (MIBI), founded by Asger Jorn in 1953. At the third conference of this movement in Alba, Italy, Constant met Debord and was struck by the convergence of their views on art, architecture, and urban space. When the SI was established in 1957, Constant would be one of the founding members. The encounter between him and Debord established the elusive notion of ambience and its correlates as the foundation of Constant's project. Thus, New Babylon turned into an attempt to make the situationist critique of the city concrete—to realize the idea of a different, unitary urbanism in a new city that would be centered around the construction of ambience and atmosphere. The idea would occupy Constant for the years to come, a project constantly changing, perpetually evolving, definitively unfinished in 1974. However, by the time Constant gave his lecture on the theme in the winter of 1960, the discussion between him and Debord had already come to an end. From then on, their paths diverged. Constant worked alone on New Babylon. Debord turned away from a constructive interest in urbanism. Their theorization of microambiences as an element of the urban tissue and a possible basis for a new architecture thus occurs in a closely circumscribed period, roughly from 1953 to 1959.

The aim of the present essay is to show how minutiae become a decisive factor in this theoretical trajectory, insofar as this understanding of ambience is linked to extremely minute nuances—humidity of the air, intensity of light, density of smell, mixture of sounds, rhythms in the streets, and textures and materials—that are conceived as gathering together to form the atmospheric fabric of a city. In fact, this interest in an aesthetics of urban ambience is partly taken up from surrealism, but it undergoes a decisive shift with the situationist theorization of urbanism. Accordingly, the main thesis to be developed here is that this fascination with minute elements of urban space, which remain beyond the threshold of intentional perception, was prepared through the surrealist poetics of the city. Yet for the surrealists, this atmospheric layer of the city revealed itself only through a highly individual, aestheticized experience of chance encounters. By contrast, the situationists try to conceive the minute elements in the city's atmospheric texture as qualities of social space and to develop an architecture that would be able to influence these minute elements—that is to say, a collective architecture of microambiences.

Hubert Damisch once wrote that "questions of the visibility of the city, or—as we would now say—of its 'legibility,' if not its 'representability,' became pressing only from the moment its image was unsettled."[4] In Constant and Debord, we witness an incisive moment where the question of the city's legibility is posed against the unsettling of its image through the functionalist and capitalist transformation of urban space that dominated the postwar reconstruction of European cities. This led to the prominence of notions such as *atmosphere, milieu, climate,* and *ambience* as the defining elements of a new syntax and lexis for the city's "legibility"—a legibility that is elusive precisely because of its dependence on minute elements. Only a decade earlier than the situationists, Leo Spitzer had, for the first time, undertaken an archaeology of precisely this semantic field, subtly lamenting its erosion. In Spitzer's analysis, the original understanding of the ancient Greek *periechon*—a protective sheltering space—passed through the Latin "ambience" and then arrived at the (seemingly deterministic) modern notions of "milieu" and "environment."[5] If Spitzer's historical study can be taken as a historical symptom of the increasing difficulty of thinking a surrounding space in terms of entanglement and interconnection, as opposed to atomistic cause-and-effect relations, then Constant and Debord's project could be read as the attempt to reinvent *ambience* and *atmosphere* as a form of being-in-relation-to and participation in urban space.

Such, at least, is the legibility that our contemporary moment allows for. Today, the expansion of various networks is increasingly recognized as the driving force that reconfigures our concepts of space and place. Faced with this phenomena, contemporary theory has recently undertaken a reexamination of hitherto neglected concepts and categories that might account for this new regime of connectivity. Arguably, the proliferating research on "ecology," "milieu," "atmosphere," "Umwelt," and "environment" punctuates a theoretical demand to rethink topological and topographical categories that may be capable of describing our heterogeneously entangled habitat. It is, therefore, astounding that the conceptual field that we saw surfacing in the debate between Debord and Constant—the semantics of ambience, atmosphere, milieu, and ecology—has received so little attention, having for the most part been dismissed as vague but self-evident metaphors. In his attempt to rethink spherical surroundings, Peter Sloterdijk, for instance, acknowledges the situationist vocabulary of atmosphere yet fails to provide any discussion of the role of minutiae for this approach.[6]

Lastly, the situationists' explicit use of the semantics of ecology invites an additional contextualization in contemporary aesthetics, architecture, and media theory. Here, the most striking aspect of the situationists' theory of microambiences is not that it anticipates a variety of theoretical developments: the elaboration of an aesthetics of atmosphere, be it as a general paradigm[7] or with precise reference to architecture,[8] and the link between the elementary notions of "media" and "environment"[9] or the understanding of air as a medium.[10] More important, rather, is the degree to which the situationist approach is located in the trajectory of ecologization and becoming environmental, which has been theorized by Erich Hörl as being divided between "general" and "restrictive" tendencies. The former refers to a paradigm that "does not turn relations into minor and derivative entities but considers them to be originary," whereas the latter describes the exploitation of relations through the "becoming-environmental" of power and capital, which culminates in contemporary "environmentalitarian" forms of surveillance.[11] The way in which the situationists targeted the organization of modern urban space—an environment that is essentially a closed circuit of work and consumption, where all mobility is determined by the flows of capital—is quite precisely a critique of a restrictive form of ecology. By contrast, the architecture of micoambiences corresponds to general ecology: a form of nonfunctional, nonutilitarian movement in urban space, attuned to a new regime of connectivity and relationality.

Benjamin's Surrealism: The Psychotopography of Atmospheres

THE LETTRIST and situationist approaches to the city are deeply influenced by their surrealist predecessors, notwithstanding the former's severe polemic against the latter. Precisely insofar as the surrealists achieved a turn away from the contemplative relation to art and explored the lived experience of the city, they provided the precondition for the lettrist and situationist projects in both a positive and a negative sense. Seen in this light, what links *and* separates their respective approaches toward the city is the notion of "atmosphere." In the literature on surrealism, Walter Benjamin's essay on the movement has long held a canonical status. If the following discussion reads surrealism through the lens of Benjaminian concepts, it is, however, not because of his essay's canonicity but because Benjamin provides the most acute theorization of surrealism's understanding of "atmosphere" as a psychotopographical view of Paris. This acuteness may, perhaps, stem from the productive tension Benjamin felt toward surrealism: an enthusiasm for the movement's political impetus combined with a skepticism about the appropriateness of its means—a tension that in many ways foreshadows the situationists' critique of their predecessors.

In his "Little History of Photography," Benjamin characterizes Eugène Atget as a precursor to surrealism insofar as his work looked "for what was unremarked, forgotten, cast adrift," meaning those minute details that are usually not deemed worthy of attention.[12] This genealogy in which Atget is the precursor of surrealism is closely linked to the notion of "atmosphere," for Atget, Benjamin writes, was the first to "disinfect the stifling atmosphere generated by conventional portrait photography in the age of decline." He goes on to note that Atget "cleanses this atmosphere—indeed, he dispels it altogether."[13] What appears here as a rejection of atmosphere is refined, in Benjamin's seminal essay on surrealism, as a further shading of the concept. In fact, one could say that the kinship between Atget and the surrealists consists, for Benjamin, in the abolishment of preceding conventions of creating ambiences and the search for new atmospheres—efforts deeply linked to the forgotten and marginalized spaces of the old Paris that Atget explored. Thus, Benjamin notes in regard to the surrealists, "they bring the immense force of 'atmosphere' [*Stimmung*] concealed in these things to the point of explosion."[14]

Crucially, Benjamin here puts the notion of "atmosphere" (*Stimmung*), which condenses the entire argument, in quotation marks, as if to signal its contrast to the preceding forms of ambience-generating conventions. The new, eminently critical, even revolutionary meaning that "atmosphere"—rendered by Benjamin no longer as the Latinate *Atmosphäre* but as *Stimmung*, which can also translate as "attunement"—receives in surrealism is, according to this analysis, deeply linked to architecture and certain marginal regions of urban space: "At the center of this world of things stands the most dreamed-about of their objects: the city of Paris itself."[15] The city's space is the very medium of atmosphere—a latent medium, however, since it only comes into view from a specific angle, carved out by surrealist aesthetics. What this aesthetic brings to light is an intimate link between the marginal and the minute, as these elements, usually beneath the threshold of attention, are constitutive of the city's atmospheric fabric. Emphatically, this atmospheric dimension of the city is the counterpoint to the routines of work, functionality, and usefulness. For the concept of "atmosphere" figures the city as charged with latent meaning, a microcosmos hidden in the city's recesses. And this atmospheric "intensity," which only certain regions in urban space possess, is linked, for Benjamin, to the fact that they are marked by destitution. That is to say, atmosphere, as conceived in surrealism, has a temporal signature. It is manifest in zones of the useless, the démodé, and the forgotten, where multiple layers of history condense into an atmospheric stratification. This temporal density functions as a resistance to the homogenization of urban space-time in accordance with the logic of utility and transparency: atmosphere is, one could say, an anachronistic remnant of capitalist modernization—whence its "revolutionary" potential.

The atmosphere of the outmoded is also at play in Benjamin's theory of the flaneur, a figure that is linked with—although not identical to—the surrealist experience of the city. Thus, in his texts on Franz Hessel—the review of *Heimliches Berlin* and "The Return of the Flâneur"—Benjamin essentially restates the theory of multiple temporal layers that defines Hessel's surrealist experience of the city. He explicitly speaks of "atmospheric resistances" that the flaneur encounters in Berlin (as opposed to Paris).[16] Similarly, in his Baudelaire essays, Benjamin delineates the nonplace proper to the flaneur: "Let the many attend to their daily affairs; the man of leisure can indulge in the perambulations of the flâneur only if as such he is already out of place."[17] What is decisive here is, again, that the flaneur is perceptive to the fact that the topographic space of the city is shot through with heterogeneous atmospheres that are at variance with the space-time of modern capitalism. Their attention lingers on minute details, be it the play of light and shadow, traces of the past that appear beneath the façade of buildings, or the chance encounter of city sounds and smells. Atmosphere denotes, then, a dimension of urban space that remains irreducible to the logic of capital—a milieu of affect, memory, and the unconscious. Alexander Gelley pointedly characterizes this attunement to the city: "The flâneur gives over to the associational currents of his environment and thus makes himself into a registering sensibility."[18] Surrealism and flânerie, one could say, explore ways of opening onto, and letting oneself be led by, these atmospheric minutiae of urban space, where the forces of desire, association, and memory interrupt the mundane experience of city life.

The language of currents, environment, and interchange aptly describes how this praxis aims at making space porous, relational, fluid—opening it up to an essentially subconscious and affective dimension that is triggered by details in the texture of the city. And this, in fact, exactly describes the surrealist experience of the city's atmosphere. In an important study, Susan Laxton has recently stressed that the attempt to extend the aesthetic exploration of the unconscious to the exploration of the city as a topography of desire emerged early on in the movement. In 1924, André Breton, Louis Aragon, Max Morise, and Roger Vitrac started from Blois, a town randomly chosen on the map, to undertake a stroll in the "absence of any goal." As Laxton succinctly notes, "From this time on, the structure and operations of the psyche would be understood to have their analogue in the Paris streets."[19] In a phrase that seems to echo the Benjaminian figure of the flaneur precisely, Breton describes their activities in these terms: "An uninterrupted quest was given free rein: its purpose was to behold and disclose what lay hidden under appearances."[20] What lies hidden under appearances are the minutiae in the urban fabric, which acquire their singular significance for the individual psyche.

For the surrealists, the Paris of work, regularity, and functionality thus conceals another layer of the city, linked to dream and desire, which can only be accessed through such aimless quests. In keeping with this desire for the psychic topography of the city, always understood as the minute texture of the city that usually goes unnoticed, Roger Caillois asks how one could fail to see that this setting "conceals another Paris, the true Paris, a ghostly, nocturnal, intangible Paris that is all the more powerful insofar as it is more secret; a Paris that anywhere and at any time dangerously intrudes upon the other one."[21] The attunement to these atmospheric minutiae—sound and smell, light and rhythm as well as the associations and affects, memories and reveries that they evoke—opens up an aesthetic dimension of the city that, as it were, escapes the prose of the world. Or, as Benjamin puts it in one note, "Flânerie is the rhythm of this slumber."[22]

Yet it is well known that Benjamin's figure of the flaneur is marked by a certain ambivalence, and it is in fact possible, as has been previously suggested, to link this ambivalence to the criticism that the situationists leveled against their surrealist predecessors. For, to the degree that flânerie remains an individual and ultimately "poetic" practice, this quest for marginalized, outmoded, and purposeless spaces remains, Benjamin argues, determined by what it implicitly critiques: capital as the governing matrix of urban space. This is because the drift through changing environments is, according to Benjamin, underwritten by an altogether different logic: "The intoxication to which the flâneur surrenders is the intoxication of the commodity immersed in a surging stream of customers."[23] Despite the fact that they may refuse the imperatives of work, the individual flaneur thrives on the commodified space of the metropolis, unable to resist it in a politically meaningful way. And we may add to this that the weak, passive agency of the flaneur, attentive as he might be to the atmospheric fabric, nonetheless remains a clearly gendered one, a point to which Benjamin remains in fact oblivious.

Without wanting to equate the surrealist poetics of the city with Benjamin's materialist deciphering of the flaneur (or even claim that Benjamin himself would equate them), their exploration of the city remains marked by comparable reliance on the individual subject, especially in Breton. Thus, to the degree that we can extrapolate from Benjamin's theory of flânerie an account of the surrealist experience of Paris, we could argue that their quest for "subversive" atmospheres is deeply intertwined with the processes of urban commodification—which they thus resist and bring about. In addition to the fact that the aimless exploration of Paris presupposes the aesthete's privilege of not having to work (and thus obey the space-time of utility and functionality), it is worth dwelling on one passage from Breton's *La Clé des champs*, which, as several commentators have already remarked, seems to anticipate the cartographic practice of the situationists. Upon closer examination, however, one can discern not only the kinship with but also the contrast to the situationist endeavor. Of the unconscious topography of Paris, Breton writes, "A walk down a single street . . . can provide . . . alternating zones of well-being and disquiet. No doubt a highly significant map should be drawn up *for each individual* which would indicate in white the places he is prone to haunt and in black those he avoids, the rest being divided into shades of grey according to the greater or lesser degree of attraction or repulsion exerted."[24] Importantly, the experience of the city is, here, an individual experience, keyed to "certain aspects," which concerns precisely the nuances in the urban fabric and can therefore produce alternating zones within the space of a single street. Thus, the way in which the atmospheric minutiae punctuate the topography of Paris is indexed to the individual's itineraries in the city, for the "exterior" space is indissociable from the "interior" space of the subject insofar as the city obtains its atmospheric significance only in relation to an individual psyche. Michael Sheringham aptly dubs this "Paris-as-subjectivity," a view of the city that ultimately figures the work of deciphering the "hidden" meanings of the city as an analog to deciphering the hidden meaning of the individual subject.[25]

Reconfiguring the Aesthetic

THE LI and SI's signal concept for the exploration of urban space, "psychogeography," readily betrays an indebtedness to surrealism, as it obviously refers to a psychic or subconscious quality of urban space. However, faced with the postwar attempts of Breton to reassemble his movement, the situationists openly declared the necessity of leaving surrealism behind. In the first issue of the SI's journal, *Situationist International Anthology*, we read, "For us, surrealism has been only a beginning of a revolutionary experiment in culture, an experiment that almost immediately ground to a practical and theoretical halt. We have to go further."[26] With regard to the elements that compose the city's atmospheric fabric, the surrealists' theoretical and practical impasse was, as will be shown in this section, the result of their focus on the individual subject. Going further than surrealism means, therefore, delinking these atmospheric minutiae from their reference to the individual psyche in order to subject them to collective control and creation. For the situationists, a new form of architecture is supposed to achieve this shift: architecture moves beyond surrealism's individualism but also beyond the avant-garde's negation of art in order to reconfigure the very meaning of the "aesthetic." Here, aesthetics no longer names the individual perception—operative in a separate sphere—of the beautiful, the sublime, or even the shockingly ugly.

Rather, aesthetics comes to name an expansive *aisthesis* of microambiences, which are both materialized and conditioned by a new collective architecture.

In a glossary in the very first issue of SI's journal, the semantic field of milieu and atmosphere takes center stage.[27] Thus, "psychogeography" is defined as "the study of the specific effects of the geographical milieu [*milieu géographique*] (whether consciously organized or not) on the affective behavior of individuals." And the notion of "unitary urbanism" receives the following gloss: "The theory of the combined use of arts and techniques as means contributing to the integral construction of a milieu [*la construction integrale d'un milieu*] in dynamic relation with experiments in behavior." Already at this level, it is evident that the pride of place given to the semantics of milieu and ambience denotes something different from the surrealist understanding of the city-as-artwork and especially of Paris-as-subjectivity, while the talk of "specific effects" quite clearly takes up the interest in minute details of the urban fabric. It is as if the surrealist sense of the potential of atmospheric minutiae is subjected to dry scientific scrutiny and to a political, almost technical application. This impression is deliberate. As Constant puts it in 1958, "To our way of thinking, the traditional arts will no longer be able to play a role in the creation of the new ambiance [*l'ambiance nouvelle*] in which we want to live.... Our conception of urbanism is thus a social one."[28] For the situationists, the ambient quality of the city is thus understood as the quality of *social* space. This affects the turn from an individual aesthetics of urban atmospheres to a political ecology of ambience: atmospheric minutiae no longer appear from the point of view of a highly aestheticized subject but as elements of social construction.

Recent scholarship has demonstrated to what degree Debord's initial interest in urban space was mediated through his encounter with Ivan Chtcheglov in 1953.[29] When Debord met Chtcheglov, he had already begun to develop the idea of constructing lived "situations" that would go beyond preceding forms of avant-garde experimentation and had been practicing aimless strolls through Paris, which would serve as the practical element in his subsequent theorization of urbanism. But it was Chtcheglov who made Debord aware of the fact that the construction of highly intense situations of lived experience cannot be abstracted from the architectural elements of the spatial environment where they take place. Henceforth, it was clear that a revolution of everyday life cannot be anything but the appropriation and reconstruction of everyday space, its architectural forms, and its atmospheric minutiae.

Steeped in the surrealist poetics of the city as well as in occult and mythological literature, Chtcheglov, in October 1953, outlined his vision of the city in a seminal essay titled "Formulaire pour un urbanisme nouveau," signed with the pseudonym "Gilles Ivan." In a lyrical vein, Chtcheglov opens this text with an elegy of the state of contemporary life in the city, which he deems to be deprived of sense and significance. Against the backdrop of the analysis in the last section, we can easily discern the echo of the surrealist aesthetics of anachronistic atmospheres as Chtcheglov evokes the presence of multiple temporal layers in the city: "All cities are geological. You can't take three steps without encountering ghosts bearing all the prestige of their legends. We move within a closed landscape whose landmarks constantly draw us toward the past."[30] What has disappeared from this metaphorical spatialization of time, however, is analogy with the psychic topography of the individual subject. That this "geological" understanding of the city was moving toward a collective pole is also underlined by the fact that Chtcheglov visualized this in his "metagraphie" from the same year, a generic subway map of Paris on which he superimposed cut-out fragments from maps of Africa, Asia, and India to suggest the spatialization of multiple temporalities and atmospheres in the city.

While Chtcheglov does credit the Dadaist and surrealist attempts to recover this dimension of urban space, he considers their impetus lost. Therefore, he introduces a new element, designating architecture as the means for an active restoration of this dimension. Agitating, like Breton before him, against what he perceives as Le Corbusier's stale, Cartesian functionalism, Chtcheglov sketches a synthetic view of architecture, centered around affect, desire, and play—an architecture, that is to say, that would be capable of modulating time and space to an as-of-yet-unknown degree: "The architecture of tomorrow will be a means of modifying present conceptions of time and space.... One can only speak of a new architecture if it expresses a new civilization."[31] Architecture, in Chtcheglov, is thus imagined as the technical facilitator of a veritable new cosmogony—shaping time, space, and collective action. Importantly, this "new architecture" is conceived as mobile, allowing—through its plasticity—for a "continuous drifting" as the habitual way of movement of the city's inhabitants. Hence is not a preexisting, static structure that is externally imposed on people; rather, it exists as a mobile structure in a dynamic, reflexive interchange between social practices and architectural form, constantly oscillating between structuration and destructuration.

Chtcheglov acknowledges that this conception remains somewhat elusive and vague, but he insists that this will necessarily remain so until "experimentation with patterns of behavior has taken place in cities specifically established for this purpose."[32] Besides this linking of architecture and practical experimentation, there are two other incisive points in the text for our investigation. First, Chtcheglov uses the concept of *ambience* in a sentence where he suggests that the works of the baroque painter Claude Lorrain offer a prefiguration of such a mobile architecture: "This ambience is provoked by an *unaccustomed architectural space*."[33] Here, "ambience" is conceived as a synaesthetic dynamic between architecture and comportment, such that architectural form provokes an incitement to drift through an affectively charged space. Instead of conceiving action and architecture as two separate pools, where one determines the other, the notion of ambience tries to think of their interrelation. To a certain degree, one could already speak here of an ecology of ambience, inasmuch as the term designates precisely the dynamic in-between space of practice and form. Secondly, Chtcheglov conceives of a correlation between different affects and partitions of the city: "The districts of this city could correspond to the whole spectrum of diverse feelings that one encounters *by chance* in everyday life."[34] Here the traditional partition of the city into quartiers—each of which has a distinct atmosphere due to minute differences in its fabric—is rethought as corresponding to the affects of chance encounters, which obviously echoes the aleatory aesthetics of surrealism. Yet in contrast to traditional quartiers, which attain their atmospheric signature in a slow historical process, Chtcheglov's vision clearly conceives this atmospheric and affective physiognomy as an object of construction, not to say manipulation. This, in a certain way, translates the surrealist aesthetics of chance encounters in the city into concrete architectural forms, if still in an extremely utopian way.

Both of these thoughts will become extremely important in Debord's appropriation of Chtcheglov's thought. The text Debord wrote as an immediate reaction to Chtcheglov's essay bears the title "Manifeste pour une construction de situations." In this "manifesto," Debord outlines the construction of situations and brings it together with Chtcheglov's thoughts about an architecture to come. The text opens with the declaration that both bourgeois and preceding avant-garde groups, such as the surrealists, have reached an impasse—a moment that is, importantly, conceived as a diagnosis about the state and fate of *aesthetics*. "Our time," Debord writes, "sees the Aesthetic die."[35] After art—that is, after aesthetics-as-art and art-as-aesthetics—Debord, following Chtcheglov, conceives of architecture as a holistic enterprise capable of changing the very meaning of the aesthetic: "The new architecture will condition everything . . . In succeeding to use the other arts . . . architecture will become this directive synthesis of the arts that marked the great epochs of the Aesthetic."[36] Architecture is thus conceived of as surpassing the avant-garde's negation of art. It literally reconstructs the "aesthetic" from art's ruins, and it does so through modeling the minute texture of the city that influences the *aisthesis* of atmosphere. Regarding this conception, Thomas Y.

Levin rightly notes that urbanism seemed to the situationists to be "at once a negation and a realisation of art."[37] But instead of conceiving this double movement as a sublation in a *Gesamtkunstwerk* of the city, one can perhaps understand it, with Agamben, as a "point of indifference." In his most subtle text on Debord, Agamben glosses the situationists' signal concept as follows: "The situation is neither the becoming-art of life nor the becoming-life of art.... The 'Northwest passage of the geography of the true life' is a point of indifference between life and art, where *both* undergo a decisive metamorphosis *simultaneously*."[38] In this zone of indifference, where art and life are transformed without simply merging together, the very notion of the aesthetic receives its new meaning: postartistic aesthetics will be an aesthetics of microambiences, or it will not be.

Further, Debord explicitly takes up Chtcheglov's partition of the city in accordance with different affective zones and specifically highlights the "ecological" aspect of this future architecture. Here, we can see how Debord wrestles with the surrealist legacy of an individual aesthetics of the city and the ambition of transforming the space of collective life: "One has to establish a specific cartography founded on the new psychogeographical data.... The beginning of this cartography will probably be subjective. Everyone would have their own map of a city. But I do not doubt that a new objectivity, perceptible to everyone, would establish itself quickly."[39] What is important here is the clear articulation of the conflict between, on the one hand, an essentially surrealist sensibility for the minute nuances of the city—which are in danger of being flattened out by gentrification and homogenization—and, on the other hand, the attempt to translate these into collective forms. To put it pointedly, we witness the transition from Breton's map of Paris-as-subjectivity to a collective map of the city's ambient effects, where the latter could guide a collective architecture. Pushing Chtcheglov's cartographic approach further, Debord also began to collaborate with Asger Jorn to produce radically fragmented "psychogeographic" maps, which show parts of Paris with particularly intense, ambient profiles separated by empty space and connected by arrows that visualize the forces of attraction and repulsion. In lieu of Chtcheglov's spatialization of time, these maps further fragment the city to disclose existent ambient zones in their spatial discontinuity as material for a new construction of urban space. If Chtcheglov held on to the quartier as the atmospheric locus, these maps effectively zero in on even smaller units: crossroads, a few interconnected streets and places, where minute elements gather together to create a particular atmosphere.

Situationist Ecology

BESIDES CARTOGRAPHIC practice, Debord, in a series of texts, sets out to define the aesthetics of ambience as a new field of research. Ambience, Debord argues, cannot be considered the result of stylistic or historical changes in architecture, nor of housing conditions, nor of other "hard" urbanist data. In his "Introduction to a Critique of Urban Geography," he explains, "The sudden change of ambience in a street within the space of a few meters; the evident division of a city into zones of distinct psychic atmospheres . . . all these phenomena seem to be neglected. . . . In fact, the variety of possible combinations of ambiences, analogous to the blending of pure chemicals in an infinite number of mixtures, gives rise to feelings as differentiated and complex as any other form of spectacle can evoke."[40] Here, the highly nuanced, minute ambient effects of the urban environment—the play of light and shadow in a street, the interaction of light and architecture, the murmur in the background—are defined as the object of an inquiry that oscillates between parascientific aspiration, playful experiment, and constructive ambition. This epistemic ambivalence is precisely reflected in a set of notes from the following year—where Debord calls for a "scientific" (in quotation marks) investigation of ambiences—and, even more forcefully, in the note for a planned exhibition, where Debord characterizes psychogeography as the "science-fiction of urbanism."[41] But science fiction, Debord notes in the text he sent to Constant, must be "understood as a part of actual life, whose proposals are destined for a practical application."[42] Thus, ambience—the name for an impossible object, the most minute nuances of the experience of urban space—turns into the horizon of a new aesthetics, a different knowledge, and the basis of a radically political architecture.

These elusive ambient effects, Debord argues in continuation of the surrealist approach, can only be traced through a "total *insubordination* to habitual influences," such as tourism or consumerism.[43] Yet this flaneur-like distance from utilitarian imperatives is now in the service of a collective apprehension, appropriation, and future construction of urban space. What is important here is the attempt to think *social* space in as nuanced a way as Breton and company wanted to consider the individual, aesthetic experience of the city. This line of thought is directly continued in Debord's important "Theory of the Dérive." For Debord, the "dérive" describes an aimless drift through the city—capable of opening up to the ambient currents of the environment—a practice he and other group members had been experimenting with since 1953. The link to the surrealist flânerie is quite evident here, as the dérive is also a form of attunement to the atmospheric texture. Exploring the highly receptive potential of this immobility in movement, Debord explicitly acknowledges and criticizes the surrealist exploration of the city in this text:

> Among the diverse Situationist procedures, the *dérive* defines itself as a technique of rapid passage through varied ambiences [*ambiances variées*]. . . . In a *dérive* one or more persons during a certain period drop their relations, their work and leisure activities, and all their other usual motives for movement and action, and let themselves be drawn by the attractions of the terrain and the encounters they find there. Chance is a less important factor in this activity than one might think: from a derive point of view cities have psychogeographical contours, with constant currents, fixed points and vortexes that strongly discourage entry into or exit from certain zones.[44]

With the *epoché* of the capitalist demands of work and its insistence on the ludic and experimental, the dérive clearly bears traces of the surrealist *errance*, yet Debord is careful to insist that it has nothing to do with "classic notions of journey or stroll" and the importance formerly attributed to chance. It is precisely against the surrealist notion of chance, understood by Breton as the knot that binds the individual subject to the "hidden" atmospheric minutiae of the city, that Debord positions the notion of ambience as a collective, quasi-objective quality of urban space, its discontinuous and heterogeneous contours notwithstanding. What was once glimpsed in an aleatory aesthetics must be turned into an objective datum for construction: "Progress means breaking through fields where chance holds sway by creating new conditions more favorable to our purposes."[45] Thus, Debord goes on to note that the greatest challenge facing psychogeography is that the "different unities of atmosphere [*unités d'atmosphère*] and of dwellings are not precisely marked off," defining the goal of the dérive to map these as-yet-unknown regions.[46]

There could be no starker contrast to Breton's idea of drawing a map "for each individual." Atmosphere has become a quality of social space, manifest in a specific "milieu," which can be traced, appropriated, and used as the basis for a different urbanism. Importantly, atmosphere is still conceived as being composed of minute nuances that are not traceable through neutral perception, but this elusiveness is not an obstacle to their conscious and collective manipulation. The resulting shift—from the individual to the collective—is, again, nothing less than a transformation of the very meaning of the "aesthetic." Aesthetics comes to be understood as a collective architecture of microambiences: "Integral art . . . can only materialize at the level of urbanism.

But it can no longer correspond with any traditional definitions of the aesthetic. In each of its experimental cities, unitary urbanism will work through a certain number of force fields, which we can temporarily designate by the standard expression *quartier*."[47] Note the ambivalent characterization of the "quartier" here, which seems to return to Chtcheglov's imagined partition of the city into affective zones. This ambivalence might, in fact, be due to Debord's second major theoretical source for his approach. Decisive for the entire question of the postartistic, parascientific determination of atmosphere but especially for the question of the district was Debord's encounter with the work of Paul-Henry Chombart de Lauwe. A student of Maurice Halbwachs, Chombart was an important postwar sociologist of the working class whose approach moved away from a deterministic understanding of the milieu to a more dynamic view of the relation between collective practices, social imaginary, and urban space, focusing particularly on the dynamics within a quartier.[48] With the introduction of new visualizations of the city, such as aerial photography, diagrammatic maps of people's daily movements, and sociopsychological charts, Chombart aimed at a new interpretation of the city that could make room for the ways people inhabiting space perceive and interpret it. In a synthesis of different disciplines, Chombart thus tried to investigate how natural, technical, economical, and architectural developments interact with the inhabitant's experience and symbolic interpretation of urban space.

To this end, Chombart also drew on the work of the Chicago School of Sociology (Burgess and Park), who developed a partition of the city in concentric zones and introduced field-work methods from ethnography to study the inhabitant's behavior. Chombart also worked with the further refinement of this concentric partition scheme in the "human ecology" of James A. Quinn.[49] When Debord (who did not read English) speaks of ecology, the concept is always mediated through Chombart's discussion and appropriation of these studies of inhabitants' interaction with the urban environment. For our investigation, there are two important concepts to highlight here, since Debord explicitly draws on them. First is the notion of the "urban tissue," adopted from urban ecology, which is structured by quartiers whose "centers of attraction" consist of shops, working places, and schools.[50] Second is the notion of "microclimates," which encompass, for Chombart, "sunlight, the circulation of air in certain streets, the presence of trees" and do have an influence not only "on the hygiene and the morbidity rate, but also on the practices and habits of the inhabitants."[51] Whereas the "center of attraction" refers to socioeconomic dynamics, the notion of "microclimates" denotes the dynamic between natural influences and urban space. It has, strikingly, escaped commentators thus far that the situationists fused these two distinct meanings in their aesthetics of ambience.

For Debord, still laboring under the influence of Chtcheglov's new vision of the city, the encounter with Chombart's vision was extremely potent.[52] As we saw, Chtcheglov had already envisioned a partition of the city according to districts with different affective registers. Now in Chombart's work, Debord encountered an explicitly "ecological" view that focuses precisely on the interaction between the structure of the quartier and its perception by the inhabitants, offering a "scientific" correlate to Chtcheglov's postsurrealist reveries. It is thus no coincidence that Debord highlights precisely this aspect of Chombart's work as useful, noting that the mapping of extremely repetitive daily movements of a young student over the course of one year is an example of a "modern poetry capable of provoking sharp emotional reactions (in this particular case, outrage at the fact that anyone's life can be so pathetically limited)."[53] If we recall Damisch's thesis about the unsettling of the city's image as the cause for the emergence of the question of its legibility, we can say that the visualization of the city in postwar urbanism provokes here an interrogation of a new phenomenal region for which no visual and theoretical language yet exists. Both the situationist theory of an aesthetics of ambiences and the cartographic practice of mapping this dimension can, in this perspective, be seen both as a reception and as a *détournement* (the situationist technique of recoding cultural signifiers) of the then existent forms of urban ecology.

Debord objects to Chombart, Burgess, and Quinn's division of the city into "unities of ambiance" inasmuch as these approaches conceive the population as confined to a district, which is essentially defined by the rhythms of consumption, work, and leisure. Against this homogenized space, where the population remains essentially "*based,* rooted," psychogeography "takes the point of view of passage."[54] Thus, the situationist partition of the city can, for Debord, coincide but also oppose the urbanist distribution inasmuch as the latter remains tied to the spatial matrix of work and leisure. For Debord, existing forms of urban ecology are right in that they refuse to consider the individual in isolation and highlight his or her interaction with the environment. But they privilege the wrong kind of relations: urban ecology only brings to light functional relations—relations that are significant from the point of view of capital. In contrast, what we may call "situationist ecology" explores nonfunctional relations, which manifest themselves in the attraction to the margin and minutiae of different ambiences: "In the margins of utilitarian relations, psychogeography studies relations through the attractions of ambiences [les relations par attirance des ambiances]."[55] Against the current understanding of urban ecology, Debord thus calls for a radical deterritorialization of ambience: "Dissociating the habitat—in the current, restricted sense—from the general milieu, psychogeography introduces the notion of uninhabitable ambiances [*d'ambiances inhabitables*] (for play, passage, . . . that is to say, for disassociating architectural ambiances from the notion of habitat-housing)."[56] Here, we encounter what might be the most ambitious statement of a "situationist ecology" that explores an entirely new concept of ambient relationality. Forces of attraction are studied not as operative in the immediate "work and leisure" environment of people but as forces that draw them *elsewhere*. Irreducible to either the socioeconomic forces of attraction in a quartier or the idea of natural "microclimates," Debord seeks to think of their entanglement in a notion of connectivity and relationality that can only be experienced from a marginal, nomadic, nonutilitarian view of the city—an experience of connectivity in urban space that is determined not by capital and leisure but by the specific atmosphere of spaces that allow for highly intense situations. However, instead of poeticizing the marginal and ephemeral, as Breton arguably did, Debord and Constant try to ground a political and architectonic project in these fleeting, minute ambiences. This is why ambience and atmosphere are the objects of both a reconstruction through the dérive—where practitioners chart the last traces of these nonutilitarian, nonreified spaces in the existing city—and a construction where new forms of architecture and urbanism aim to create the conditions under which these marginal spaces can be turned into qualities of social space.

In Constant and Debord's "Declaration d'Amsterdam," the complex play between the minute and the general as well as the transitory and durable, which is required for this new general ecology, is pointedly defined: "The construction of a situation consists in the edification of a transitory micro-ambiance and a play of events for a unique moment in the life of some persons. It is inseparable from the construction of a general ambiance, relatively more durable, in contemporary urbanism."[57] Note how the authors here repurpose the Chombartian concept of "microclimates," blending the cultural and the natural. The construction of ambiences is thus conceived as the modulation of the most minute elements in experience, encompassing the manipulation of air, light, smell, and sound. Constant goes so far as to conceive this multisensory architecture—to be executed through artificial lighting, air-conditioning, and Plexiglas constructions—as an overcoming of nature: "Far from a return to nature . . . we see in such immense constructions the possibility of overcoming nature and regulating at will the atmosphere, lighting, and sounds in these various spaces."[58] It is a long way from the highly individual experience of the city's atmosphere in surrealism to this vision of a natureless ecology, where every detail of ambience could be modulated through architecture to make room for the highest intensity in experience for the people moving through it. And yet in the course of this essay, it clearly emerged how the interest in the most minute elements of urban space, understood as the constituents of the atmospheric tissue, runs like a guiding thread from the surrealist aesthetics to Constant and Debord's plans for reconstructing social space.

Recently, commentators such as Rem Kohlhaas and Mark Wigley spoke of the topsy-turvy anticipation of developments through the situationist project.[59] From this perspective, the total surveillance and manipulation of the atmosphere in shopping malls, integrated offices, and the new emerging forms of environmental monitoring through AI could appear as the catastrophic materialization of a once emancipatory project. But if the situationist city proves so very topical, this might mean something more, or at least something else, than these old, if not ancient, narrative patterns allow for. Like the process of ecologization as a whole, the vision of an architecture of microambiences might be precisely inscribed on a fault line, oscillating between critique and capture. This, then, might be the minute, if decisive, difference of the allegedly necessary reversal of a utopian vision into dystopian normalcy—leaving us, still, with the question about which forms of architecture and urbanism would redeem the promise of different ambiences.

Notes

1. Guy Debord, "Écologie, psychogéographie et transformation du milieu humain" (1959), in *Œuvres*, ed. Jean-Louis Rançon (Paris: Gallimard, 2006), 457–61, esp. 457.
2. Constant and Guy Debord, "Declaration d'Amsterdam," *Internationale Situationniste* 2 (1958): 31–32, esp. 32.
3. Constant, "Unitair Urbanisme" (unpublished manuscript of a lecture at the Stedelijk Museum, Amsterdam, December 20, 1960), cited in Mark Wigley, *Constant's New Babylon: The Hyper-architecture of Desire* (Rotterdam: Witte de With, 1998), 9.
4. Hubert Damisch, *Skyline: The Narcissistic City*, trans. John Goodman (Stanford, CA: Stanford University Press, 2001), 14.
5. Leo Spitzer, "Milieu and Ambiance: An Essay in Historical Semantics," *Philosophy and Phenomenological Research* 3 (1942): 1–42, 169–218.
6. Peter Sloterdjik, *Spheres*, vol. 3, *Foams*, trans. Wieland Hoban (South Pasadena, CA: Semiotext(e), 2016), 611–23, esp. 615–16.
7. Gernot Böhme, "Atmosphere as a Fundamental Concept of a New Aesthetics," trans. David Roberts, in *The Aesthetics of Atmospheres*, ed. Jean-Paul Thibaud (London: Routledge, 2017), 11–24.

8. Prominently by Peter Zumthor, *Atmospheres: Architectural Environments, Surrounding Objects* (Basel: Birkhäuser, 2006).
9. John Durham Peters, *The Marvelous Clouds: Toward a Philosophy of Elemental Media* (Chicago: University of Chicago Press, 2015).
10. Eva Horn, "Air as Medium," *Grey Room* 73 (2018): 6–25.
11. Erich Hörl, "Introduction to General Ecology: The Ecologization of Thinking," in *General Ecology: The New Ecological Paradigm*, ed. Erich Hörl and James Burton (London: Bloomsbury, 2017), 1–75, esp. 7.
12. Walter Benjamin, "Little History of Photography," trans. Edmund Jephcott and Kingsley Shorter, in *Selected Writings*, ed. Michael W. Jennings, Howard Eiland, and Gary Smith (Cambridge, MA: Belknap Press of Harvard University Press, 2006), vol. 2, 507–30, esp. 518.
13. Ibid., 518.
14. Walter Benjamin, "Surrealism: The Last Snapshot of the European Intelligentsia," trans. Edmund Jephcott, in Jennings, Eiland, and Smith, *Selected Writings*, 207–21, esp. 210.
15. Ibid., 211.
16. Walter Benjamin, "Die Wiederkehr des Flaneurs," in *Gesammelte Schriften*, vol. 3, ed. Rolf Tiedemann and Gerhard Schweppenhäuser (Frankfurt/Main: Suhrkamp, 1991), 194–98, esp. 198; "Franz Hessel, *Heimliches Berlin* [Review]," in Tiedemann and Schweppenhäuser, *Gesammelte Schriften*, 82–84.
17. Walter Benjamin, "On Some Motifs in Baudelaire," trans. Harry Zohn, in Jennings and Eiland, *Selected Writings*, vol. 4, 313–55, esp. 326.
18. Alexander Gelley, *Benjamin's Passages: Dreaming, Awakening* (New York: Fordham University Press, 2015), 119–20.
19. Susan Laxton, *Surrealism at Play* (Durham: Duke University Press, 2019), 93. The following two citations from Breton and Caillois are taken from Laxton's meticulous account (93–94).
20. André Breton, *Conversations: The Autobiography of Surrealism*, trans. Mark Polizzotti (New York: Paragon House, 1993), 106.
21. Roger Caillois, "Paris, a Modern Myth," in *The Edge of Surrealism: A Roger Caillois Reader*, ed. Claudine Frank (Durham, NC: Duke University Press, 2003), 177–89, esp. 180.
22. Walter Benjamin, *Das Passagen-Werk*, in Tiedemann and Schweppenhäuser, *Gesammelte Schriften*, vol. 5, 1054.
23. Walter Benjamin, "The Paris of the Second Empire in Baudelaire," trans. Harry Zohn, in Jennings and Eiland, *Selected Writings*, vol. 4, 3–92, esp. 31.
24. André Breton, *La Clé des champs* (Paris: Jean-Jacques Pauvert, 1967), 280. This is Michael Sheringham's translation, who cites the passage in his excellent article "City Space, Mental Space, Poetic Space: Paris in Breton, Benjamin and Réda," in *Parisian Fields*, ed. Michael Sheringham (London: Reaktion Books, 1996), 85–114, esp. 90.
25. Ibid., 96.

26. Situationist International, "The Sound and the Fury," in *Situationist International Anthology*, trans. and ed. Ken Knabb (Berkeley, CA: Bureau of Public Secrets, 2006), 47–49, esp. 48.
27. The glossary, where the following two definitions are taken from, can be found as "Definitions," in Knabb, *Situationist International Anthology*, 51–52 (translations modified).
28. Constant, "A Different City for a Different Life," trans. John Shepley, in *Guy Debord and the Situationist International*, ed. Tom McDonough (Cambridge, MA: MIT Press / October Books, 2002), 95–101, esp. 96 (translation amended).
29. See the groundbreaking study by Jean-Marie Apostolidès and Boris Donné, *Ivan Chtcheglov: Profil perdu* (Paris: Editions Allia, 2006), 55–67.
30. Ivan Chtcheglov, "Formulary for a New Urbanism" (1953), in Knabb, *Situationist International Anthology*, 1–8, esp. 2. (translation modified). For an important alternative account of the relation between surrealism, Chtcheglov, and Debord, see Boris Donné, "Debord & Chtcheglov, Bois & Charbons: La dérive et ses sources surréalistes occultées," *Mélusine* 28 (2008): 109–24.
31. Chtcheglov, "Formulary for a New Urbanism," 3.
32. Ibid., 5.
33. Ibid., 5.
34. Ibid., 6.
35. Guy Debord, "Manifeste pour une construction de situations" (1953), in *Œuvres*, ed. Jean-Louis Rançon (Paris: Gallimard, 2006), 105–12, esp. 105.
36. Ibid., 109.
37. Thomas Y. Levin, "Geopolitics of Hibernation: The Drift of Situationist Urbanism," in *Situacionistas: Arte, política, urbanismo* [Situationists: Art, politics, urbanism], ed. Libero Andreotti and Xavier Costa (Barcelona: Museu d'Art Contemporani, 1996), 111–46, esp. 111–12.
38. Giorgio Agamben, "Marginal Notes on Commentaries on the *Society of the Spectacle*," in *Means without End*, trans. Vincenzo Binetti and Cesare Casarino (Minneapolis: University of Minnesota Press, 2000), 73–89, esp. 77.
39. Debord, "Manifeste pour une construction de situations," cited in Apostolidès and Donné, *Ivan Chtcheglov*, 63. This decisive passage is not included in the version published in Debord's *Œuvres*.
40. Guy Debord, "Introduction to a Critique of Urban Geography" (1955), in Knabb, *Situationist International Anthology*, 8–11, esp. 10.
41. Guy Debord, "Programme de travaux concrets" (1956), in Rançon, *Œuvres*, 268–69, esp. 269. For the notes on the exhibition, see Guy Debord, "La psychogéographie, c'est la science-fiction de l'urbanisme" (1957), in Rançon, *Œuvres*, 283.
42. Guy Debord, "Écologie, psychogéographie et transformation du milieu humain" (1959), in *Œuvres*, ed. Jean-Louis Rançon (Paris: Gallimard, 2006), 457.
43. Debord, "Introduction to a Critique of Urban Geography," 11.
44. Guy Debord, "Theory of the Dérive" (1956), in Knabb, *Situationist International Anthology*, 62–66, esp. 62 (translation modified).
45. Ibid., 63.

46. Ibid., 66.
47. Guy Debord, "Report on the Construction of Situations and on the Terms of Organization and Action of the International Situationist Tendency" (1957), trans. Tom McDonough, in McDonough, *Guy Debord and the Situationist International*, 29–50, esp. 44 (translation modified).
48. Cf. Rosemary Wakeman, *The Heroic City: Paris, 1945–1958* (Chicago: Chicago University Press, 2014), 170–76, esp. 171–72.
49. Ernest W. Burgess and Robert E. Park, *The City: Suggestions for Investigation of Human Behavior in the Urban Environment* (Chicago: Chicago University Press, 1925); James A. Quinn, *Human Ecology* (New York: Prentice Hall, 1950). These works are discussed in Paul-Henry Chombart de Lauwe, *Paris: Essais de sociologie 1952–1964* (Paris: Éditions ouvrières, 1965), 45–47.
50. Chombart de Lauwe, *Paris*, 65–67.
51. Ibid., 33–34.
52. There are excellent studies on Debord's relation to Chombart de Lauwe that, however, do not address the importance of "ecology." See especially Anthony Vidler, "*Terres Inconnues*: Cartographies of a Landscape to Be Invented," *October* 115 (2006): 13–20; Tom McDonough, "Situationist Space," *October* 67 (1994): 58–77; Tom McDonough, "The Dérive and Situationist Paris," in Andreotti and Costa, *Situacionistas*, 54–66, which develops crucial observations on the relation between the situationists and the Chicago school.
53. Debord, "Theory of the Dérive," 62–63.
54. Debord, "Écologie, psychogéographie et transformation du milieu humain," in Rançon, *Œuvres*, 459.
55. Ibid., 459.
56. Ibid., 460–61.
57. Constant and Debord, "Declaration d'Amsterdam," 32.
58. Constant, "A Different City for a Different Life," in McDonough, *Guy Debord and the Situationist International*, 96 (translation amended).
59. Rem Koolhaas, "The Topsy-Turvy as Utopian Architecture," in *New Babylon: To Us, Liberty*, by Constant, exh. cat. (Den Haag: Gemeentemuseum, 2016), 64–67, esp. 64–65; Mark Wigley, "Extreme Hospitality," in Constant, *New Babylon*, 38–49.

Bibliography

Agamben, Giorgio, "Marginal Notes on Commentaries on the *Society of the Spectacle*," in *Means without End*, trans. Vincenzo Binetti and Cesare Casarino (Minneapolis: University of Minnesota Press, 2000), 73–89.
Apostolidès, Jean-Marie, and Donné, Boris, *Ivan Chtcheglov: Profil perdu* (Paris: Editions Allia, 2006).
Benjamin, Walter, *Das Passagen-Werk*, in *Gesammelte Schriften*, vol. 5, ed. Rolf Tiedemann and Gerhard Schweppenhäuser (Frankfurt/Main: Suhrkamp, 1991).

———, "Die Wiederkehr des Flaneurs," in *Gesammelte Schriften*, vol. 3, ed. Rolf Tiedemann and Gerhard Schweppenhäuser (Frankfurt/Main: Suhrkamp, 1991), 194–98.

———, "Franz Hessel, *Heimliches Berlin* [Review]," in *Gesammelte Schriften*, vol. 3, ed. Rolf Tiedemann and Gerhard Schweppenhäuser (Frankfurt/Main: Suhrkamp, 1991), 82–84.

———, "Little History of Photography," trans. Edmund Jephcott and Kingsley Shorter, in *Selected Writings*, vol. 2, ed. Michael W. Jennings, Howard Eiland, Gary Smith (Cambridge, MA: Belknap Press of Harvard University Press, 2006), 507–30.

———, "On Some Motifs in Baudelaire," trans. Harry Zohn, in *Selected Writings*, vol. 4, ed. Michael W. Jennings and Howard Eiland (Cambridge, MA: Belknap Press of Harvard University Press, 2006), 313–55.

———, "The Paris of the Second Empire in Baudelaire," trans. Harry Zohn, in *Selected Writings*, ed. Marcus Bullock, Howard Eiland, and Gary Smith (Cambridge, MA: Belknap Press of Harvard University Press, 2006), 3–92.

———, "Surrealism: The Last Snapshot of the European Intelligentsia," trans. Edmund Jephcott, in *Selected Writings*, vol. 2, ed. Michael W. Jennings, Howard Eiland, Gary Smith (Cambridge, MA: Belknap Press of Harvard University Press, 2005), 207–21.

Boehme, Gernot, "Atmosphere as a Fundamental Concept of a New Aesthetics," trans. David Roberts, in *The Aesthetics of Atmospheres*, ed. Jean-Paul Thibaud (London: Routledge, 2017), 11–24.

Breton, André, *Conversations: The Autobiography of Surrealism*, trans. Mark Polizzotti (New York: Paragon House, 1993).

———, *La Clé des champs* (Paris: Jean-Jacques Pauvert, 1967).

Burgess, Ernest W., and Park, Robert E., *The City: Suggestions for Investigation of Human Behavior in the Urban Environment* (Chicago: Chicago University Press, 1925).

Caillois, Roger, "Paris, a Modern Myth," in *The Edge of Surrealism: A Roger Caillois Reader*, ed. Claudine Frank (Durham, NC: Duke University Press, 2003), 177–89.

Chombart de Lauwe, Paul-Henry, *Paris: Essais de sociologie 1952–1964* (Paris: Éditions ouvrières, 1965).

Chtcheglov, Ivan, "Formulary for a New Urbanism" (1953). In *Situationist International Anthology*, trans. and ed. Ken Knabb (Berkeley: Bureau of Public Secrets, 2006), 1–8.

Constant, "A Different City for a Different Life" (1959), trans. John Shepley, in *Guy Debord and the Situationist International*, ed. Tom McDonough (Cambridge, MA: MIT Press / October Books, 2002), 95–101.

Constant, and Debord, Guy, "Declaration d'Amsterdam," *Internationale Situationniste* 2 (1958): 31–32.

Damisch, Hubert, *Skyline: The Narcissistic City*, trans. John Goodman (Stanford, CA: Stanford University Press, 2001).

Debord, Guy, "Écologie, psychogéographie et transformation du milieu humain" (1959), in *Œuvres*, ed. Jean-Louis Rançon (Paris: Gallimard, 2006), 457–61.

———, "Introduction to a Critique of Urban Geography" (1955), in *Situationist International Anthology*, trans. and ed. Ken Knabb (Berkeley, CA: Bureau of Public Secrets, 2006), 8–11.

———, "La psychogéographie, c'est la science-fiction de l'urbanisme" (1957), in *Œuvres*, ed. Jean-Louis Rançon (Paris: Gallimard, 2006), 283.

———, "Manifeste pour une construction de situations" (1953), in *Œuvres*, ed. Jean-Louis Rançon (Paris: Gallimard, 2006), 105–12.

———, "Programme de travaux concrets" (1956), in *Œuvres*, ed. Jean-Louis Rançon (Paris: Gallimard, 2006), 268–69.

———, "Report on the Construction of Situations and on the Terms of Organization and Action of the International Situationist Tendency" (1957), trans. Tom McDonough, in *Guy Debord and the Situationist International*, ed. Tom McDonough (Cambridge, MA: MIT Press / October Books, 2002), 29–50.

———, "Theory of the Dérive" (1956), in *Situationist International Anthology*, trans. and ed. Ken Knabb (Berkeley, CA: Bureau of Public Secrets, 2006), 62–66.

Gelley, Alexander, *Benjamin's Passages: Dreaming, Awakening* (New York: Fordham University Press, 2015).

Hörl, Erich, "Introduction to General Ecology: The Ecologization of Thinking," in *General Ecology: The New Ecological Paradigm*, ed. Erich Hörl and James Burton (London: Bloomsbury, 2017), 1–75.

Horn, Eva, "Air as Medium," *Grey Room* 73 (2018): 6–25.

Internationale Situationniste, "The Sound and the Fury," in *Situationist International Anthology*, trans. and ed. Ken Knabb (Berkeley, CA: Bureau of Public Secrets, 2006), 47–49.

Koolhaas, Rem, "The Topsy-Turvy as Utopian Architecture," in *New Babylon: To Us, Liberty*, by Constant, exh. cat. (Den Haag: Gemeentemuseum, 2016), 64–67.

Laxton, Susan, *Surrealism at Play* (Durham: Duke University Press, 2019).

Levin, Thomas Y., "Geopolitics of Hibernation: The Drift of Situationist Urbanism," in *Situacionistas: Arte, política, urbanismo* [Situationists: Art, politics, urbanism], ed. Libero Andreotti and Xavier Costa (Barcelona: Museu d'Art Contemporani, 1996), 111–46.

McDonough, Tom, "The Dérive and Situationist Paris," in *Situacionistas: Arte, política, urbanismo* [Situationists: Art, politics, urbanism], ed. Libero Andreotti and Xavier Costa (Barcelona: Museu d'Art Contemporani, 1996), 54–66.

———. "Situationist Space," *October* 67 (1994): 58–77.

Peters, John Durham, *The Marvelous Clouds: Toward a Philosophy of Elemental Media* (Chicago: University of Chicago Press, 2015).

Quinn, James A., *Human Ecology* (New York: Prentice Hall, 1950).

Sheringham, Michael, "City Space, Mental Space, Poetic Space: Paris in Breton, Benjamin and Réda," in *Parisian Fields*, ed. Michael Sheringham (London: Reaktion Books, 1996), 85–114.

Sloterdijk, Peter, *Spheres*, vol. 3: *Foams*, trans. Wieland Hoban (South Pasadena, CA: Semiotext(e), 2016).

Spitzer, Leo, "Milieu and Ambiance: An Essay in Historical Semantics," *Philosophy and Phenomenological Research* 3 (1942): 1–42, 169–218.
Vidler, Anthony, "*Terres Inconnues*: Cartographies of a Landscape to Be Invented," *October* 115 (2006): 13–20.
Wakeman, Rosemary, *The Heroic City: Paris, 1945–1958* (Chicago: Chicago University Press, 2014).
Wigley, Mark, *Constant's New Babylon: The Hyper-architecture of Desire* (Rotterdam: Witte de With, 1998).
———. "Extreme Hospitality," in *New Babylon: To Us, Liberty*, by Constant, exh. cat. (Den Haag: Gemeentemuseum, 2016), 38–49.
Zumthor, Peter, *Atmospheres: Architectural Environments, Surrounding Objects* (Basel: Birkhäuser, 2006).

Cell and Cosmos

Siegfried Ebeling's Environmental Architecture

MARGARETA INGRID CHRISTIAN

✦ ✦ ✦

SIEGFRIED EBELING'S tract *Space as Membrane* (1926, *Der Raum als Membran*), reviewed by Bruno Taut, owned by Walter Gropius, and read by Ludwig Mies van der Rohe, was translated and published for the first time in English in 2011.[1] Ebeling (1894–1963), a little-known Bauhaus student and architectural utopian, argues that a house should function like a miniature environment within the world at large. He draws on cell biology, the work of the microbiologist Raoul Heinrich Francé, and theories of the microcosm to suggest that architectural space should function like a biological milieu. This analogy between the space of dwelling in architecture and the space of living in the life sciences was not new. For instance, Paul Scheerbart's *Glass Architecture* (*Glasarchitektur*, 1914) begins by implicitly applying Darwin's notion of evolution to a concept of architectural space: just as the environment influences natural selection and, thus, the evolution of organisms, architectural space determines the evolution of culture. Culture is the "product" of our built environment, and in order to change our culture, we must transform our habitational milieu.[2] Yet whereas for Scheerbart, this transformation consists in divesting our spaces of their sense of enclosure,[3] for Ebeling, this transformation involves the opposite maneuver. Ebeling develops a counterintuitive theory of architectural milieu—one that rests not on influence and interaction (as implied by the concept of milieu) but rather on separation and enclosure. Furthermore, whereas Scheerbart's architectural vision rests on a planetary and cosmic consciousness, Ebeling's model of the built environment incorporates the minutiae of existence as well: cells and their constitutive parts, such as nucleus, plasma, and membrane.

It was only in 2016, after the historian of design, Walter Scheiffele, brought attention to Ebeling's work that *Space as Membrane* was reprinted in the original German (in an edition that incorporated Mies's underlinings). It is surprising that Ebeling remains an obscure figure given the context in which his work is being discussed today: his manifesto of what he terms "biological architecture" points forward to our own time's interest in biomimetic architecture, "living buildings,"[4] and sustainable design. In a remarkably prescient moment, Ebeling writes about "designing the house as its own energy source."[5] When contemporary architects and engineers want to build "biological houses" and propose buildings that are "self-sustainable and responsive structures infused with a bit of biology,"[6] they echo Ebeling's *Space as Membrane* almost word for word. Finally, as the title suggests, Ebeling's tract prefigures many later ideas associated with lightweight construction practices, especially those that center on the concept of a membrane (for instance, Frei Otto's roof of the Olympic Stadium in Munich).

Indeed, much has been made in the existing literature on Ebeling of the idea of the membrane and of notions of exchange and permeability implied by it. For architectural scholars, Ebeling "began to re-imagine and valorize the periphery" of a building and thus fits into a longer line of thinkers interested in reconceptualizing the architectural margin, such as Heinrich Tessenow or Bruno Taut.[7] Ebeling's text is celebrated for the "reenergizing of the building façade,"[8] for conceiving the wall as a "medium for energy flow,"[9] and for understanding walls as "breathing, regulating, selectively permeable entities."[10] No longer "static and impermeable,"[11] Ebeling's buildings posit the wall as an "open boundary."[12] However, this emphasis on permeability, this stress on the exchange between the house organism (*Hausorganismus*) and its environment, misses an important point in Ebeling's utopian manifesto—indeed, it misunderstands its ultimate purpose.

Despite the concept of the "wall skin" (*Wandhaut*),[13] which *Space as Membrane* borrows from the phenomenon of cellular osmosis, the logic animating Ebeling's architectural manifesto is that of isolation. He wants to undo the idea of milieu and enclose dwellers in themselves. Ebeling uses biology as a model for engineering, yet he does so in order to free humans from nature. He posits nature as a paradigm for the built environment in order to neutralize nature for the dweller. Even though *Space as Membrane* deals with cells, plants, and nature broadly understood, it is concerned even more with spiritual cleansing and freedom from the vagaries of nature. Although several concepts with which the text operates—"membrane," "wall skin"—suggest an architecture of openness, Ebeling's vision is not, in fact, in line with Sigfried Giedion's famous description of modern buildings that open themselves in every possible way and seek "connection" and "interpenetration."[14] Ebeling's house seeks out its environment to delimit itself from it.

To understand Ebeling's paradox of biological architecture,[15] as he calls it, one must examine the different cultural forces acting on his text, ranging from cell biology and theories of macrocosm and microcosm to his understanding of Nietzsche's *Lebensphilosophie*. Indeed, one of the aims of this essay is to untangle the varied influences that come together in *Space as Membrane*. The surfeit of influences is significant in a text like Ebeling's—a text that is concerned precisely with a mode of dwelling that can annul external, foreign influences on dwellers. In Ebeling's writing, the architectural mitigation of influence is built on the textual mobilization of influence. His program of architectural neutralization extends even to the influence of space. The neutralization of space—its consignment "below the threshold of consciousness"[16]—is the major task of Ebeling's architectural utopia. His vision of the built environment is resolved to nullify the very idea of environment.

Ebeling led an unsettled life. Born in 1894, he studied philology and theology before volunteering in the First World War. After being released from war captivity in 1920, he took up his studies again, this time in Christian archaeology and art history alongside experimental physics. To support himself, he worked as a casual laborer, a house painter, and a welder until he joined the Bauhaus in Weimar as well as the propaganda department of the aircraft manufacturer Junkers Aircraft and Motor Works in Dessau in 1922.[17] After sporadic studies in Weimar, he left the Bauhaus in 1925. He worked for Junkers by writing articles that popularized global air transportation and by researching metallic house construction possibilities until 1927, when he left the company.[18] Ebeling took dance classes with Rudolf Laban in Berlin,[19] stayed with Rudolf Steiner in Dornach, worked in a wagon factory, became the private secretary of a sugar factory owner, and remained an architectural theorist and visionary yet a failed architect.[20] Scheiffele hypothesizes that the achievement of the rank of officer in the Luftwaffe served as a compensation of sorts for Ebeling's failures in his career as an architect. His tract *Space as Membrane* was published in 1926. It shows a neglected, utopian side of the Bauhaus at a time when the latter had turned to pragmatic serial production and industrial manufacturing.[21]

Microcosmologies of Architecture

IN SPACE *as Membrane*, a building functions like an organism, such as a plant, or a biological unit, such as a cell. In his short article "Cosmological Space-Cells" ("Kosmologe Raumzellen," 1924), which introduces many of the ideas he goes on to develop two years later in *Space as Membrane*, Ebeling conceives of the micromilieu of a house in terms of a "house organism" (*Hausorganismus*) or a "space cell" (*Raumzelle*).[22] In line with this biological view of architecture, he argues that the time is ripe to "adapt three-dimensional space, as crudely defined by physics, into a three-dimensional membrane—biologically defined—between our body (as a plasmatic weak substance) and the latent minute forces of the spheres."[23] In this sentence, Ebeling conceives of the human organism as an entity that needs a biologically functioning membrane both to protect it from the immediate exterior and to mediate its interactions with a distant, cosmic exterior ("forces of the spheres"). Elsewhere, he conceives of a building's edge in terms of a "wall skin" and of the inhabitant in terms of a cell's nucleus. By relying on the idea of a biological cell (German: *Zelle*), Ebeling activates the etymology of the word that goes back to the Latin *cella*, meaning a "small room."[24] Resorting to a parallel between habitational and cytological environments, Ebeling argues that built space should resemble the membrane of a cell. Ebeling's theory of dwelling thus stretches between two extreme domains: the basic structural unit of an organism (the cell) and the largest environment conceivable (the universe). As we will see, it is by harnessing the micrologic mechanisms of the cell that Ebeling's house is meant to shelter the dweller from the macrodomains of local weather, global climate, and cosmic force.

Ebeling's vision of the house spans, thus, biology and cosmology. The convergence of these two scales is well illustrated by his claim that a house is a "relatively rigid, multi-celled spatial entity"[25] that acts as a meeting point for forces coming from the earth and the sky. This idea of the house as a place of encounter between microscopic and cosmic milieus is also suggested in the title of his earlier text, "Cosmological Space-Cells." Ebeling's attentiveness to the relation between the human microcosm and macrocosmic forces fits into the cosmological interests of his era. Rudolf Steiner lectured on the analogies between microcosm and macrocosm in 1920; the botanist, microbiologist, and popular science writer Raoul Heinrich Francé (whom Ebeling cites in *Space as Membrane*) wrote on concepts of harmony between microcosmic and macrocosmic domains (e.g., molecular and celestial harmony, intracellular and planetary harmony) in his book *Bios: Die Gesetze der Welt* (Bios: The laws of the world, 1921).[26] In 1926, the same year that Ebeling published his manifesto, Fritz Saxl held a lecture on depictions of "Microcosm and Macrocosm" in art from antiquity to the Renaissance.[27] The list could continue with Ebeling's dance teacher, Rudolf Laban, and the latter's views on the correspondences between microcosmic and macrocosmic movements in dance; Ebeling's fellow architects' "constructivist cosmology of architecture";[28] or Gropius's interest in relating architecture to a "totality of space" (*Allraum*).[29] While these cosmological perspectives on affinities between the parts and the whole have a long history dating back to Greek antiquity,[30] at the beginning of the twentieth century, they were inflected by Ernst Haeckel's monist natural philosophy. Ebeling's architectural cosmology is indebted to Haeckel's monism popularized by *Art Forms in Nature* (*Kunstformen der Natur*, 1904). Haeckel's work relied on the microscopic study of radiolarians, single-celled organisms that create complex mineral skeletons.

Many artists and architects of the time recognized in these skeletal minutiae fundamental structural units present not only in all living creatures but also in crystals or the movement of astral bodies.[31] Ebeling's architectural cosmology sees (to quote Detlef Mertins) an "isomorphism in organic and inorganic materials as well as microscopic and macroscopic events."[32] On the one hand, this early twentieth-century cosmological view is obliged to the technological advances of potent microscopes and telescopes.[33] On the other hand, the invention of photography in the nineteenth century opened new possibilities for the employment of older optical instruments. The photographic plate, which stood, in the words of the French astronomer Pierre Janssen, for the "true retina of the scientist," brought formerly inaccessible realms within human vision's reach and opened new cosmological perspectives.[34]

Ebeling's micrological approach to architecture is particularly beholden to the work of Francé.[35] Indeed, Francé's work influenced not only Ebeling but a number of Bauhaus figures, including Mies van der Rohe, László Moholy-Nagy, and El Lissitzky.[36] They were especially enthused by Francé's concept of *Biotechnik*, which proposed that human engineering should learn from plants' technological mechanisms. Francé cofounded the German Micrological Society (Deutsche Mikrologische Gesellschaft) in 1907. This society, which exists to this day in a slightly different iteration as the Microbiological Association of Munich (Mikrobiologische Vereinigung München), set as its purpose the advancement of the study of nature through the use of the microscope.[37] The initiative for the founding of the Micrological Society came from Francé's book, *Forays into the Water Droplet* (*Streifzüge im Wassertropfen*, 1907), which dealt with unknown worlds that only the microscope could reveal. It was in this book that Francé exclaimed, "Let us establish a micrological society, that aims to encourage the popular use of the microscope and that seeks to make more accessible the consolidation of the new science of the fine build and life of plants and animals."[38] The aim was to make the newest developments in the life sciences in the period accessible to the wider public. The intended vehicle was to be the microscope—a natural choice for Francé, the naturalist, naturist, and plant protector, whose work focused on micrological life-forms. In line with this society, Francé also cofounded the magazine *Microcosm* (*Mikrokosmos*) in 1907, directed primarily at the amateur microscopist.[39] Although he started out studying protozoans (single-celled organisms), his work soon left the strictly scientific domain and wandered into the realm of natural philosophy. Today he is remembered especially for his work on the *edaphon* (the community of organisms living in the soil)[40] and for his contributions to soil ecology.

Telluric and Atmospheric Architecture

EBELING MENTIONS Francé's book *The Technical Accomplishments of Plants* (*Die Technischen Leistungen der Pflanzen*, 1919) in a paragraph in which he also discusses a Japanese bacteriologist's experiments with a glass house. Both the work of Francé and that of the bacteriologist serve Ebeling as models for an architecture of the future that applies cells' and plants' expertise to the building process. Ebeling describes the bacteriologist's research into a glasshouse in which "individual glass blocks are filled with a saline solution that absorbs the heat of the sun."[41] The solution helps keep the glass rooms cooler during the day and warmer during the night when it releases the sun's heat.[42] However, Ebeling's building is not only a point of convergence between the bacteriological and the solar, the microscopic and the cosmic—his building is also the site of an elemental encounter between telluric and atmospheric forces that act upon it. His house is not a milieu that acts upon the dweller but rather a miniature milieu that is itself acted upon by the large-scale cosmic environment.

"The house," claims Ebeling, "is to be perceived as a conducting medium [*Durchgangsmedium*] channeling a continuous stream of forces ... flowing from a ground surface that in geophysical terms is variously defined, through hollow space [that of the building], to an open space that is also variously defined, and then back again in the reverse direction. In the center of this play of forces, in each instance, are organisms subject to both physiological *and* psychological laws."[43] Although the "play of forces" remains somewhat vague in Ebeling's prose, it must be understood in the context of the notion of *Helio-Biozönose* (with the cumbersome meaning of "sun-earth-life correspondence"). This is a notion that Ebeling likely encountered in his readings of articles by Raoul Hausmann, published in the avant-garde magazine G.[44] Ebeling relates this vague doctrine of correspondence between earthly and solar forms of life to another related correspondence between telluric and atmospheric rays acting on the house organism. For the latter correspondence, he draws on the work of Wilhelm Ostwald, a physical chemist who argued in 1895 for the "overcoming of scientific materialism" by replacing physics' central concept of matter with a doctrine of energy.[45] If plants could collect energy from the sun, perhaps Ebeling's energy-efficient house could emulate plants by learning to instrumentalize the environmental energies coming from the earth and the sky.[46]

Finally, Ebeling draws on the work of the physician Willy Hellpach, a founder of environmental psychology, in particular his text *The Geo-psychic Phenomena: The Human Soul under the Influence of Weather and Climate, Soil and Landscape* (*Die geopsychischen Erscheinungen: Die Menschenseele unter dem Einfluß von Wetter und Klima, Boden und Landschaft*, 1911).[47] In his footnotes, Ebeling quotes from Hellpach's work, which claims that an organism partakes in the equalization of influence between the field of the earth and that of the air.[48] Hellpach goes so far as to claim that, since humans are a part of the electrostatic field of the earth, some individuals manifest perceptible electric discharges, and he hypothesizes that these might have considerable psychophysical effects. Others' psychophysical states might be dependent on the composition of the earth (on ore, coal, water, rock).[49] In view of Hellpach's foundational work in environmental psychology, it is no wonder that Ebeling will come to see the neutralization of environmental factors—of various agents of influence acting on the human psyche—as the main aim of buildings and their walls. For Ebeling, interior spaces and walls become the agents of this neutralization. They become, to use a phrase of Le Corbusier, *murs neutralisants* (neutralizing walls).[50] Whereas Hellpach's neo-Hippocratic factors of *Geo-psychic Phenomena* constitute a large-scale milieu that affects the minutiae of living organisms, Ebeling's building constitutes a small-scale milieu meant to nullify the effects of the large-scale one.

Cutaneous Enclosures

IN SPACE *as Membrane*, Ebeling blurs the difference between the organism of the dweller and the organism of the dwelling because, for him, the interior space is meant to be an extension of the human skin.[51] Ebeling is not the first to conceive of a building's wall in terms of skin. His vision is part of a longer history of organic architecture that stretches from Vitruvius and Alberti to

Gottfried Semper and Frederick Kiesler.[52] According to Didem Ekici, in the nineteenth century, the idea of the house as skin arose in the context of the public hygiene discourse in Germany, in particular, in the work of the physician Max von Pettenkofer, who "conceptualized the dwelling as a skin that envelops its inhabitants."[53] (It was around the same time that Semper transposed Pettenkofer's ideas into architectural notions of *Bekleidung*, or "dressing."[54]) For Ebeling, a house should fulfill a cutaneous function of enclosure; like the skin that contains us, it should separate us from the outside. Ebeling is at times inconsistent in the way he suggests this. Even so, throughout his text, he returns to the idea of self-enclosure.

Consider, for instance, the following excerpt:

> Innumerable measurable (or as yet immeasurable) minute flows stemming from the breathing earth or the radiating space of the sky bombard the walls of our houses and either bounce off unharnessed ... or are *neutralized* in their cavities. Thus, while the so-called "breathing" wall skin [*Wandhaut*]—made of wood, mud, stone or substitutes—may be able to regulate in a very crude way the relation between the natural climate and the artificial interior climate, it does nothing to prevent the human occupants from being exposed to the detrimental effects of subtler atmospheric fluctuations (such as thunderstorms, blizzards or the Foehn or Scirocco winds).[55]

Initially, it sounds as if Ebeling were lamenting the fact that buildings' walls are not sufficiently porous insofar as various telluric and celestial currents bounce off them. The problem appears to be that houses do not utilize the natural phenomena around them; they do not make them productive for the inhabitants. The desire to solve this problem is what makes Ebeling into a pioneer of self-sustainable building. But then comes a twist in his argument: the ultimate aim of self-sustainable architecture is to enable corresponding "self-sustainable" inhabitants. His "energetically autonomous"[56] dwelling is a vehicle for the creation of the psychologically autonomous dweller. In this latter sense, in the second part of the previous quote, Ebeling laments that walls are *too* porous and not isolating enough. They cannot keep out the finer atmospheric fluctuations (such as those before storms, snow, foehn, sirocco); they allow inhabitants to be at the mercy of these meteorological variations, which, in line with the environmental psychology of Hellpach, affect the psychological household of the inhabitants. Too much mobility between inside and outside hinders the dweller's psychic mobility. Ebeling envisions the house as a cocoon that should protect its resident from meteorosensitivity

(*Wetterfühligkeit*). His conception of the membrane is therefore less osmotic than homeostatic: "The, as yet, very porously coated space will become, through the creation of new structural conditions ... a membrane between our body as nucleus and the plasmatic energies of the wide world."[57]

Space as a Negative Factor

CONVERGING IN Ebeling's views here are both Hippocratic tenets or early environmental psychology and a Wölfflinian psychology of architecture. Throughout his text, Ebeling blurs the boundaries between dweller and dwelling as organism, between psychological and architectural "interior," and between built and natural environment. The purpose of the house is not only to free inhabitants of the natural milieu without but also to protect them from the man-made milieu within. In this sense, Ebeling's text is directed against Wölfflin's *Prolegomena to a Psychology of Architecture* and its interest in architecture's ability to elicit spiritual effects (*seelische Wirkungen*).[58] Instead, the purpose of architecture should be the transformation of space into a negative factor, into a nonagent: "Negative space becomes the preliminary state of spatial activation. The more we try to separate this space from our body—conceiving of it as an environment that defines our mind—the more it will become a veritable function of the 'psychological' world that is expanding within us."[59] Furthermore, Ebeling proposes "a reevaluation of architecture. Space should no longer be perceived as a positive agency that exerts a certain psychological influence on the people who inhabit it.... Rather, space has to be perceived more as a *negative*."[60] The aim of Ebeling's environmental architecture is to combine the natural resources around a building with the built-in technological features to render the building's *milieu intérieur* innocuous. He wants to instrumentalize a building's external environment precisely to neutralize its variable effects in the building's internal environment. In other words, he wants to free humans from the environment (with all its contingencies and variabilities such as changing conditions of light or heat) precisely by utilizing the environment. In a sense, this emphasis on autonomy is fitting for the modernist Ebeling insofar as modernism was, as Mertins pointed out, very preoccupied with self-generation (of new forms versus historical styles and mass production).[61]

Perhaps counterintuitively, the emphasis on the cosmic scale in Ebeling's writing already hints at the way in which his biological science of building is intent on omitting nature: albeit modeled on botanical units, his house is not conceived in relation to natural phenomena such as landscapes, trees, gardens, or the fresh air. Instead of viewing the house in relation to these immediate natural parameters the way Bruno Taut does when he conceives of *Alpine Architecture* or Paul Scheerbart does when he thinks about continuities between house and garden in *Glass Architecture*—that is, instead of seeing the house as embedded in the nature surrounding it—Ebeling views it as embedded in abstract physical and chemical "forces," "rays," and "energies" that come from the earth and the sun.

The Paradoxes of Biotechnik

EVEN AS Ebeling conceives of the house in terms of a biological structure, in true "bioconstructivist" manner, this structure can only be achieved by harnessing the forces of engineering, science, and technology. The very notion of bioconstructivism relies, of course, on this productive conjointment of nature and technology—on the possibility of approximating nature by resorting to technology. Ebeling's house will ultimately be a feat of technology and engineering that, in line with Francé's tenets of *Biotechnik*, emulates nature and borrows engineering techniques from plants' technical accomplishments.[62] Ebeling, under the influence of Francé, regards natural organisms as "prototypes of human technology."[63] However, unlike Francé, who studies bacteria and microscopic parasites and notes that they are well adjusted to penetrating surroundings (*das Durchdringen der Umgebung*),[64] Ebeling draws on various cellular mechanisms to conclude that buildings must serve for the dweller as encasements of impenetrability. Ebeling's selective understanding of the minutiae of cellular activities (focused on aspects of homeostasis over osmosis) reinforces the idea of the building as an impermeable milieu. Indeed, the stable internal environment of Ebeling's building makes it resemble a *milieu intérieur* in the sense of the nineteenth-century French physiologist Claude Bernard. Francé frequently stresses that all living beings are united together in communities of animals and plants, in biocenosis, associations, and formations.[65] However, it is precisely this all-encompassing association among living forms that disquiets Ebeling and pushes him to envision an architecture that can assure the individual freedom from cosmic interrelation and ecological symbiosis. As we will see in what follows, the basis for his uneasiness is racist, biological, and technological. Thus he worries about telepathic connections with far-off nations, ecological oneness, and the technologically induced shrinkage of distances as seen, for instance, in aviation. Consider the following passage in which he writes of the dangers of the "parapsychological interdependence" of Earth's inhabitants: "A psychology based on the principle of symbiosis and the parapsychological relationships of the living environment is aware that the dull lethargy of an Eskimo in winter, for example, may well affect people living in our environment, under particular [atmo]spheric conditions. It is therefore in everyone's greater interest if humanity ... not only joins battle against the dangers of a single environment [*einzelnen Umweltraum*] ... but also fights against environmental threats as a whole [*Umweltraum überhaupt*], taking into account the parapsychological interconnectedness of all creatures on earth."[66] In Ebeling's racist vision, far-off climatic effects (and affects: the Eskimo's lethargy) act on a house dweller in Germany. In *Space as Membrane*, he emphasizes the significance of a German idea of architecture, frets about foreign influences (the effect of foreign climates reaching the German dweller, the affect of foreign nations reaching the German spirit), and finds correspondences between Germany and Japan in terms of a shared concern for what sounds like *Lebensraum* ("that young, intelligent, spatially constrained nation [*raumbeengte Volk*] in the Pacific that we so much resemble!"[67]). These deeply problematic viewpoints are in line with similar pronouncements in other texts. For instance, in his manuscript titled "The Airplane in the Context of Cultural Events" ("Das Flugzeug im Sinnzusammenhang des kulturellen Geschehens," 1923), Ebeling writes about Germans' special (superior) relationship to flight; he wonders which nation will retain spiritual world leadership and declares that Germany must lodge itself in the metaphysical ground of its "nationness" (*Volkheit*) in order to better fly and better mechanically overcome and spiritually master the spaces.[68]

Ebeling's house is meant to ensure the autonomy of the dweller's psychological household by protecting it from both the immediate, local environmental conditions and the distant, global ones. The built environment is thus the encasement that protects from progressively wider circles of environment. "Millions of people," he writes, "are exposed to appalling climatic conditions that could at least be ameliorated, if not wholly eradicated, using scientific means."[69] Ebeling wants to create a "climatologically differentiated architecture"[70] in order to eradicate the influences of climatic conditions on the inhabitant. Just as he seeks an architecture that learns from nature in order to dampen nature's effect on the inhabitant, he seeks to instrumentalize cosmic forces ("chemical-physical forces of the ground and the radiation forces present in open space"[71]) in order to seal the inhabitant from them. And while his architecture wants to exploit technological advances, one cannot help but feel that this architecture is also meant to protect against a new world enabled by technological possibilities. Aviation, in particular, must have made the interconnectedness of people on the globe explicit for Ebeling. This interconnectedness was after all one of the selling points of aviation—and one that Ebeling emphasized in his texts written for the propaganda department of the Junkers Aircraft and Motor Works. In a sense, then,

Ebeling's architectural utopia sought to mitigate the technological forces it was built on.

His understanding of the relation between nature and technology is also in line with his reception of Nietzsche's *Lebensphilosophie*. Ebeling's view of the house, which emulates nature in order to better learn to neutralize its effects,[72] is in accordance with his view of the body: humanity must return to the body, to physiology, but precisely in order to overcome it. His aim is to "raise the physical onto a mental plane,"[73] to "draw out the spiritual in man by raising the body by several orders of magnitude."[74] In his understanding of Nietzsche, humanity must resort to the body in order to reach the heights of pure spirit.[75] Although in Ebeling's agriarchitectural vision building houses should amount to cultivating the earth, his ultimate purpose is the cultivation of the new type of human (*den neuen Menschentypus anzubauen*) in a Nietzschean sense.[76]

Conclusion

ALTHOUGH THE influence of Haeckel and Francé on Ebeling cannot be overstated, their evolutionary and ecological perspective is nevertheless superseded in *Space as Membrane* by Goethe's vision of biological envelopment. Ebeling's futuristic architecture invokes an anachronistic representative of biological knowledge at the time—namely, Goethe's *The Metamorphosis of Plants* (*Die Metamorphose der Pflanzen*, 1817). Rather than Francé's work on symbioses, it is Goethe's writing on encasement that serves as the ultimate guiding principle for Ebeling's architecture of isolation. "All things," Ebeling quotes from *The Metamorphosis of Plants*, "if they are to have a vital effect, must be enveloped. Thus all things that are turned towards the outside yield prematurely to a gradual decay."[77] Ebeling's architectural vision is driven by a fear of the outside; it is animated by the desire to turn the house into a protective husk. For him, architecture's potential "does not go beyond the principle of the tree-bark or, framed in terms of the cell nucleus, the principle of the membrane."[78] The architect Walter Curt Behrendt argued in *The Victory of the New Building Style* (*Der Sieg des Neuen Baustils*), published one year after *Space as Membrane*, that "the biological principle of self-generation has been subsumed ... in theories of evolution (Darwin) and ecology (Uexküll).... At the turn of the century, scientists such as the zoologist Ernst Haeckel and later the botanist Raoul Francé argued that knowledge from the life sciences should be used to guide human works and ways of life."[79] *Space as Membrane* shows how Ebeling employed ideas from botany and microbiology and applied them to architecture in a way that turned these ideas against themselves. He invoked ecological perspectives to argue for an architecture of self-generation. And he delineated a futuristic mode of dwelling that rests on an atavistic mode of biological being.

The paradox of Ebeling's work lies in the fact that he builds his architectural theory on notions of evolutionary biologists such as Haeckel and Francé, yet he does so in order to create a vision of a biological architecture in which self-enclosure and spirit (not permeability and nature) reign supreme. He invokes the membrane that regulates processes of exchange between a house organism and its external milieu, he envisions a site-specific building that relies on the natural elements surrounding it, and he projects technological innovations that make the house react and adjust to external climatic changes. However, these ecological factors are not about the house's openness toward nature or the symbiosis between urban and natural environment; instead, these factors are marshaled to support Ebeling's ultimate architectural aim, which is to free the human of nature, to enable the human as a sealed microcosm: space should be something that "merely creates the physiological preconditions under which the individual, in accordance with his psychological make-up, can develop in complete autonomy, free from all external influences, into a self-contained Being-for-oneself (*Fürsichsein*)—a microcosm."[80] In *Space as Membrane*, the house as microcosm and the dweller as microcosm merge into one. Thereby Ebeling inverts the usual function of the milieu as well as the logic of scale customarily associated with it. Ebeling's architectural milieu does not determine and influence minutiae within it (in the sense that a space conditions its inhabitants, for instance); instead, Ebeling's milieu wards off determinants and influences. Furthermore, the house no longer signifies a container for something smaller within it such as a dweller; rather, house and dweller fuse into a single entity, into a micromilieu, that is itself contained within a world at large that must be kept at bay.

Notes

I am grateful to Nathan Modlin and Marina Resende Santos for their editorial and bibliographic assistance respectively.

1. Wolfgang Thöner, introduction to *Das leichte Haus: Utopie und Realität der Membranarchitektur*, by Walter Scheiffele, Edition Bauhaus 44 (Leipzig: Spector Books, 2016), 7.
2. Paul Scheerbart, *Glasarchitektur*, accessed online: https://www.projekt-gutenberg.org/scheerba/glas/glas.html (my translation). See the first section, titled "The Milieu and Its Influence on the Evolution of Culture."

3. Scheerbart, *Glasarchitektur*: "Und dieses wird uns nur dann möglich sein, wenn wir den Räumen, in denen wir leben, das Geschlossene nehmen."
4. For living buildings, see Katherine Guimapang, "A Living Breathing Building: How Biology and Architecture Will Change Construction and the Built Environment," Archinect, July 24, 2019, https://archinect.com/news/article/150147904/a-living-breathing-building-how-biology-and-architecture-will-change-construction-and-the-built-environment.
5. Siegfried Ebeling, *Space as Membrane*, ed. Spyros Papapetros, trans. Pamela Johnston (London: Architectural Association, 2010), 16.
6. Guimapang, "Living Breathing Building."
7. See Gerald Adler, "Energising the Building Edge: Siegfried Ebeling, Bauhaus Bioconstructivist," in *Peripheries: Edge Conditions in Architecture*, ed. Ruth Morrow and Mohamed Gamal Abdelmonem, Critiques: Critical Studies in Architectural Humanities 8 (Abingdon: Routledge, Abingdon, 2013), 186, 193.
8. Adler, 193.
9. Adler, 194.
10. Matina Kousidi, "Breathing Wall Skins: Theorizing the Building Envelope as a Membrane," *MD Journal* 1 (2016): 85.
11. Walter Scheiffele, *Bauhaus, Junkers, Sozialdemokratie: Ein Kraftfeld der Moderne* (Berlin: Form + Zweck, 2003), 185.
12. Christoph Asendorf, "Grenzfläche—Übergang," *Neue Bildende Kunst*, no. 2 (1996): 48.
13. Siegfried Ebeling, *Der Raum als Membran*, Edition Bauhaus 43 (Leipzig: Spector Books, 2016), 10.
14. Sigfried Giedion, *Building in France, Building in Iron, Building in Ferroconcrete*, trans. J. Duncan Berry (Santa Monica, CA: Getty Center for the History of Art and Humanities, 1995), 91. This passage is cited by Walter Benjamin in his *Arcades Project*.
15. Ebeling, *Space as Membrane*, 17. Ebeling writes about the "paradoxical turns towards biological architecture."
16. Ebeling, 18.
17. Scheiffele, *Das leichte Haus*, 157–58.
18. Scheiffele, 180. For more information on why Ebeling left, see Scheiffele, *Bauhaus, Junkers, Sozialdemokratie*, 193.
19. I have written elsewhere on the parallels between Ebeling's and Laban's works. See Margareta Ingrid Christian, *Objects in Air: Artworks and Their Outside around 1900* (Chicago: University of Chicago Press, 2021).
20. Scheiffele, *Bauhaus, Junkers, Sozialdemokratie*, 222.
21. Walter Scheiffele, afterword to Ebeling, *Der Raum als Membran*, 43 (my translation).
22. Siegfried Ebeling, "Kosmologe Raumzellen: Ideen zur Ethik des konstruktiven Denkens," in "Junge Menschen," special issue, *Weimar Bauhaus* 5, no. 8 (November 1924): 173–74. Quotes from this text are my translations. Ebeling thinks in

terms of the human "species" or "genus"; he constructs similes between human cultural phenomena and the natural processes of "decomposition" and "decay"; he ponders the opposition between the spiritual and the organic; he resorts to biological vocabulary such as "bloods" (*Geblüt*), "phalanx," "inner growth," and "excitability" (*Reizbarkeit*); and he mentions the ability of life to self-generate. This terminology continues in *Space as Membrane*, in which Ebeling puts forth an agri-architectural vision: building dwellings should amount to cultivating the earth ("die Erde anbauen," 5) and to making cities into "true urban landscapes" (9). He also declares the living plant to be "the new primal symbol of architecture" (18).

23. Ebeling, *Space as Membrane*, 16.
24. Deutsches Wörterbuch von Jacob Grimm und Wilhelm Grimm, s.v. "ZELLE, f.," accessed on February 14, 2021, www.woerterbuchnetz.de/DWB/zelle.
25. Ebeling, *Space as Membrane*, 8.
26. Raoul H. Francé, *Bios: Die Gesetze der Welt*, vol. 2 (Munich: Hanfstaengl, 1921), see 255–80.
27. Spyros Papapetros, "MICRO/MACRO: Architecture, Cosmology, and the Real World," *Perspecta* 42 (2010): 114.
28. Detlef Mertins, "Bioconstructivisms," in *Departmental Papers* (City and Regional Planning, 2004), 363. Mertins is referring here to the work of H. P. Berlage.
29. Scheiffele, *Bauhaus, Junkers, Sozialdemokratie*, 186.
30. M. Gatzemeier, "Makrokosmos/Mikrokosmos," in *Historisches Wörterbuch der Philosophie*, vol. 5, ed. Joachim Ritter and Karlfried Gründer (1980), 640–42.
31. See, for example, the Dutch architect H. P. Berlage, who writes, "I need only remind you . . . of the stereometric-ellipsordic forms of the astral bodies, and of the purely geometrical shape of their courses; of the shapes of plants, flowers, and different animals, with the setting of their component parts in purely geometrical figures; of the crystals with their purely stereometrical forms, even so that some of their modifications remind one especially of the forms of the Gothic style; and lastly, of the admirable systematicalness of the lower animal and vegetable orders, in latter times brought to our knowledge by the microscope." Quoted in Mertins, "Bioconstructivisms," 363.
32. Mertins writes here about Paul Weidlinger's "Form in Engineering," but his phrasing is very appropriate for Ebeling as well. Mertins, 365.
33. See also Mertins, 363–65.
34. Quoted in Corey Keller, "Abbilder des Unsichtbaren," in *Brought to Light: Photography and the Invisible, 1840–1900*, ed. Corey Keller (New Haven: Yale University Press, 2008), 30. In his writings, Ebeling adds to the cosmological perspectives a Hegelian phenomenological dimension, insofar as he writes about an abstract historical spirit that expresses itself in the concrete phenomena of architecture, about a general spiritual direction that manifests itself in the sensuous singularities of design. This phenomenological thrust is

present in both *Space as Membrane* and *Cosmological Space-Cells*. In the latter he writes, for instance, about the "development of the creative spirit" ("Gang des schöpferischen Geistes," my translation) and in the former about "the enormous potential of a purely spiritual movement to seed a new style.... The whole of the Gothic, including proto-Gothic, is a single *form*-giving process for this spiritual crystallization of the core" (Ebeling, *Space as Membrane*, 26; emphasis in original).

35. For a biography of Francé, see Martin Müllerott, "Francé, Raoul," *Neue Deutsche Biographie* 5 (1961), https://www.deutsche-biographie.de/pnd118692453.html#ndbcontent.

36. Detlef Mertins, "Where Architecture Meets Biology: An Interview with Detlef Mertins," *Departmental Papers (Architecture)* 7 (2007), http://repository.upenn.edu/arch_papers/7. As Müllerott points out, Francé's work was received by *völkisch* sympathizers as well.

37. "Vereinszweck ist die Förderung der Naturkunde mit dem Mikroskop," Impressum der Webseite der Mikrobiologischen Vereinigung München e.V. (MVM), https://www.mikroskopie-muenchen.de/impressum.html.

38. Cited in "Vereinsgeschichte der MVM": https://www.klaus-henkel.de/geschichte.html: "Gründen wir eine mikrologische Gesellschaft, die den Gebrauch der Mikroskope volkstümlicher machen will und die ganze große Vertiefung der neueren Wissenschaft vom feinen Bau und Leben der Pflanzen und Tiere dem Verständnis näher rücken wird."

39. Francé's microcosmic interests paralleled macrocosmic ones: in 1904, he also cofounded the Kosmos Society of Nature Lovers (Kosmos-Gesellschaft der Naturfreunde), which capitalized on the public's upsurge of interest in the natural sciences and also published the popular science magazine titled *Kosmos*. This still exists today in a new iteration as an *Umweltmagazin* (environmental magazine). See https://www.wissenschaft.de/natur/. See also Mertins: "Francé presented an entire cosmology—his publisher was even called Kosmos—which someone like Lissitzky was very sympathetic with, since he was oriented towards a new cosmology of world reconstruction. Moholy-Nagy and other constructivists of the 1920s all wanted to have that kind of comprehensive, scientific worldview as a platform for their experimental work"; Mertins, "Where Architecture Meets Biology."

40. Francé's definition of the "edaphon" in *Bios: Die Gesetze der Welt* is "die Lebensgemeinschaft der im Erdboden lebenden Organismen" (275).

41. Ebeling, *Space as Membrane*, 24.

42. Ebeling, 25.

43. Ebeling, 32–33 (emphasis in original). Cf. Walter Benjamin's claim in "The Return of the Flâneur" that Le Corbusier turns human residences into spaces of transit for all manner of forces as well as light and air waves ("Durchgangsraum aller erdenklichen Kräfte und Wellen von Licht und Luft"); quoted in Asendorf, "Grenzfläche—Übergang," 48. Yet whereas this spells architectural transparency for Benjamin, for Ebeling, it necessitates architectural impermeability.

44. Scheiffele, *Das leichte Haus*, 184–85: "Helio-Biozönose heisst etwas umständlich ausgedrückt: Sonnen-Erden-Lebensgesamtkorrespondenz, durchgeführt für alle Gebiete, gleichgültig ob Physik, Medizin, Psychologie, Soziologie, Geologie." Here Scheiffele is quoting Hausmann giving credit to Hanns Fischer for the term.
45. Scheiffele, *Bauhaus, Junkers, Sozialdemokratie*, 29.
46. See Scheiffele, 30.
47. For deeply problematic aspects of Hellpach's work, see Horst Gundlach, "Willy Hellpachs Sozial- und Völkerpsychologie unter dem Aspekt der Auseinandersetzung mit der Rassenideologie," in *Rassenmythos und Sozialwissenschaften in Deutschland: Ein verdrängtes Kapitel sozialwissenschaftlicher Wirkungsgeschichte*, ed. Carsten Klingemann, Beiträge zur sozialwissenschaftlichen Forschung, vol. 85. (Opladen: Westdeutscher Verlag, 1987), 242–76.
48. Ebeling, *Space as Membrane*, 9n1.
49. Ebeling, 9n2.
50. Le Corbusier, quoted in Kousidi, "Breathing Wall Skins," 86.
51. See Spyros Papapetros, "Future Skins," afterword to *Space as Membrane*, by Siegfried Ebeling (London: Architectural Association, 2010), xiii.
52. For more on this history, see Caroline van Eck, *Organicism in Nineteenth-Century Architecture: An Enquiry into Its Theoretical and Philosophical Background* (Amsterdam: Architectura & Natura Press, 1994); and Didem Ekici, "Skin, Clothing, and Dwelling: Max von Pettenkofer, the Science of Hygiene, and Breathing Walls," *Journal of the Society of Architectural Historians* 75, no. 3 (September 2016): 281–98, https://online.ucpress.edu/jsah/article/75/3/281/60879/Skin-Clothing-and-DwellingMax-von-Pettenkofer-the.
53. Ekici, "Skin, Clothing, and Dwelling," 281.
54. Ekici, 281, 293.
55. Ebeling, *Space as Membrane*, 8 (emphasis in original).
56. Scheiffele, *Bauhaus, Junkers, Sozialdemokratie*, 194.
57. Ebeling, "Kosmologe Raumzellen": "Der heute noch massiv-porös umkleidete Raum wird durch die Schöpfung neuer Strukturverhältnisse ... zu einer Membran zwischen unserem Körper als Kern und den plasmatischen Energien der Grosswelt werden" (my translation).
58. Heinrich Wölfflin, "Prolegomena zu einer Psychologie der Architektur," in *Kleine Schriften*, ed. Joseph Gantner (Basel: Schwabe, 1946), 13, accessed through Universitätsbibliothek Heidelberg Digital Library, https://digi.ub.uni-heidelberg.de/diglit/woelfflin1946/0017. Cf. Ebeling arguing for the "overthrow of every kind of aesthetically defined building that is based on an ossified conception of the relationship between building material and body matter (plasma) and an ultimately dualistic world-view ('pictorial,' 'plastic')"; Ebeling, *Space as Membrane*, 20.
59. Ebeling, *Space as Membrane*, 17.
60. Ebeling, 10.
61. See Mertins, "Bioconstructivisms."

62. See Raoul Heinrich Francé, *Die Pflanze als Erfinder* (Stuttgart: Kosmos, Gesellschaft der Naturfreunde, 1920), 8. Francé introduces the term *Biotechnik*.
63. Mertins, "Where Architecture Meets Biology."
64. Francé, *Bios*, 260, caption to image 115.
65. Francé, *Bios*, 275.
66. Ebeling, *Space as Membrane*, 15. I made a modification to the original English translation, which translates "einzelnen Umweltraum" as "individualised environment." In my opinion, it should be "single" or "particular environment." Ebeling writes about the dangers of an individual environment (of a city, a particular climate) vs. those of the environment as a whole (of the global, the planetary).
67. Ebeling, *Space as Membrane*, 24 (my emphasis).
68. Ebeling's manuscript, found among documents of the Junkers propaganda department, is reproduced in Scheiffele, *Bauhaus, Junkers, Sozialdemokratie*, 233–34. Consider also his misogynistic claims in "Kosmologe Raumzellen" that the new movement of design must overcome the female phase and come into its own as a "male eros of the present," as a "male design [*Gestaltung*] worthy of us" (my translation).
69. Ebeling, *Space as Membrane*, 15.
70. Ebeling, 30.
71. Ebeling, 29.
72. See also "Natural light should be blended with the house-machine's artificial light . . . so as to create a profound unity in the interior with an automatic regulation of the quality of light (avoiding twilights, dawn and greyish light on dull days, particularly in the sub-Arctic)"; Ebeling, 31.
73. Ebeling, 18.
74. Ebeling, 17.
75. Ebeling, 10n** [notes not numbered but rather marked by asterisk]: "The 'soul,' as it grows more distant from the mother earth of eros-bound animality, will . . . attempt to understand with eyes torn open, with a relentless consciousness, the true reason of the world, not in order to collapse in front of its image, but rather to overcome it as purified spirit" (my translation).
76. Ebeling, *Der Raum als Membran*, 33 (my translation).
77. Ebeling, *Space as Membrane*, 18.
78. Ebeling, 18.
79. Quoted in Adler, "Energising the Building Edge," 196.
80. Ebeling, *Space as Membrane*, 10–11.

Literature as a Milieu of the Small
Kleist/Kafka

MARIANNE SCHULLER

✦ ✦ ✦

THIS ESSAY seeks to read literature as a milieu of the "small." The orientation provided by the title does not so much pursue the question of literature as a scene of the small or the minute in the sense of tiny objects and small forms but rather sketches the "small" as a microstructural moment of literature significant for modernity, its processes, and its becoming. Just as the small is first brought forth in the literary process, it is also true that the contour of the literary milieu, as it appears in the following exploration, is not simply given but is formed in a retroactive volte-face through the performance of the small. In this respect, there is less recourse here to a more or less established arsenal of concepts and knowledge; instead, an attempt is made to design the minutiae of literature as that which designs itself within a space of literary knowledge. The broad field of minutiae is, here, however, unfolded in a micrological reading. The micrological reading can perhaps overcome any rather assertive gestures by paying attention to the particularity and the unforeseen nuances of the literary performance per se.

The following reading focuses on Kleist and Kafka. In doing so, it takes up a much-discussed constellation. Kafka himself repeatedly expressed his admiration for Kleist; his drastic remark in a letter to his friend Max Brod in January 1911 has become almost proverbial: "Kleist bläst in mich, wie in eine Schweinsblase" (Kleist blows into me as if into a pig's bladder).[1] The references between the two modernist authors are manifold and explored in many ways. I will first turn to a text by Kleist that, most famously, thematizes this procedure of breaks and undercuts on the level of minutiae: the short text "Über die allmähliche Verfertigung der Gedanken beim Reden" ("On the Gradual Production of Thoughts Whilst Speaking," circa 1805–6), which comprises only a few pages. In so doing, I make use of a punchline that consists of the fact that Kleist's text not only forms the beginning for the readings to be unfolded here but begins itself with the question of beginning, with a question that inexorably entails that of the end. In contrast, Kafka's text "Josefine die Sängerin, oder Das Volk der Mäuse" ("Josefine the Singer, or The People of the Mice," 1924) not only is his last text, completed and published shortly before his death, but is itself carried after the end as a problem of literature. The constellation of the two texts thus reveals a great range (from beginning to end, as it were) brought about by minutiae, which at the same time encompasses elements of literary modernism.

"The Twitching of an Upper Lip": On Kleist's "On the Gradual Production of Thoughts Whilst Speaking"

I begin with Kleist's "Über die allmähliche Verfertigung der Gedanken beim Reden"[2] because it presents the beginning as a paradoxical figure right from the start. When the text begins, something will always have preceded it that makes the beginning possible and at the same time makes it seem impossible. Because the beginning of a text cannot be founded in itself, it is, therefore, always the consequence, as Kleist says, of "good luck." If he is lucky, sentences and further texts emerge from this setting that subsequently appear to justify the beginning.[3]

This paradox of the beginning starts with the fact that one does not know where the text begins. With the first sentence? In the paratextual address "To R. v. L."? If this abbreviation refers to Otto August Rühle von Lilienstern, a childhood friend of Kleist's, there remains an uncertainty that concerns not only the addressee but also the identity of the text genre: Is it a record of a conversation with a counterpart addressed as a "dear meaningful friend"? Is it a letter? Or should the reader of a potential public[4] feel equally and individually addressed?

These questions, which seem of no great importance but necessarily arise with the beginning, cannot be decided later in the course of the text. In their undecidability, they seem to initiate a move that is essential to the text. The questions must remain undecided to a certain extent because the beginning can only be postulated, retroactively through the text, as being well founded. The same applies to the conclusion. The text ends with the note "to be continued," in brackets, to which the abbreviation—signaling authorship—is added (32). Naturally, this minimal reference here may refer to the medium of the magazine but can also be read as a performative anticipation of the text process: just as the "continuation" corresponds to the setting, the conclusion also indicates that the production of thoughts is not completely fulfilled in a finished text object and thus comes to a conclusion. Rather, the text becomes a place or scene that performatively brings about a shift in a common understanding of the text. The text is not fulfilled in an object constituted by it but vice versa: it realizes itself and its desire, the production of thoughts in speech, by coproducing the incompleteness of language with every incremental step of its production. Far from the theorems of progress, there is a structural "more" of which Kleist's textual desire is in pursuit. With each further step of its production, an absence emerges that reproduces the text's passage in the sense of a "to be continued." The lack constituting desire is not attached to an existing but still indistinct object that could fill it but rather arises from its subversion: from the breaking up of the symbolic consistency of language.[5] This process takes place in Kleist's text insofar as it continues from interruptions, from gaping blanks, from pauses, stretches, and other techniques that do not subordinate themselves to the direction of any grand scheme of a given meaning. In these linguistic microprocesses, however, an "excess of desire" emerges that discovers a nonknowledge in knowledge and thus opens a path to modernity.

From the first sentence, the subject of knowledge and language is on the agenda. Here the narrative-I takes on the role of an advisor/mentor, whose advice is not conveyed as a finished message but is staged as a search process. "If you want to know something and cannot find it through meditation, I advise you, my dear, meaningful friend, to talk about it with the next acquaintance who you encounter (*aufstößt*)" (27). The sought-after knowledge, thus the starting point, is not to be found in a methodical-rational procedure for which the allusion to the Cartesian "meditations" stands. It requires quite different steps. The narrative-I sketches the scene of a chance encounter with a fleeting acquaintance who, as the saying goes, "bumps into" him. Just as the word *encounter* resonates with the opposite, the word *burp* (*aufstoßen*), which is inappropriate if not offensive in this context, triggers not only the association of a bump (*aufstoßen*) from outside but also that of an unintentional and usually unpleasant process taking place, invisibly and on the level of smallest perceptions, inside the body. Not a dialogical situation but the collisions with another and with something other than language, something of minutest (if any) importance, provide the impetus to speak, whereby the object remains below perception—indeed, it does not (yet) seem to exist at all. The ego, as if it were longing for knowledge, does not seem to be oriented toward the extension or completion of previously unknown objects of knowledge; rather, its desire is for another knowledge and for something other than knowledge, which is realized in the search for the metonymically unattainable.

This fleeting scene is joined, without transitions, by another, which in turn provides the impetus (*Anstoß*) for the search for an unknown object postulated by the desire for knowledge. This scene creates a space that is used, by the I-narrator and his sister, as a workspace. At the center of the scene is not the object of their respective work but the topographical arrangement of the places taken by brother and sister.[6] While the brother looks into the light in order to clarify the legal and algebraic tasks that, for him, obviously remain in the dark of ignorance or the nonperceptual, the sister gives other speeches that rather miss the relevant discourse. And these speeches, the subjects of which are not mentioned, hit the brother, as it were, from behind; they are issued behind his back. So while the brother is turned to the bright light of enlightenment without achieving a solution to the problem, the sisterly speeches come from the darkness, behind his back, triggering a kind of invisible enlightenment, a cognition on some undetermined microlevel: "And behold when I talk to my sister, who sits behind me and works, I find out what I would not have been able to get out by brooding for perhaps hours. Not as if she had told me, in the true sense of the word, for she knows neither the Code of Law, nor has she studied Euler or Kästner" (27).

The small scene is constructed simply; it appears like the reproduction of a negligible everyday event. And yet it is highly condensed and compact. The compactness is created by the fact that in it, in the depicted everyday scene, even the questions are already effective, which move the text noticeably but almost imperceptibly, without being written out or even conscious—a knowledge that, unlike the epoch and the procedure of the Enlightenment, does not presuppose it can be explained but deconstructs this presupposition and thus opens up a nonknowledge in knowledge. While the narrative-I still awaits the Enlightenment and thus the unlocking of the closed, Kleist's text conversely, in the course of his performative procedure, unlocks a closedness[7] that incalculably expands and fundamentally changes knowledge via the dimension of nonknowledge. Thus the sister appears as the one who, with her speeches based on the illumination of a still dark discussion of the facts, not only turns around and makes absurd speeches behind his back but also attempts to interrupt the brother's speech with small gestures. If the theater man Kleist, who enigmatizes the scene by means of the exteriority of the gestures, can be seen here, then the enigmatic character of the scene lies not least in the fact that it presents the exterior as an unfathomable element of language, as an Other belonging to it: "Nothing is salutary to me than a movement of my sister, as if she wanted to interrupt me; for my already strained mind is only more excited by this attempt from outside to snatch from it the speech in whose possession it finds itself" (28).

According to the I-narrator, the sister's unbidden attempts to disturb the brother's thoughts are intended to deprive him of his inner possessions.

But it is precisely in this deprivation that the salvation lies. This thrust from outside, which causes a rupture in the inner structure of the thoughts, simultaneously triggers a dark, indistinct microconception that, as dark and indistinct as it is, is connected in an indeterminate proximity with that which the ego seeks. Here, as with Kant, the "mind" functions as a capacity for individual notions, perceptions, feelings, which the elements of language set into a relationship of the alien self: "But because I do have some dark idea that is in some connection with what I am looking for from afar, if I boldly make the beginning, the mind, while the speech progresses, in the necessity of finding an end in the beginning as well—that confused idea is chosen for complete clarity in such a way that the knowledge, to my astonishment, is finished with the period" (27–28).

Beginning and end are made and have retroactively created a consistent unity of thought and made it recognizable. It is not by chance that the astonishment that the events trigger in the narrator is mentioned, for instead of a possible explanation for the production of the thought by the sentence, the sequence contains a sign for its lack: a dash, the merest of punctuation marks, indicating a loss. This abysmal emptiness in the word sequence, this absence of meaning, can become the starting point for the leap to other elements of language that, by entering into new relationships, gather other meanings. This small trait of language requires certain techniques, all of which exhibit the moment of break as the moment of linguistic production: "I mix in inarticulate tones, extend the connecting words, use an apposition where it is not necessary, and make use of other, more expansive speech, tricks for the fabrication of my idea in the workshop of reason, to gain the necessary time" (28).

So it is not least the interruptions to the speech that produce the speech and the thoughts that the speaker does not have. Thus, the figure of a paradox emerges: the interruption, the loss of language, the not-knowing produces an "excess" of language, thoughts, perceptions, feelings, and sensations. This "excess," however, is not oriented toward the principle of perfection or progress that has prevailed since the Enlightenment, which follows the scheme of continuity, but rather this "excess" that breaks off is the result of another movement inherent in language: a path without a predetermined direction and without a reassuring sense that can enter into a relationship with other ideas. This "excess" without a definable goal, without a substantial substrate, this "excess" in the sense of "to be continued" is realized by seeking the I-narrator in speech.[8]

Just as Kleist's text continues and moves through minutiae, through interruptions and gaps, from voids to pauses and further interruptions, so too does it do so through the collision of highly different narrative segments.

The stories that are triggered are not told out but rather broken off by the incursion of another. This creates a mosaiclike text, composed of numerous and ever-smaller elements, that brings together different small genres such as anecdote and fable as well as different fields of knowledge such as natural science, politics, and history. Only the longer final sequence, which deals with "school knowledge," using the example of an exam, is distinguished by a coherent narrative. This difference once again characterizes the knowledge that the text realizes as a search: while "school knowledge" presents itself as that which is and remains substantial, knowledge—as it becomes legible in Kleist's text—is constituted in a linguistic movement that is open to determination,[9] in which knowledge realizes itself as metonymic-unattainable.

Kleist's narrator now tells a story about the outbreak of the French Revolution, which was initiated by the "Donnerkeil Mirabeau" (28–29). According to the narrator's account, it was, in Mirabeau's view, an inappropriate instruction from the king to the last meeting of the estates that prompted Mirabeau to resist. After the king's request was repeated by the "Ceremonienmeister" with the question of whether the assembled had heard the king's order, it reads,

> "Yes," replied Mirabeau, "we have heard the king's order"—I am sure that when he [Mirabeau] made this humane beginning, he [Mirabeau] did not yet think of the bayonets with which he concluded: "Yes, my lord," he repeated, "we have heard him"—you can see that he doesn't quite know what he wants yet. "But what justifies you," he continued, and now suddenly a source of tremendous ideas comes to his mind—"to suggest orders to us here? We are the representatives of the nation."—That was what he needed! "The Nation gives orders and receives none"—to swing straight to the summit of presumption. "And to explain myself clearly to you"—and only now does he find what expresses the whole resistance to which his soul is equipped: "so tell your king that we will not leave our places other than by the force of bayonets."—Whereupon he sat down, self-satisfied, on a chair. (28–29)

This scene finds a sense of satisfaction at its end. There is, however, a small, rather inconspicuous postscript, which seems like the narrator's attempt at a résumé: "Perhaps that the last thing that happened was the twitching of an upper lip, or an ambiguous play on the cuff, which in France caused the overthrow of the order of things" (28–29). Why does this postscript seem strange? Because there is a small—even minute—discrepancy between the story and the "summary." First of all, the pleasant impression arises that we are dealing here with the suggestion of a rule: every movement, however slight, and every gesture, no matter how random it may appear, can be understood as a sign or indication whose meaning is or can be understood—for, as a sign that functions via repeatability, it is not, consciously or unconsciously, integrated into the symbolic system of language that determines our social order.

The postscript, for all its similarity to the narrated scene, is a little perpendicular to it, for the simple reason that the scene itself does not show the "twitching of the upper lip" or the "plucking of the cuff" as an object. Rather, it lists new, different events that, undocumented as they remain, are (so to speak) unique. This positioning in the text deprives them of the sign character as it was brought about by the previous narrative. The inconspicuous postscript, however, which can of course also be the occasion for small stories, anecdotes, fables, plays, and so on, remains here, although postscribed, "before" the "empire of signs" (Barthes) as the realm of the symbolic. Things stand there in the absence of meaning and without suggesting a new area of definition but can perhaps accumulate new meanings. The slightest shearing out of facial expressions and the smallest gesture are not reinterpreted into something "big," but an incalculable "gap in the symbolic" (Dolar) opens up in the text, which, even a revolution can become the site of the production of an object that did not exist until then, the revolution. And so Kleist's little text ends with Mirabeau's outlined prospect of the enforcement of the republic—of the enforcement of an institution, of an Other.

Almost Nothing: On Kafka's Last Story, "Josefine the Singer, or The People of the Mice"

FRANZ KAFKA wrote the story "Josefine the Singer, or The People of the Mice" and also edited it for print shortly before his death on June 3, 1924. He assigned it the *final position* in his *A Hunger Artist: Four Stories*, which deals primarily with the subject of "art" and the "artist figure" in connection with the motifs of disappearance and endings, up to and including death.[10] The motif of disappearance also dominates Kafka's last story and gains even more significance through its placement in the *Hunger Artist*. It is also hard to avoid the assumption that the story, which falls in the period shortly before Kafka's death, is in a sense a testamentary gesture.

Initially, Kafka had intended the story to have the title "Josefine the Singer," which he then changed on short notice to "Josefine the Singer, or The People of the Mice." He commented on this change to his friend Robert Klopstock, who accompanied Kafka in the last phase of his life: "The story gets a new title Josephine, the Singer / or The People of the Mice. / Such or-titles are not very pretty / but here it perhaps has a special / sense, it has something of a scale" (*KKAD*, 462). The change of title first of all establishes a connection between the singer Josefine and a world of mice described as a "people," to which the singer also belongs. Yet it remains unclear how the "or" organizes the connection: Is it to be understood disjunctively or in a nonexclusive sense? In the first case, the narrative would occupy the space opened up by the disjunction, an empty interstice, as it were, between the singer and the people; in the second case, an elliptical arrangement would be designed that does not define Josefine without the people and the people without Josefine. In this way, the two poles would be more closely related to each other, while the space opened up by the disjunction would make them a step apart. In a sense, they lose themselves in the space of the narrative.

The narrative begins with the naming of the singer Josefine. She is the only character in the story who is given a name. This distinction makes her an exception among the people from the outset. However, the exceptional situation takes on ambiguous features right at the beginning, for, as the mouse narrator reports, the people of the mice have no musical inclinations whatsoever—music does not occupy an ancestral or desired place in the life of the people at all. For the narrator, it is indisputable that the people of the mice are unmusical. One can hardly help but think of Kafka's recurring statement about his own unmusicality, so the opening of the narrative allows a certain ambiguity of this configuration to shine through.[11] It is the narrator's excessive use of the linguistic form that produces this effect:

> Our singer is called Josefine. Those who have not heard her do not know the power of song. There is no one who is not swept away by her singing, which is all the more significant because our sex does not love music as a whole. Quiet peace is our favourite music … we consider a certain practical cleverness, which we admittedly also need extremely urgently, to be our greatest advantage, and with the smile of this cleverness we are used to comfort ourselves over everything, even if we should once—which does not happen—have the desire for the happiness that perhaps emanates from music. Only Josefine is an exception; she loves music and knows how to communicate it; she is the only one; with her passing, music will disappear from our lives—who knows how long? (*KKAD*, 350)

The cascade of negations not only undermines the setting of the beginning but also opens up a space suspended between presence and absence like the narrative. The negation symbol itself marks an object as absent and at the same time keeps it present as absence. In this way, negation in its sliding paradoxes calls up a mode of being that transcends the polar opposition of present/absent, which, according to Freud, allows "thinking a first degree of independence" from the results of repression and from the compulsion of the pleasure principle.[12] The narrator exposes the nonmusicality of the mouse people as a fact—but one that neither erases nor preserves as an identical object that which it denies. Rather, an in-between space emerges—like the space of the narrative itself—taking in something outstanding, other, unknown, or barely noticeable (for example, the question of music and unmusic that runs through Kafka's writing). In other words, the negation doubts not the statement—the unmusicality of the mice—but its unequivocal validity. It undermines the judgmental speech.

The opening sequence ends with the word *disappear*. According to the narrator, music will disappear from the lives of the mice with the singer's death. If this already hints at the motif of the end at the beginning, the narrator leaves the question open as to what the nonmusicality of the people, on the one hand, and the disappearance of music from the people's lives, on the other, could mean for his narrative. The two remarks put together remain semantically unconnected. But what about the singing of Josefine in the first place? With a heavy heart, the narrator asks himself whether Josefine's announcement is singing at all: "Is it not perhaps only a whistle after all? And, as we all know, whistling is the real skill of our people, or rather not a skill at all, but a characteristic expression of life. We all whistle" (351–52). Josefine whistles like all those who belong to the mouse people. There is only one difference, however minimal and barely detectable: her whistle is a little weaker and more attenuated than those of the other mice. Nevertheless, her barely distinguishable proclamations and performances, which she calls for without a fixed schedule but with authoritarian force, have great effects. Silence reigns in the auditorium; it is "as quiet as a mouse, as if we had become partakers of the longed-for peace" (354). This arrangement of nonsong, faint whistling, and great effect presents the brooding narrator with insoluble questions: "So if it were true that Josefine does not sing, but only whistles, and perhaps even, as it seems to me at least, hardly goes beyond the limits of the usual whistling—indeed, perhaps her strength is not even quite sufficient for this usual whistling ... if all this were true, then Josefine's alleged artistry would be disproved, but the riddle of her great effect would then be even more to be solved" (352). The narrator uses an example to address the riddle both posed and embodied by Josefine. Perhaps, as this riddle is so difficult to crack, he invokes the idea of a nutcracker who is characterized by the peculiarity that he transforms the least peculiar, the usual par excellence—namely, the cracking of a nut—into an artistic act. It is about the peculiarity that someone "solemnly stands up to do nothing but the usual. Cracking a nut is truly not an art, so no one will dare to call an audience together and crack nuts in front of them to entertain them" (353). But when he does and the entertainment function kicks in, the action attracts an audience, then puzzling questions arise again. The initially rather placating assumption that, in the case of the scene, it could not have been "mere nut cracking" is not satisfactory. It is immediately transformed into a more far-reaching contemplation: "Or it may be nut cracking, but it turns out that we have overlooked this art because we have mastered it smoothly and this new nutcracker only shows us its real nature, in which case it might even be useful for the effect if he is somewhat less proficient at nut-cracking than the majority of us" (353). This scene is funny not least because it takes a crack at the "everyday action" versus "art action." This crack makes the more or less common distinction between art and everyday action virtually invalid. Rather, there is another movement: a difference is introduced into the everyday action itself, albeit minimally and rather inconspicuously, that transforms the action into a scene. In this case, it is solemn gestures that the nutcracker presents and that, by bringing about an interruption in the everyday action, transform the surroundings into an audience. Both the decontextualization of the action created by the distance and something else accomplish the transformation of the everyday action: a certain deprivation of the usual perfection. Nut cracking, which is usually "smoothly" mastered, is transformed into the strange not by an increase in perfection but by a certain weakness in the execution. By the fact that the cracking is not perfect, a small breach is made in the well-known process that exhibits the breaking "itself"—the breakage as a moment of the artistic process, a kind of nothingness that lends the everyday object an incomprehensible, almost solemn dignity.

Mladen Dolar has related this gesture of withdrawal, the introduction of lack, to the invention of the ready-made (1913) by Marcel Duchamp.[13] As with Kafka's nutcracker, Duchamp transformed the ready-made into an art object by taking an ordinary object and stripping it of its ordinary context. Duchamp's work is about an ordinary bicycle as part of an ordinary, mass-produced bicycle that, stripped of its usual milieu and function, he has transformed into an art object. Similar to Kafka's nutcracker, Duchamp does not introduce something different, foreign, exotic, unusual as an additional element to a finished object—here, it is an industrial mass-produced object; rather, he adds an almost-nothingness to the object that "elevates the ordinary object to strange sublimity."[14] This reference, which places Kafka's nutcracker scene within the horizon of modern art, simultaneously reveals striking differences: whereas in Duchamp's work, a scene is created that programmatically and with resounding success redefines the art of modernism—that is to say, a large scene—and institutionalizes it as art to a certain extent, in Kafka's case, the programmatic dimension also becomes obsolete. The scene of nut cracking as art is, as it were, a street scene that can take place anywhere and at any time because (and insofar as) it is removed from institutional recognition as art. Art becomes, to use Kafka's reflections on "small literature," an "affair of the people."[15]

The enigmatic nutcracker scene perhaps bears a similarity to the enigma of the Josefine story. Both stories are about the transformation of a common act—nut cracking, mouse whistling—into an object traded as art. But the comparison also highlights the difference. While the nut cracking introduces the break as an object, "Josefine" is about an object that already functions as art: singing, which gains its meaning as art from a harmony of some kind with the object's "voice." The only catch is that the singer does not sing; she, as one might assume, cannot even sing. She has, as the narrator puts it in a rarely apodictic statement, a "nothing of voice" (362). This statement—the nothing of voice—seems to refer to the singing voice as the voice that has emerged from the cultivation of the nonstructured, presymbolic object of voice, for Josefine does have vocal proclamations: she whistles like all mice do. Whistling is the "expression of life" of mice, which as such is not subject to the cultivation of the object voice. The (unexplained) nothingness of voice would then be a voice from which the translations into the symbolic and meaningful are, as it were, subtracted or expressed from a different perspective, a voice before the transformation of the physiological part of a process of sound formation that transforms the voice into the symbolically expressive—one could also say, a voice before art.[16]

When Josefine's singing and its great effect are promoted by the narrator's stated nothingness of voice, then it is not only a curious moment that proves the nonmusicality of the mice. Rather, in connection with Josefine's undisputed artistry, a moment that is essential for the question of art—for the question of the artistic object—emerges: it is about the zone of connection between something and nothing. To the extent that the voice, especially the voice of singing, is determined by the intersection between barely perceptible physiological processes and symbolic expressive procedures, the voice is the object that subverts the opposition of "nature" and "culture." Even the pronouncement of the whistle, which all mice possess, is not simply "nature" but is, as the narrator points out, an utterance of life, an utterance that manifests life. In other words, life is not without the voice, without the whistle and the peep, because life needs the utterance, even in its minutest instantiations, in order to be life.

If this also undermines the strict opposition of life and symbolic utterance, the observation of Josefine's nothingness of voice as an element of her artistry points to the connection between nothingness and her performances. They seem to derive their efficacy not from the fact that they do not substitute something for nothingness—that is to say, replace it—but in the fact that nothingness dwells in their pronouncements and thus becomes audible. Just as the life of mice is not without whistles, so Josefine's art is not without nothing. Part of this is that Josefine knows nothing about it and wants to know nothing about it; indeed, on the contrary, she is convinced that her artistic creation knows no lack. According to her conviction, her art is an incomparable, unique specialty because she, like no other artist, possesses all the requirements associated with the art of singing in the highest, flawless perfection.

While the people of the mice enjoy the events, characterized by feeble whistling, as a place of togetherness that briefly dampens the distress of their lives, for Josefine, the performances are the place where she can flaunt her self-expression as an artist: her delicacy and her fragility, her exhausting of nerves, her sole dominion over the audience of mice, her claim to the support and protection of the sycophants surrounding her. All her airs and graces, taken from the repertoire of the big-city divas of the early twentieth century, are part of her vaudevillian spectacle, whose outsized pretensions are comically at odds with the diminutiveness of her appearance and the slightness of her expressions.

Now, when the narrator follows all these efforts of the singer to stage herself as the embodiment of art, he does so without a critical or even arrogant gesture that exploits the comedy of the scene. Rather, he seems to be of a porous receptivity to the airs and graces of grandeur. Perhaps because he perceives in it the homage to an art that, in Josefine's case, is only evident in the fact that it is missing from her or in the exiguity, the almost-nothingness, of her performances. And it is this lack, this almost-nothingness as a marker of a breaking point between nothing and something, that the narrator perceives and that affects him and his narration itself. The narration seems almost like an echo of Josefine's lovingly delivered performances. But while Josefine tries to deny the almost nothing in her oversized allure, the narrator tries to bring out the almost nothing in the small, the scurrying, the moving, the minimal in the narration itself. The narrative balances between something and nothing on the level of the micrological, which indicates the breaking point in speech.

The tide turns, however, at the moment when Josefine puts forward the claim for recognition of her performance as art (the claim for institutionalization of art as an exception). This claim to recognition as an exception is in no way connected to Josefine's refusal to help with the toil and labor of the mouse's life; she is concerned with the purely symbolic act of recognition as an artist by the people. She belongs to the people and at the same time wants to win recognition as an exception to the people. In a tireless effort to gain recognition as a great artist, which is heightened to the heartbreakingly comic by her tininess, she wants to achieve the uniqueness of art and of herself as an artist. She wants, in other words, to institutionalize the "small rupture" that has occurred in each case through her performances, through her lack of voice, and has brought about an unspeakable disruption of the same in the same. What Josefine strives for, then, "is only the public, unequivocal recognition of her art that transcends time, that rises far above anything previously known" (370).

If the institutionalization of the rupture is itself already an irresolvably comic and impossible thing, the same applies to its capers of recognition: they aim at a structure of power that is represented in the figure of the sovereign. Just as the sovereign stands inside and outside the law,[17] Josefine also wants to position herself and her art as the authority that takes the place of the outside for "das Volk." By establishing this structural reference and breaking it into tiny increments, the world of the comic is once again stretched across the narrative.

It is precisely this step, however, that is not taken by the people of the mice. The people are unimpressed by all the threats she makes in case they refuse to recognize her as an artist. They are even unimpressed that Josefine is considering reducing or even eliminating the coloratura. As the narrator suggests, the people's indifference is due not least to the fact that there were no coloraturas anyway. Without an argument, without a for and an against, the people "calmly reject the demand" (369). All the arguments that Josefine presents in the struggle for recognition "the people listen to and pass over. This people, so easily moved, is sometimes not moved at all" (368). Far from making an aesthetically justified judgment, the silent refusal makes it inexorably clear that recognition as an artist could at the same time mean that Josefine is "almost outside the law" (368). For the people of the mice, then, there is no outside the law, not even in the mode of art. This attitude, as the narrator suggests, brings about a kind of reversal of the relationship between the "people" and the individual with regard to the question of the message: it is not the exceptional individual who addresses a message to the people but the permeable, thin, minutest whistle that appears like the "message of the people to the individual." In Josefine's events, the crowd "is withdrawn into itself. Here ... the people dream, it is as if the limbs were loosened, as if the restless one could stretch and stretch at will in the great warm bed of the people. And now and then Josephine's whistle sounds in these dreams. . . . Something of the poor short childhood is in it, something of lost happiness that can never be found again, but something of the active life of today is also in it, of its small, incomprehensible and yet existing and not too lively life" (366–67). As much as the narrative opens up references to the history and life of the Jews, as much as it addresses the related question of law and the sovereign formation of power, these references are themselves readable as the justification of a poetological contour. Just as the nutcracker introduces a difference as an interruption of the ordinary in the mode of a property accessible to all, just as Duchamp's gestures of withdrawal into the existing object of the ready-made produce an Other Unknown, so the comedy—as well as the tragedy—of the singer Josefine, lies in establishing the nothingness of voice qua recognition as something under the name of art, in assigning it an institutionalized place in society. While Duchamp makes something out of nothing, Josefine wants to make something out of nothing, which lacks nothing. But with this, with the institutionalization of the rupture, it would be, in Dolar's felicitous formulation, "the end of art."[18]

If the small, the minimal, that which necessarily appears as minutiae, is already comical in view of the great goal and the prospect of eternal fame, undimmed by time, then—at the same time—an abysmal, testamentary seriousness breaks out in the comedy: not as criticism, not as a program, but in the gesture of rupture. With Hamacher's reading of Benjamin, the gesture can be marked as a kind of breaking point in language: "The gesture marks in language that which is no longer language."[19] And therefore Kafka's narrative must move in the milieu of the people of the mice, trembling with tenderness, the minimally small, tiny, because it is the movement factor of the narrative. The teeming nation of mice is minute not only in shape but also in terms of language: it communicates through a "childish whisper," through "admittedly innocent gossip that merely moves the lips" (358). What appears as sheer minutiae is neither that which is outside nor that which is inside the order. Like Odradek, it is a figure of the in-between, the figure of an unfixable transience between nothing and something, which comes about through the minimal but does not show itself in the minimal.[20] In a journal entry from February 15, 1920, Kafka traces this figure of the connection between nothing and something, which he also wishes for the "View of Life":

> As the most important or most wish was to gain a view of life (and—this was necessarily connected—to be able to convince others of it in writing) in which life preserved its natural heavy falling and rising, but at the same time was recognised with no less clarity as a nothing, as a dream,

as a suspension. Perhaps a beautiful wish, if I had wished for it correctly. For example, as a wish to hammer a table together with scrupulously neat craftsmanship and at the same time to do nothing, and not in such a way that one could say: "To him, hammering is nothing" but "to him, hammering is real hammering and at the same time also nothing," whereby hammering would have become even bolder, even more determined, even more real and, if you like, even more insane.[21]

The desire, then, goes to the nonaction in action, to the point of difference between speech and muteness as that which constitutes the reality of action, of speech, of speaking. If the point of difference, the nothingness "itself," can never come to appear because it is what produces the appearance, it establishes the connection to the minimal, to the infinitesimal, to the almost-nothingness that runs through Kafka's literature as a motif and procedure. Just as in Josefine's meager whistling as an almost nothing, the difference opens up in the same—all mice whistle—so in Kafka's small literature, the difference opens up as the object of the "most real reality," which, like Odradek, dwells immeasurably in the reality of the symbolic. And if the master of the house in which Odradek roams worries that he might outlive him and his symbolic world, this worry is justified.

Josefine does not survive, at least not as an artist. Things must go downhill with her. First, she neglects the events and finally disappears from the scene altogether. The attempt to save art through the act of its symbolic recognition and institutionalization as an exception to the law—as an exception to the rule, which consumes its strength—fails. It fails in a double sense: on the one hand, because, structurally speaking, the exception to the rule is necessarily included in the rule procedure; on the other hand, because the people reject the exception. One could say that the attempt to save art is its downfall. But it is the people who survive. They will, as the narrator says and knows, "overcome the loss" (376).

Where does this knowledge come from? Not from an appeal to aesthetic judgment on the part of either the people or the narrator. Nor is it to be taken as an indication of the people's indifferent or generous toleration of Josefine's inability in the matter of singing, but it owes itself to a kind of nonperception of what Josefine is trying to assert as art. The prospect of overcoming the loss caused by disappearance and silencing can be overcome because it was present in Josefine's art itself. While Josefine seeks to exclude the almost nothing, the hollow, through symbolic recognition, through the act of institutionalization, the story tells us that this would be the end of art. The weakness of Josefine's real whistling is effective in that it adds a loss to reality that changes the presence of the real whistling. It appears itself as a kind of memory that does not refer to a given, perhaps recurring object; rather, the thin and tiny whistle evokes a memory that cannot be caught up by any object, of an almost-nothingness that, like the life of the people, passes into an unlosable oblivion that is always already and never lost. Also, because the rupture, the difference, the in-between, the gap cannot be recorded, archived, preserved, or institutionalized in any files of history or as history, it is what blows through Kafka's little literature. And Josefine, too, "will happily lose herself in the countless multitude of our people and soon, as we do not do history, will be forgotten in heightened redemption like all her brothers" (377).

Notes

1. Franz Kafka, *Schriften Tagebücher Briefe, Kritische Ausgabe*, ed. Gerhard Neumann, Malcolm Pasley, and Jost Schillemeit, Frankfurt am Main: S. Fischer, Briefe 1900–1912, ed. Hans-Gerd Koch, 1999, vol. 2, 275 (in the following section on Kafka, all quotes will be from this edition). See also the title of an essay by Claudia Liebrand, "Kafka's Kleist: Pig's Bladders, Broken Jugs and Protracted Trials." *Textverkehr: Kafka und die Tradition*, ed. Claudia Liebrand and Franziska Schössler, Würzburg 2004, 73–100.
2. As quoted from Heinrich von Kleist, Brandenburger Ausgabe (BKA) II/9: *Sonstige Prosa*, ed. by Roland Reuß, Frankfurt a.M. / Basel: Stroemfeld Verlag 2007, 25–32. In the following, the page numbers are given in brackets. The editorial-historical information can be found in Kleist, BKA, II, 25. For further references, see Marcel Schmid, *Autopoiesis und Literatur: Die kurze Geschichte eines endlosen Verfahrens*, Bielefeld, transcript Verlag 2016, esp. 13–28. For dating, see BKA, II, 25. The text was probably written in Königsberg in 1805–6; the original manuscript is missing.
3. See Bettine Menke, "Anfangen—zur 'Herkunft der Rede,'" in Barbara Thums et al. (eds.), *Herkünfte: Historisch, ästhetisch, kulturell*, Heidelberg, Winter Verlag 2004, 13–37.
4. The text was intended for publication in a journal, most likely for Cottas *Morgenblatt für gebildete Stände* or for *Phöbus*; see Schmid, 17n18.
5. I refer here to Slavoj Zizek, "Mehr-Geniessen: Lacan und die Populärkultur," in *Wo es war* 1 (Wien: Turia & Kant 1992), esp. 7–12.
6. The reference here to Kleist's favorite sister, Ulrike, is quite apparent.
7. On the figure of opening up as closure in the realm of literary speech, see Werner Hamacher, *Entferntes Verstehen* (Frankfurt am Main: Suhrkamp), 98.
8. See also here Zizek, op. cit.
9. Hamacher, op. cit., 85.
10. The volume was published by *Die Schmiede*, Berlin, at the end of August 1924—i.e., after Kafka's death. Further details to be found in Kafka, Kritische

Ausgabe, *Drucke zu Lebzeiten Apparateband*, ed. Wolf Kittler, Hans–Gerd Koch and Gerhard Neumann, Frankfurt am Main: 1996, 386. In the following notes, cited as KKAD plus app., page references in parentheses in the text.
11. See Marianne Schuller, "Verschwinden ohne Ende, Ein Schluss-Licht," in Günther Ortmann and Marianne Schuller (eds.), *Kafka: Organisation, Recht und Schrift*, Weilerswist, Velbrück Wissenschaft 2019, 423–35.
12. See Sigmund Freud, "*Die Verneinung*," in *Freud, Gesammelte Werke*, ed. Anna Freud, with the collaboration of Marie Bonaparte among others, Frankfurt am Main 1991, 7. Edition, Bd. XIV, 11–15, here 15. See Schuller, op. cit., 426.
13. Mladen Dolar, *His Masters Voice: Eine Theorie der Stimme*, Frankfurt am Main: Suhrkamp 2007, chap. 7, 217–38, esp. 233.
14. Ibid., 233.
15. See Kafka's notes from his journals and their recording by Deleuze/Guattari, *Kafka: Für eine kleine Literatur*, Frankfurt am Main: Suhrkamp Verlag 1976, esp. chap. 3, 24–39.
16. An examination of Roland Barthes's concept of the voice would be appropriate here as it tends toward a "physiological reduction" of the voice. I can but give the general reference here to Roland Barthes, "Die Rauheit der Stimme," in ders. *Was singt mir, der ich höre in meinem Körper das Lied*, Merve Verlag Berlin 1979, 19–336.
17. I am hinting here at a connection to Giorgio Agamben that is important for Kafka readings, especially to his book *Homo sacer: Die souveräne Macht und das nackte Leben*, Frankfurt am Main: Suhrkamp 2002; I am referring here to the pertinent discussion of the text by Günther Ortmann, "Der Ausnahmezustand: *Homo sacer*," in ders. *Regel und Ausnahme: Paradoxien sozialer Ordnung*, Frankfurt am Main: Suhrkamp 2003, 88–90.
18. Dolar, 237.
19. Hamacher, "Die Geste im Namen. Benjamin und Kafka," in *Entferntes Verstehen*, a.a.O., 280–323, here 313.
20. See "Die Sorge des Hausvaters," in KKAD, app., 282–84.
21. KKAT (=Tagebücher), 855.

CONTRIBUTORS

ELIZABETH BROGDEN is a writer and editor based in Cambridge, Massachusetts. She holds a PhD in English from Johns Hopkins University and a BA in English from Barnard College. Her scholarly articles and reviews have appeared in NOVEL: A Forum on Fiction, Studies in American Fiction, Studies in the Novel, Journal of Victorian Culture, and Modern Language Notes.

MARGARETA INGRID CHRISTIAN is assistant professor in Germanic studies at the University of Chicago. She works at the intersection of literature, art writing, and the history of science. She is the author of *Objects in Air: Artworks and Their Outside around 1900* (University of Chicago Press, 2021).

ELENA FABIETTI is assistant professor of German literature at the University of Regensburg (Germany). She completed her doctoral studies in comparative literature in Italy (Siena) and the United States (Johns Hopkins University). She has published a book on images in Baudelaire's and Rilke's poetry (2015) and is working on a second book entitled *Bodies of Glass: A Cultural and Literary History of Transparent Humans*.

CHRISTIANE FREY is associate professor of German literature at Johns Hopkins University and codirector of the Max Kade Center for Modern German Thought. She previously taught at the University of Chicago and Princeton University. Among her recent publications are *Laune: Poetiken der Selbstsorge von Montaigne bis Tieck* (Fink, 2016) and the coedited volume *Säkularisierung: Grundlagentexte zur Theoriegeschichte* (Suhrkamp, 2020).

CHRISTOPHER D. JOHNSON is associate professor of Spanish, German, and comparative literature at Arizona State University. He previously taught at Harvard University; University of California, Los Angeles; and Northwestern University and was a research associate at the Warburg Institute, London. He is the author of *Memory, Metaphor, and Aby Warburg's Atlas of Images* (2012) and *Hyperboles: The Rhetoric of Excess in Baroque Literature and Thought* (2010).

DANIEL LIU is currently a research associate (*Wissenschaftlicher Mitarbeiter*) at the Chair for the History of Science in the Historical Seminar of the Ludwig-Maximilians Universität München, where he is the Principal investigator on a three-year Deutsche Forschungsgemeinschaft (DFG)–funded project, Living Matter under the Microscope: Protoplasm Theory, "*wissenschaftliche Mikroskopie*," and the Molecularization of Cell Structure, 1840–1940. He received his PhD in the history of science, medicine, and technology from the University of Wisconsin–Madison in 2016.

ANDREAS MAHLER is professor of English literature at Freie Universität in Berlin. His main areas of research are early modern epistemology, Elizabethan satire, early modern genres, and London as an early modern center. His publications include *Moderne Satireforschung und elisabethanische Verssatire* (1992) and edited collections on images of the city (1999) and Shakespeare's subcultures (2002).

ROGER MAIOLI is associate professor of English at the University of Florida. Besides articles and reviews published in a variety of journals, he is the author of *Empiricism and the Early Theory of the Novel: Fielding to Austen* (Palgrave, 2016).

MALTE FABIAN RAUCH is a philosopher and art theorist based in Berlin. He received his PhD from the Universität der Künste and also holds an MA in philosophy from the New School, New York.

HANS-JÖRG RHEINBERGER is senior professor of the history of science at the Technische Universität in Berlin. A biologist by training, he was a director of the Max-Planck-Institut for the History of Science Berlin (1997–2014). Among many renowned publications, he is the author of *Toward a History of Epistemic Things* (Stanford University Press, 1997).

MAREIKE SCHILDMANN is assistant professor of German literature at the Universität Bremen and was previously a research associate at the *Zentrum für Literatur- und Kulturforschung*, Berlin. She holds a PhD in German literature and is the author of *Poetik der Kindheit: Literatur und Wissen bei Robert Walser* (Wallstein, 2021).

CARMEN SCHMECHEL is a historian of medicine and science, focusing on the early modern constellations, particularly knowledge transfers between natural sciences and philosophical thought. Currently a DFG postdoctoral fellow at the Institute for Philosophy of the Freie Universität in Berlin, she is writing a book on the extensive cultural history of obsolete theories of fermentation.

MARIANNE SCHULLER was senior professor of German literature at the University of Hamburg and coauthored, among many other books, *Mikrologien: Literarische und philosophische Figuren des Kleinen* (2003).

CYNTHIA WALL is the William R. Kenan Jr. Professor of English at the University of Virginia. She is the author of *Grammars of Approach: Landscape, Narrative, and the Linguistic Picturesque* (2019); *The Prose of Things: Transformations of Description in the Eighteenth Century* (2006); and *The Literary and Cultural Spaces of Restoration London* (1998), as well as editor of works by Alexander Pope, Daniel Defoe, and John Bunyan.

INDEX

Abbe, Ernst, 10, 132, 133–34
Addison, Joseph, 187
aestheticism, 22–23, 73–74, 86, 240, 241, 243, 246–47, 249–52
Agamben, Giorgio, 214, 250
ambience: architecture and, 248–50, 255–56; definitions, 241, 251; social construction aspects of, 247, 252, 254–55; urbanism and, 239–40. *See also* architecture; atmosphere; milieu; minutiae
Ambronn, Hermann, 142–43
analogism, 172–74, 263
Ansatzpunkte, 209, 224–25, 226–27
anthropology, 48
Aragon, Louis, 244
architecture: biological architecture, 13–14, 263–64, 266, 267, 269–70, 274–75; collective architecture, 246–47, 251–52; cultivating isolation in, 264–65, 273; psychological aspects of, 271–72; role in ambience, 248–50, 255–56. *See also* ambience; atmosphere; milieu
Aristotle, 21, 26–27, 30, 34, 99–100, 101
artificial intelligence (AI), 144
Atget, Eugène, 242–43
atmosphere: artistic medium in, 239; surrealist interpretation, 242–44, 255–56; urbanism, 13, 241–42, 244–45, 246–47, 249, 252–53. *See also* ambience; milieu; minutiae
atomism, 6, 49, 50, 56, 97, 139
Auerbach, Erich: critiques of Curtius, 222–23, 224–25; microcosmic departure points, 209, 211; *Mimesis*, 226–29; "Philologie der Weltliteratur," 223–24, 225–26
avant-garde, 246–47, 249

Bachelard, Gaston, 6
Bacon, Francis, 25, 26, 27–28, 28, 31, 36
bacteria, 151–54

bacteriophages, 151, 153–54
Baker, Henry, 80
Baker, Richard, 165
Balzac, Honoré de, 5
Banchetti-Robino, Marina, 103, 109–10
Barry, Edward, 31
Batteux, Charles, 33
Beattie, James, 33
Behrendt, Walter Curt, 274
Bender, John, 23, 34
Benjamin, Walter, 214–15, 242–43, 243–45
Bernard, Claude, 6, 149, 150
Bigg, Charlotte, 140
biology: architecture and, 264–65, 266, 274–75; bacteriophages, 151, 153–54; enzymology, 155–56, 159; history of, 150–51, 153, 158–59; plaques, 151–52, 153, 155; protein synthesis, 158. *See also* micropreparations; microscopy; natural philosophy
Blair, Hugh, 23, 36–37, 184, 187
Blake, William, 21–22
Bligh, William: *A Narrative of the Mutiny*, 12, 182–83, 190, 195–201
Breton, André, 244, 245–46, 252
Britain: eighteenth-century novel in, 22, 24–25, 29–30, 36, 195; Elizabethan era culture, 165–66, 171; empiricist philosophy in, 30, 33
Brown, John, 52, 53–54, 58, 79, 140
Brown, Lancelot "Capability," 181, 182–83, 201
Buffon, George Louis Leclerc, 78–79
Burnet, Thomas, 30

Caesar, Michael, 74
Caillois, Roger, 245
Campana, Andrea, 79
Canguilhem, Georges, 6
capitalism, 242, 244–45, 247–48, 256
Chambers, Ephraim, 183, 184
Chapone, Hester, 29

Chombart de Lauwe, Paul-Henry, 253–54
Christian, Fletcher, 195–96
Chtcheglov, Ivan, 247–49, 250, 254
city, the: ambience and, 239–40, 241–42, 251–52; atmosphere and, 242–44; capitalist structure of, 242, 244–45, 247–48, 256; collective architecture, 246–47, 253; legibility, 240–41, 254; social architecture, 12–13, 248–50, 255–56
Cocker, Edward, 182
Cole, Lucinda, 102
Coleridge, Samuel Taylor, 187
Collier, Jeremy, 31
Constant (artist), 13, 239–40, 247, 255
contagion: physical factors of, 97, 104, 105–6, 107–8, 111–12; spiritual factors of, 100–101, 109, 110. *See also* fermentation; miasma theory; putrefaction
Cowell, Philip: *This Is Me, Full Stop*, 189
Croce, Benedetto, 213, 214
Curtius, Ernst Robert: criticisms of, 223, 224–25; *European Literature and the Latin Middle Ages*, 216–18, 219–20; on synchronic phenomenology, 210; Warburg meeting, 215–16

Dacier, André, 26, 32
d'Alembert, Jean le Rond, 5, 25
Damisch, Hubert, 240, 254
da Monte, Giambattista, 106
Daston, Lorraine, 28, 167
Davidson, Jenny, 23
Day, Angel, 184
Debord, Guy: on architecture, 13, 249–52; collaboration with Chtcheglov, 247; collaboration with Constant, 239–40; on urban ecology, 253–55
Defoe, Daniel, 11, 31, 182–83, 190–94, 195
Delbrück, Max, 152–53
Dennis, John, 26
Dibdin, Charles, 37
Diderot, Denis, 5, 25
Diodorus of Sicily, 100
disease: contagion, 97, 100–101, 104, 105–6, 107–8, 109, 110, 111–12; fermentation, 97–99, 107–8, 113–15; miasma theory, 9, 98, 99–102, 102, 103, 104–5, 113; putrefaction, 99–100, 101–2, 106, 111, 112–14
DNA, 125, 156–57
Donne, John: "Break of Day," 167; cultural milieu, 166–67, 170–71; "The Flea," 11, 167, 168–70, 172–74; portrait of, 165–66
Dove, John, 30
Drury, Joseph, 23, 38
Duchamp, Marcel, 290–91, 294
Dürer, Albrecht, 213
dynamism, 50

Ebeling, Siegfried: background, 13–14, 265; biological architecture, 263–64, 266, 268, 269–70; cosmic considerations, 271–72; influences, 267–68, 269; isolation in architecture, 270–71, 272–73, 274–75; nationalist sympathies, 273
Efal, Adi, 213
Ekici, Didem, 270
empiricism: bodily dimension of, 74–75, 85; critiques of fiction, 23, 25, 31–33, 34, 39–40; history and, 28–30, 35–37, 126; inductive thinking, 171, 173, 210. *See also* particularity
enzymology, 155–56, 159
Escherichia coli (bacteria), 151–53, 158–59
Euclid, 211
Evelyn, John, 186–87
experimental methodology. *See* in vitro experimentation; micropreparations; microscopy

Ferguson, James or Adam, 36
fermentation, 97–99, 107–8, 113–15. *See also* contagion; miasma theory; putrefaction
Fielding, Henry, 33, 34, 40
flânerie, 243–45, 251
Fordyce, David, 36
Foucault, Michel, 171, 172
Fracastoro, Girolamo: on contagion, 97, 104, 105–6, 107, 110–11; physical factors of minutiae, 103; spiritual factors of minutiae, 104–5
Francé, Raoul Heinrich, 266, 267–68, 272, 274–75
Freud, Sigmund, 289

Frey-Wyssling, Albert, 143
Friedmann, Herbert, 150

Galen of Pergamon, 98, 100, 103, 110, 113–14
Gelley, Alexander, 244
genetics, 151–52, 153–55, 158–59
German Micrological Society (Deutsche Mikrologische Gesellschaft), 267
Germany, 270, 273
Giedion, Sigfried, 265
Gilbert, Geoffrey, 40
Glenn, John, 181
Godwin, William, 33, 38
Goethe, Johann Wolfgang von, 58, 274
Guillemin, Anna, 213

Haeckel, Ernst, 267
Haller, Albrecht, 53
Hamacher, Werner, 11, 294
Hamilton, Elizabeth, 39–40
Harris, John (Mr. Stops): *Punctuation Personified*, 185
Hausmann, Raoul, 269
Hawkesworth, John, 21
Hays, Mary, 38–39, 40
Hellpach, Willy, 269
Helmholtz, Hermann, 10, 132, 133
Hessel, Franz, 243–44
Hildebrand, Caz: *This Is Me, Full Stop*, 189
Hippocrates, 99, 100, 110
Homer: *Odyssey*, 226
Hooke, Robert, 79–80, 126, 128–29, 134, 212
Hörl, Erich, 242
Hume, David, 29
Hurd, Richard, 27

imagination, 26, 35–36, 56, 75, 81–82, 85–86, 110
imaging. *See* microscopy; optics (physics)
in vitro experimentation, 9–10, 149–50, 151–52, 158. *See also* micropreparations

Johnson, Samuel: *Rasselas*, 21
Jones, William (of Nayland), 31
Jorn, Asger, 240, 250

Kafka, Franz: *A Hunger Artist*, 287; "Josefine die Sängerin, oder Das Volk der Mäuse," 282, 287–90, 291–94, 295–96; on Kleist, 281
Kant, Immanuel, 49–50, 285
Keck telescopes, 131
Keymer, Thomas, 186
King, Stephen, 189–90
Kircher, Athanasius, 98, 110–13
Kleist, Heinrich von: "Über die allmähliche Verfertigung der Gedanken beim Reden," 282–87
Klopstock, Robert, 287
Knox, Vicesimus, 32
Köhler, August, 124, 139

Laban, Rudolf, 265, 267
Lamarck, Jean-Baptiste, 5
Lamb, Jonathan, 192, 198
Laxton, Susan, 244
Leeuwenhoek, Antony van, 80, 128
Leibniz, Gottfried, 2, 49, 55, 64, 65, 211, 214
Lennard, John, 183, 186, 188
lenses, 122, 126, 131–32, 133, 135, 138. *See also* microscopy
Leopardi, Giacomo: aesthetic philosophy of, 73–74; background, 78–79, 80–81; empiricism of, 74–75, 85; function of detail, 72–73, 81–82; poetics of, 81, 86; theory of pleasure, 82; *Zibaldone di pensieri*, 8, 71–72, 76–78, 83–85, 85–86
L'Estrange, Roger, 186
Lettrist International, 239
Lilienstern, Otto August Rühle von, 282
Linfert, Carl, 215
linguistics, 210–11, 213, 220–21, 226–27, 283. *See also* philology
Linnaeus, Carl, 78
Lister, Joseph Jackson, 131
literary theory: *locus amoenus*, 78; narrative structure, 283, 285–87, 292–94; nomenclature, 287–88; point of view, 283–85, 288–90, 292, 295–96; virtual witnessing, 7, 23–24, 34–35, 36–37, 40. *See also* novel (literary form); poetics
Locke, John, 29, 75
Lorrain, Claude, 249

Lucretius: *De rerum natura*, 102–3
Lukács, Georg, 48, 57
Luria, Salvador, 152–53

Manutius, Aldus, 183–84
Matthaei, Heinrich, 159
medicine, 53–54, 85–86, 101–2. *See also* disease; micropreparations; microscopy
Merleau-Ponty, Maurice, 2
miasma theory: physical factors in, 9, 97–98, 99–102, 102, 113; spiritual factors in, 103, 104–5. *See also* contagion; putrefaction
microambience, 13, 239, 240–41, 247, 252–53, 256. *See also* ambience; milieu
Microbiological Association of Munich (Mikrobiologische Vereinigung München), 267
micropreparations: DNA sequencing, 155–57; experimental environments, 157–60; in vitro experimentation, 149–50; phages, 150–55
microscopy: history of, 2, 10, 80, 126–29, 267–68; imaging, 122–24, 143, 144; limitations, 130–31, 135, 140; magnification, 124–25, 126, 132–33, 144, 154, 157; resolution, 125–26, 144; role of light, 131–33, 137–38, 141, 142–43; technical developments, 129–30, 131–32, 133–35, 137–39, 141–42; types, 124, 135, 137, 139, 140–41, 142–43. *See also* optics (physics)
milieu: ambience and, 12–13, 59, 61, 293–95; architecture and, 13–14, 263–64, 266, 268, 275; biology and, 108, 109, 150–51, 157–60; contagion and, 9, 105, 106–7, 110–11, 111–12; contextualizing events, 196, 197–98, 200–201; contextualizing literature, 282–83; contextualizing objects, 166–67, 290–91; culture and, 166–67, 170–71; definitions, 5–6, 181, 184, 185–86, 247; human body as, 51–53, 54–55, 58, 111; individuality and, 60–61, 62, 83–84; linguistics and, 213, 226–27; literary studies and, 7, 11, 14–15, 34, 48, 56, 57, 210, 212, 216–18, 219–20, 222, 224, 225–26, 228–29, 281, 289–90, 292–93; miasma theory and, 97, 98, 99–100, 102; natural philosophy and, 6, 48, 51–53, 54–55, 61–62; in negotiation with minutiae, 4–5, 8, 72–73, 75, 82, 84–85, 86, 168, 191, 272. *See also* ambience; atmosphere; city, the; novel (literary form)
minutiae: absence as presence, 283–84, 288–89, 292, 294, 295–96; ambience and, 240, 241, 254–55; analogy and, 11, 168, 169–70, 171, 172–74; architecture and, 13–14, 267; atmosphere and, 12–13, 244, 246, 250, 252; biology and, 9, 10, 97, 102, 103, 110, 112–13, 156; definitions, 48, 181; as embodiment of the whole, 50–51, 55–56, 75, 80–81, 84–85, 86, 159–60; literary studies and, 14–15, 81–82, 166, 167, 191–92, 224, 227, 281, 286–87, 291; meaningful details, 5, 74, 76–77, 82, 149–50, 153–54; miasma theory and, 98, 100–101, 103, 104; microscopy and, 2, 122–24, 139, 143–45, 212; overwhelm in, 8, 219–20, 222–23, 224, 228, 285–86; perception of, 2–3, 8, 54, 58–60, 83–84, 283; philology and, 10–11, 12, 209–10, 211–12, 214–15, 216, 219–20, 221; punctuation as, 11–12, 181–82, 190, 193–94, 198–200, 201, 285; relationality, 4, 72–73, 155; role in creating milieus, 196–98, 290–91; seeds of disease, 105–6, 107, 108–9, 111, 114; seeds of Novalis, 47–48, 50–51, 61–62; spiritual dimensions, 104–5; visual arts and, 165–66. *See also specific fields of study*
molecular genetics, 151, 153, 158. *See also* biology; genetics; micropreparations
Molière (Poquelin, Jean-Baptiste), 33
Montagu, Elizabeth, 37
Montagu, Mary Wortley, 31–32
Montaigne, Michel de, 227
More, Hannah, 181
Moretti, Franco, 229
Morise, Max, 244
Mr. Stops (John Harris): *Punctuation Personified*, 185
Mullis, Kary, 156

natural philosophy: economy of nature, 79; milieu and, 6, 51–53, 54–55, 61–62; minutiae and, 8, 47–48, 50–51, 55–57, 77–78
New Babylon, 239, 240

Newman, William R., 109
Nietzsche, Friedrich, 11, 274
Nieuwenhuys, Constant. *See* Constant (artist)
Nirenberg, Marshall, 159
Novak, Maximilian, 195
Novalis (Friedrich von Hardenberg): background, 8, 49; *Henry of Ofterdingen*, 57–60; milieu as organic environment, 48, 51–53, 60–61; minutiae as seeds, 47–48, 50–51, 56–57, 61–62; natural philosophy of, 49–50, 53–55
novel (literary form): empiricism and, 23, 25, 31–33, 34, 39–40; faculty of imagination, 26, 35–36; as milieu, 14–15, 34, 56, 57, 60; narrative structure, 283, 285–87, 292–94; as organism, 48, 61; point of view, 283–85, 288–90, 292, 295–96; role of setting, 22, 24; virtual witnessing, 7, 23–24, 34–35, 36–37, 40. *See also* literary theory

optics (physics), 122–24, 131, 135–36, 144. *See also* microscopy
Ostwald, Wilhelm, 269

parenthesis: history, 183–84; in *A Narrative of the Mutiny* (Bligh), 198–200; in "Parentheses: A Bestiary," 188–89, 191, 201; in *Robinson Crusoe* (Defoe), 190–94; usage, 11–12, 184–87, 188–90, 201
Paris, 243, 244–45, 246
Parker, Robert, 101
Parkes, M. B., 188
particularity: aesthetics of, 23; Aristotelian criticism, 26; empiricism and, 7, 74–75, 85; history of, 28–30, 31, 35–37; literary theory and, 24–25, 37–38; natural philosophy and, 27–28; neoclassical poetics, 21–22, 25. *See also* empiricism
Pasquier, Étienne, 167
Peirce, Charles Sanders, 154
Pelling, Margaret, 104
Pemberton, Henry, 37
Perrin, Jean, 139, 140
Pettenkofer, Max von, 270
philology: definition, 209; historical perspective, 222, 224, 225–26, 228, 229; meaningful details, 211–12, 213–14, 216–17, 221–22, 226–27; minutiae in, 10–11, 12, 209–10; overwhelming details, 219–20, 222–23; stylistic expression, 213, 221, 222, 226–27
plaques, 151–52, 153, 155
Plato, 26–27, 33, 284
poetics: milieu and, 61, 86; minutiae and, 61–62, 73, 81–82, 84; particularity and, 21–22, 25–26, 32–33; universals and, 21. *See also* literary theory
Pollock, Sheldon, 209
Poquelin, Jean-Baptiste (Molière), 33
Pound, Ezra, 210
Priestley, Joseph, 29, 36, 37–38
psychogeography, 239, 242, 246, 247, 250, 251–52, 254–55
punctuation, 11–12, 181–84, 190, 193–94, 198–200, 201, 285
putrefaction, 99–100, 101–2, 106, 111, 112–14. *See also* contagion; miasma theory

Quinn, James A., 253

Rabelais, François, 227
Reynolds, Joshua, 21–22, 33
Richardson, Samuel, 187–88, 201
Robertson, Joseph, 182, 183, 184
Romanticism, 49, 52, 56, 57, 86
Rousseau, Jean-Jacques, 77

Said, Edward, 209, 227
Sanger, Frederick, 156
Scheerbart, Paul: *Glass Architecture*, 263, 272
Scheiffele, Walter, 264, 265
Sedlmayr, Hans, 214–15
Shaffer, Simon, 34–35
Shapin, Steven, 34–35
Sheringham, Michael, 246
Sidney, Philip: *Defense of Poesy*, 25–26
Simon, Maurya: "Parentheses: A Bestiary," 188–89, 191, 201
situationism, 240–42, 245–46, 250, 254–55
Situationist International, 239, 246
Slayter, Elizabeth M., 139
Sloterdijk, Peter, 241
Smith, Adam, 31

Smith, John, 183
Solmi, Sergio, 78
Spitzer, Leo: critique of Curtius, 218–20, 223; "Linguistics and Literary History," 12, 220–22; "Milieu and Ambiance," 209, 220; semantic changes in language, 210–11, 241
Stafford, Barbara, 80
Stanhope, Philip (Earl of Chesterfield), 29
Steele, Richard, 187
Steiner, Rudolf, 265, 266
Stent, Gunther, 153
Sterne, Laurence: *Tristram Shandy*, 4–5, 80
Stewart, Dugald, 29, 31
Stewart, Susan, 3
Stoicism, 103
Stower, Caleb, 184
surrealism, 240, 242–43, 244–46, 247, 249–50, 252

Taine, Hippolyte, 220
Taylor, Thomas, 30
telescopes, 126, 130–32
Thermus aquaticus (bacteria), 155–56
Thompson, Helen, 23–24, 35
Trifler (pseudonym), 32
Tristram Shandy (Sterne), 80

universals, 21, 26–27, 30, 32–33, 74, 82
urbanism, 239–40, 241–42, 243–45, 246–47, 249, 251–52

van Helmont, Jan Baptista: contagion theory of, 108–9; ferments and fermentation, 97–98, 107–8, 114; spiritual agent of disease, 110, 113; spontaneous generation, 101–2
virtual witnessing, 23–24, 34–35, 36–37, 40
Vitrac, Roger, 244
von Hardenberg, Friedrich (Novalis): background, 8, 49; *Henry of Ofterdingen*, 57–60; milieu as organic environment, 48, 51–53, 60–61; minutiae as seeds, 47–48, 50–51, 56–57, 61–62; natural philosophy, 49–50, 53–55
Vossler, Karl, 213, 214

Warburg, Aby, 210, 213–14, 215–16, 218
Weber, Samuel, 214
Whately, Richard, 33–34, 38
Wheare, Degory, 28–29
Wilson, Catherine, 2, 126
Wölfflin, Heinrich, 271
Wollaston, William, 30
Woolf, Virgina, 181–82

Young, William, 29–30

Zamecnik, Paul, 158
Zola, Émile: *The Experimental Novel*, 5–6

www.ingramcontent.com/pod-product-compliance
Ingram Content Group UK Ltd.
Pitfield, Milton Keynes, MK11 3LW, UK
UKHW040824291224
452713UK00019B/21